GUIDE TO MUTATION DETECTION

GUIDE TO MUTATION DETECTION

Edited by
GRAHAM R. TAYLOR, Ph.D.
Yorkshire Regional DNA Laboratory
St. James's University Hospital
Leeds, United Kingdom

IAN N. M. DAY, Ph.D.
Division of Human Genetics
Southampton University Hospital
Southampton, United Kingdom

Published in association with the Human Genome Organization

WILEY-LISS

A JOHN WILEY & SONS, INC., PUBLICATION

Published by John Wiley & Sons, Inc., Hoboken, New Jersey.
Published simultaneously in Canada.

Library of Congress Cataloging-in-Publication Data:
Guide to mutation detection / edited by Graham R. Taylor, Ian N. M. Day.
p. ; cm.
"Published in association with the Human Genome Organization."
Includes bibliographical references and index.
ISBN 0-471-23444-3 (pbk. : alk. paper)
1. Mutation (Biology)—Laboratory manuals.
[DNLM: 1. DNA Mutational Analysis—instrumentation. 2. DNA Mutational Analysis—methods. 3. Gene Dosage. 4. Polymorphism, Single Nucleotide. QU 58.5 G946 2004]
I. Taylor, G. R. (Graham R.) II. Day, Ian N. M. III. Human Genome Organisation.
QH461.G85 2005
576.5′49—dc22 2004004692

Printed in the United States of America

10 9 8 7 6 5 4 3 2 1

CONTENTS

CONTRIBUTORS

Mohammed A. Aldahmesh, Division of Human Genetics, Southampton University Hospital, Southampton, United Kingdom

Khalid K. Alharbi, Division of Human Genetics, Southampton University Hospital, Southampton, United Kingdom

Norbert Arnold, Clinic for Gynecology and Obstetrics, University of Kiel, Kiel, Germany

David E. Barton, National Centre for Medical Genetics, Department of Medical Genetics, University College of Dublin, Dublin, Ireland

Christophe Beroud, Laboratoire de Génétique Moléculaire, Centre Hospitalier Universitaire (CHU) Montpellier, Institut Universitaire de Recherche Cliniquie (IURC), Montpellier, France

James Bonfield, Medical Research Council, Laboratory of Molecular Biology, Cambridge, United Kingdom

Patrick G. Buckley, Department of Genetics and Pathology, Uppsala University, Uppsala, Sweden

Ruth Charlton, Molecular Genetics Laboratory, Institute of Child Health, London, United Kingdom

Xiao-he Chen, Division of Human Genetics, Southampton University Hospital, Southampton, United Kingdom

Mireille Claustres, Laboratoire de Génétique Moléculaire, IURC, Institut Universitaire de Recherche Clinique, Montpellier, France

David Cockburn, Regional DNA Laboratory, St. James's University Hospital, Leeds, United Kingdom

Richard G. H. Cotton, Genomic Disorders Research Centre, St. Vincent's Hospital, The University of Melbourne, Victoria, Australia

Darren Cuthbert-Heavens, Mutation Detection Facility, Cancer Research UK, St. James's University Hospital, Leeds, United Kingdom

Ian N. M. Day, Division of Human Genetics, Southampton University Hospital, Southampton, United Kingdom

Gert de Voss, Ingeny International BV, Amundsenweg 17, Goes, The Netherlands

Jan P. Dumanski, Department of Genetics and Pathology, Uppsala University, Uppsala, Sweden

Diana Eccles, Division of Human Genetics, Southampton University Hospital, Southampton, United Kingdom

Kim Flintoff, Regional DNA Laboratory, St. James's University Hospital, Leeds, United Kingdom

Ivo Glynne Gut, Centre National de Génotypage, Evry, France

Lema Haddad, Division of Human Genetics, Southampton University Hospital, Southampton, United Kingdom

John F. Harvey, National Genetics Reference Laboratory, Salisbury District Hospital, Salisbury, Wiltshire, United Kingdom

Rania Horaitis, Human Genome Variation Society, Genomic Disorders Research Centre, Carlton South, Australia

Steve E. Humphries, Division of Human Genetics, Southampton University Hospital, Southampton, United Kingdom

Lucy Jenkins, Regional DNA Laboratory, St. James's University Hospital, Leeds, United Kingdom

David Judge, Department of Genetics, University of Cambridge, Cambridge, United Kingdom

Ulrika Liljedahl, Department of Medical Sciences, Uppsala University, Uppsala, Sweden

Katarina Lindroos, Department of Medical Sciences, Uppsala University, Uppsala, Sweden

Sue Malcolm, Clinical and Molecular Genetics Unit, Institute of Child Health, London, United Kingdom

Hamid Rassoulian, Division of Human Genetics, Southampton University Hospital, Southampton, United Kingdom

Mark D. Robinson, Regional DNA Laboratory, St. James's University Hospital, Leeds, United Kingdom

Jan Schouten, MRC-Holland, Amsterdam, The Netherlands

Snaevar Sigurdsson, Department of Medical Sciences, Uppsala University, Uppsala, Sweden

Helen Snowden, Regional DNA Laboratory, St. James's University Hospital, Leeds, United Kingdom

Emmanuel Spanakis, Division of Human Genetics, Southampton University Hospital, Southampton, United Kingdom

Ann-Christine Syvänen, Department of Medical Sciences, Uppsala University, Uppsala, Sweden

Claire F. Taylor, Cancer Research UK, Mutation Detection Facility, Leeds, United Kingdom

Graham R. Taylor, Regional DNA Laboratory, St. James's University Hospital, Leeds, United Kingdom

Mario Tosi, Institut National de la Santé et de la Recherche Médicale, Faculté de Médecine et Pharmacie, Rouen, France

PREFACE

From the early days of chromosome-specific workshops through international scientific meetings, and by developing a standardized human gene nomenclature, the Human Genome Organization (HUGO) has been instrumental in enabling open access to the Human Genome Project. Completion of "a" human genome sequence is, however, merely a foundation from which to develop a new approach to human biology and medicine. Knowledge of one consensus sequence assembled from clones derived from relatively few individuals immediately raises the questions of how representative this sequence is of actual genomes carried by individuals, of how much variation there is between individuals, and of what biological significance such variations may have. The term "mutation" or "variation" can be used to describe a DNA sequence change compared to an arbitrary reference sequence. Clearly, the understanding of genetic variation requires experimental and informatics tools to detect and document mutations. In 1991, Dick Cotton and colleagues organized the first HUGO Mutation Detection Workshop in Oxford, England. This was a small, informal meeting designed to bring together inventors and users of mutation detection technology, both academic and commercial. These meetings have continued on a biennial basis ever since and attract many distinguished participants for the exposition and review of leading-edge technologies in the field.

In 1997, Professor Cotton and a steering group discussed the idea of alternating these meetings with a mutation detection practical course. These courses were to be aimed at postgraduate or postdoctoral scientists and comprise a combination of lectures and practical sessions covering a wide range of mutation detection techniques. Restricted to 50 participants (now increased to 60 to meet the demand), they provide a survey of currently available and emerging technologies for mutation detection. They have proved popular with teachers and participants alike. As we prepare to deliver the fourth course in Newcastle in 2004, demand for the courses is still high, attracting participants from many nations, in keeping with the international outlook of HUGO. Lectures, given by leading researchers in their fields, are complemented by demonstration and practical sessions to provide hands-on experience with many of the techniques. The lecture and practical material from these courses has formed the basis for this handbook. A degree-level understanding of the vocabulary of modern biology and biochemistry is assumed. This book is intended to provide the reader with a broad overview of available and emerging technology for mutation detection together with reliable protocols currently in use. It will have been successful if it facilitates wider access to the benefits of human genome research.

A great many people have contributed to the success of these courses and to this handbook. Cathy Pole, the executive secretary of HUGO, and the HUGO staff have coordinated the course administration. The scientific advisors, Dick Cotton, Johan Den Dunnen, Ulf Landegren, Mireille Claustres, Ann-Christine Syvänen, and Mario Tosi, laboratory organizers, Claire Taylor and Anne-Françoise Roux, and many others are responsible for the content and direction of the courses and the book.

We thank the authors for permission to adapt their teaching material to the printed page and Kristen Hauser and Luna Han at John Wiley & Sons for their patience and efficiency.

Graham R. Taylor
Ian N. M. Day

Part I

REVIEWS

1

GENOTYPE TO PHENOTYPE

Sue Malcolm
Clinical and Molecular Genetics Unit, Institute of Child Health, London, UK

INTRODUCTION

One might expect that close study of an undergraduate textbook would be all that is needed to understand this topic. If a codon is mutated to a stop codon, then full-length protein cannot be produced. If the start codon is mutated, then translation cannot take place, and if the third nucleotide of a codon is mutated so that the genetic code predicts no change in amino acid, then there will be no mutation. In reality, the situation is far from this simple, and exceptions to all the above exist and will be presented. The incredible complexity involved in splicing and the central importance of splicing mutations will also be discussed and provide a uniting theme.

It is also important to appreciate that specific mutations will have very different effects on the function of the protein that one expects to be reflected in a range of disorders corresponding to mutations in a particular gene. This is well illustrated in the craniosynostoses, which are autosomal dominant disorders in which there is premature fusion of the cranial sutures. These are associated with mutations in the fibroblast growth factor receptor (FGFR) gene family. The mutations found in Crouzon and Pfeiffer syndromes predominantly create or remove a cysteine to leave an unpaired cysteine in one of the three extracellular immunoglobulin-like domains. This results in cross-linking and covalent dimerization of subunits leading to constitutive activation of the receptor molecules, which become independent of the need for a ligand (fibroblast growth factor, FGF). On the other hand, when a specific mutation occurs in the linker region between the second and third loop a distinct clinical variant, Apert syndrome, occurs. In contrast, the specific amino acid

Guide to Mutation Detection, Edited by Graham R. Taylor and Ian N. M. Day.
ISBN 0-471-23444-3

substitutions cause enhanced or changed binding affinities to FGFs. The majority of cases are caused by either a Ser252Trp or a Pro253Arg mutation in FGFR2. The only other missense change leading to Apert syndrome is Ser252Phe. This will occur much less often than Ser252Trp because, as a result of the constraints of the genetic, code a double substitution, CG_TT, is required, but it is significant that both result in replacement of serine by a large hydrophobic residue. In one family where Ser252 mutated to leucine, which has completely different properties as an amino acid and is achieved by a single nucleotide substitution, the phenotype was so mild that some members were considered to be unaffected (reviewed in 1).

NONSENSE MUTATIONS

Contrary to intuition, the predominant consequence of nonsense mutations is not the synthesis of truncated proteins. The majority of nonsense transcripts are recognized and efficiently degraded by the cell via a pathway known as nonsense-mediated mRNA decay (NMD). This mechanism is thought to protect the organism from potentially harmful dominant-negative or gain-of-function effects of truncated proteins that could result if nonsense transcripts were stable (2).

Exon skipping often results when either the intronic g or t at a splice donor site is mutated. However, Dietz et al. (3) were the first to describe the skipping of constitutive exons in vivo induced by nonsense mutations. Maintenance of an open reading frame provides an additional level of scrutiny during splice site selection. A deletion found at the cDNA level should always be confirmed at the genomic level. It remains to be seen what proportion of these skipped exons actually arise because of disruption of an exonic splice enhancer (see below).

CHANGES INVOLVING SPLICING

Nucleotide changes within and around splice donor and acceptor sites have long been recognized as a major source of mutation in humans and are routinely screened for. However, they will seriously underrepresent the proportion of mutations that are in fact operating via the splicing system. A systematic study of mutations in ataxia-telangiectasia showed that 30/62 (48%) resulted from defective splicing (4), but fewer than half involved the canonical AG splice acceptor site or GT splice donor site. Others occurred at less stringently conserved sites, including silent mutations of the last nucleotide of exons, mutations elsewhere in the consensus splice sites, and creation of splice-acceptor or splice-donor sites in either introns or exons. These splicing mutations led to a variety of consequences, including exon skipping and, to a lesser degree, intron retention, activation of cryptic splice sites, or creation of new splice sites. Several cases of exon skipping in both normal controls and patients for whom no underlying defect could be found in genomic DNA were also observed, suggesting other so far undetected sites of mutation. Another cDNA-based study of neurofibromatosis identified splice mutations in 50% of patients for whom a mutation was found (5). Again, most of the mutations did not involve the conserved AG/GT motif. Several mutations within both exons and introns resulted in the activation of cryptic splice sites. For example, a mutation 332 bases into IVS30 resulted in the change of AGACataagt to AGACgtaagt, which gives a Shapiro and Senapathy (6) score of 86.49.

A recent interesting case involves two missense mutations in the Krit1 gene in families with cerebral cavernous malformations. The point mutations were predicted to lead to the changes D137G and Q210E, but further RNA analysis revealed that both point mutations actually activate cryptic splice-donor sites, causing aberrant splicing and leading to a frameshift (7).

More recently it has been demonstrated that some mutations within coding regions may in fact be exercising their effect through splicing by disrupting exonic splice enhancers (ESEs). Accurate splicing is a complicated business requiring an array of small nuclear ribonucleoproteins and other factors that are components of the spliceosome. ESEs are present in constitutive and alternatively spliced exons and are required for efficient splicing of those exons. The ESEs in pre-mRNAs are recognized by serine/arginine-rich (SR) proteins, a family of essential splicing factors that also regulate alternative splicing. ESEs contain a wide spectrum of sequences of approximately 6–8 nucleotides, but they are hard to detect because of their degeneracy. When missense mutations are identified in genomic DNA, particularly in a diagnostic laboratory, the usual interpretation is that the affected amino acid is crucial for the function of the protein, particularly if the change is nonconservative or the residue changed is highly conserved in evolution. But if these mutations disrupt an ESE, the consequence may be incorrect splicing of the transcript. Experimental evidence for this comes from studies in the breast cancer-causing gene BRCA1 (8). A Glu1694Ter nonsense mutation in exon 18, which causes skipping of the entire exon, has been found in a number of families with a breast cancer predisposition. Although it is difficult to search directly for ESE disruption because of the sequence degeneracy, the G_T change was predicted to disrupt a high scoring motif in a matrix analysis. Minigene constructs were tested for splicing in a HeLa cell nuclear extract. Exon jumping correlated with the disruption of high-scoring motifs rather than the presence of nonsense codons. It follows that the high-scoring motifs may also be disrupted by changes leading to missense mutations or silent third base changes. and because of the degeneracy of the ESEs these events may be frequent.

The effects on phenotype may be varied and subtle. The molecular basis for the distinction between Duchenne and the clinically less severe Becker muscular dystrophy is generally explained by the frameshift hypothesis. Both disorders can result from deletions in the dystrophin gene. but the outcome depends on the exact combination of exons removed. Deletions resulting in Duchenne have been shown to shift the translational open reading frame of triplet codons (because not all exons contain an exact multiple of three nucleotides), leading to in-frame stop codons and a truncated protein. Deletions identified in Becker result from a combination of exons that maintain the open reading frame and predict a shorter but partially functional protein. In a minority of Duchenne cases nonsense mutations are observed, again predicting a truncated protein. A Japanese patient was reported with a nonsense mutation in exon 27 of the dystrophin gene but a phenotype of Becker. It was shown by functional studies using an in vitro splicing system that the mutation led to exon skipping of exon 27 by interruption of an exon-specific enhancer.

Exon splice enhancers also provided the explanation for initially puzzling correlations of genotype and phenotype in spinal muscular atrophy (SMA). Two closely related survival motor neuron genes, predicted to encode identical proteins, map to the SMA locus. SMA is caused by mutations within the telomeric copy of the gene (SMN1), with 96% of patients showing a homozygous deletion of SMN1. Not only is SMN2 not able to compensate for loss of SMN1, but homozygous absence of

SMN2 is found in 5% of control individuals with no phenotypic effect. SMN1 produces full-length mRNA, but SMN2 expresses dramatically reduced full-length mRNA and abundant levels of incorrectly spliced transcript lacking exon 7. The splicing of exon 7 is determined by a single nucleotide difference at position +6 in exon 7 (C in SMN1 and T in SMN2), which disrupts an exon splice enhancer (9). A most exciting development for possible therapy has been the identification in *Drosophila* of a splicing factor, Htra2-β1, which promotes the inclusion of exon 7 into full-length SMN2 transcripts and works across species in human and mouse cells carrying an SMN2 minigene (10).

■ TROUBLE IN INTRONS

As mutation detection becomes more exhaustive, an increasing number of mutations are being reported deep within introns. This must pose enormous problems for diagnostic laboratories as the introns are in general many times longer than the exons and because functional tests for the significance of the change will be difficult.

A recent study of patients with the cardiac version of Fabry disease found a novel midintronic base substitution that causes an insertion of a "pseudo exon" of 57 base pairs between the normal exons 4 and 5 (11). The base change is not within the canonical GT/AG but instead increases the activity of a rare alternative splice site that normally only contributes <5% of the α-galactosidase mRNA levels but in this case produces >70% of the total α-galactosidase mRNA. Consequently, residual activity of the normal product is reduced to about 10% of normal, leading to a cardiac phenotype. The base change occurs within the additional exon, 4 base pairs before the 3′ end of it, and has been suggested to act by exon splice enhancement.

Exon splice silencers are even less well understood in humans, but a fascinating example was recently reported (12). A 4-base pair deletion within intron 20 of the ATM gene results in the incorporation of an additional pseudo-exon into the mRNA. The deletion was shown to occur in a new splice proccssing element that is essential for accurate intron removal and is part of a new U1 small nuclear ribonucleoprotein (snRNP) binding site.

■ HOW MANY HITS ARE NEEDED?

Increasingly, our simple models of autosomal dominant, recessive or, X-linked modes of inheritance are inadequate. Despite heroic efforts to unravel the molecular genetics of common disorders, progress has been slow. Two recent success stories give us some clue to why it has been so difficult. An association has been found between an insertion mutation in NOD2 gene and Crohn disease (inflammatory bowel disorder) in German and British populations (13). The variant is a C-insertion mutation in exon 11 in a leucine-rich region leading to a premature stop codon at amino acid 1007 of the protein and was initially submitted to a public database as a polymorphism. The nature of the change would imply that it is almost certainly pathogenic. This is indeed true, as a strong association was observed with the allele occurring at 4.6% in the general population and at 19% in the cases. However, even in such an apparently clear-cut case the extra risk conferred by carrying the heterozygous genotype is less than 3-fold, although the homozygous extra risk is 42-fold.

A second example has arisen during the search for minor genes contributing to breast cancer (14). CHEK2 is a cell cycle checkpoint kinase directly connected to the DNA damage response pathway. A previously described protein-truncating mutation in *CHEK2*, 1100delC, was found to segregate with breast cancer in a branch of a large family EUR60 that mapped to neither of the known high-penetrance breast cancer genes, *BRCA1* or *2*. An association between breast cancer and 1100delC was found in *BRCA1/BRCA2*-negative families compared with controls. The variant was found in 1.2% of the healthy population and 5.1% of the breast cancer patients. They estimated that the *CHEK2**1100delC variant results in an approximately 2-fold increase of breast cancer risk in women and a 10-fold increase of risk in men. Even in EUR60, it was carried by only 8 of the 17 affected women in the family.

Both these examples are proven examples of pathogenic mutations, and both are functionally nonsense mutations. However, the remarkably low penetrance shows how little we understand of the modifying factors, including environmental factors and modifying genes, that control the genotype-phenotype correlations.

A tantalizing glimpse of such interactions comes from studies showing the need for three mutant alleles that are emerging in the case of Bardet–Biedl syndrome (15).

■ ACKNOWLEDGMENTS

I thank Adrian Krainer for stimulating my interest in RNA splicing.

■ REFERENCES

1. Wilkie AOM, Morris-Kay GM (2001) Genetics of craniofacial development and malformation. Nat Rev Genet 2: 458–468.
2. Mendell JT, Dietz HC (2001) When the message goes awry: disease-producing mutations that influence mRNA content and performance. Cell 107: 411–414.
3. Dietz HC, Valle D, Francomano CA, Kendzior RJ, Pyeritz RE, Cutting GR (1993) The skipping of constitutive exons in vivo induced by nonsense mutations. Science 259: 680–683.
4. Teraoka SN, Telatar M, Becker-Catnia S, Liang T, Onengut S, Tolun A, Chessa L, Sanal O, Bernatowska E, Gatti RA, Concannon P (1999) Splicing defects in the ataxia-telangiectasia gene, ATM: Underlying mutations and consequences. Am J Hum Genet 64: 1617–1631.
5. Ars E, Serra E, Garcia J, Kruyer H, Gaona A, Lazaro C, Estevill X (2000) Mutations affecting mRNA splicing are the most common molecular defect in patients with neurofibromatosis type I. Hum Mol Genet 22: 237–247.
6. Shapiro MB, Senapathy P (1987) RNA splice junctions of different classes of eukaryotes: Sequence statistics and functional implications in gene expression. Nucleic Acids Res. 15: 7155–7174.
7. Verlaan DJ, Siegel AM, Rouleau GA (2002) Krit1 missense mutation lead to splicing errors in cerebral cavernous malformation. Am J Hum Genet 70: 1564–1567.
8. Liu H-X, Cartegni L, Zhang MQ, Krainer AR (2001) A mechanism for exon skipping caused by nonsense or missense mutations in BRCA1 and other genes. Nat Genet 27: 55–58.

9. Lorson CL, Androphy EJ (2000) An exonic enhancer is required for inclusion of an essential exon in the SMA-determining gene SMN. Hum Mol Genet 22: 259–265.
10. Hofmann Y, Lorson CL, Stamm S, Androphy EJ, Wirth B (2000) Htra2-β1 stimulates an exonic splicing enhancer and can restore full-length SMN expression to survival motor neuron 2 (SMN2). Proc Natl Acad Sci USA 97: 9618–9623.
11. Ishii S, Nakao S, Minamikawa-Tachino R, Desnick RJ, Fan J-Q (2002) Alternative splicing in the α-Galactosidase A gene: Increased exon inclusion results in the Fabry cardiac phenotype. Am J Hum Genet 70: 994–1002.
12. Pagani F, Buratti E, Stuani C, Bendix R, Dork T, Baralle FE (2002) A new type of mutation causes a splicing defect in ATM. Nat Genet 30: 426–429.
13. Hampe J, Cuthbert A, Croucher PJ, Mirza MM, Mascheretti S, Fisher S, Frenzel H, King K, Hasselmeyer A, MacPerson AJ, Bridger S, van Deventer S, Forbes A, Nikolaus S, Lennard-Jones JE, Foelsch U, Krawczak M, Lewis C, Mathew CG (2001) Association between insertion mutation in NOD2 gene and Crohn's disease in German and British populations. Lancet 357: 1925–1928.
14. Meijers-Heijboer et al. (2002) Low penetrance susceptibility to breast cancer due to CHEK2*1100del C in non-carriers of BRCA1 or BRCA2 mutations. Nat Genet 31: 55–59.
15. Badano JL, Kim JC, Hoskins BE, Lewis RA, Ansley SJ, Cutler DJ, Castellan C, Beales PL, Leroux MR, Katsanis N (2003) Heterozygous mutations in BBS1, BBS2 and BBS6 have a potential epistatic effect on Bardet–Biedl patients with two mutations at a second BBS locus. Hum Mol Genet 12: 1651–1659.

2

FREQUENCY AND NATURE OF MUTATIONS AND THE METHODS TO DETECT THEM

Mireille Claustres
Laboratoire de Génétique Moléculaire, IURC, Institut Universitaire de Recherche Clinique, Montpellier , France

INTRODUCTION

Main Classes of Mutations Occurring in Human DNA

The DNA of the human genome is subject to a variety of different types of heritable change (mutation). Any change in a specific DNA sequence is a mutation; however, the term "mutation" is generally used to mean disease-causing sequence change, whereas "allelic variant" means non-disease-causing change, although it may modulate the phenotype of a disease. Allelic sequence variation is usually defined as a *polymorphism* if more than one allele at a locus occurs with a frequency greater than 0.01. The mean heterozygosity for human genomic DNA is in the order of 0.001–0.004 (this means that 1/250 to 1/1000 bases are different between allelic sequences and these changes are mostly inherited because the mutation rates are low). Alternatively, *single nucleotide polymorphisms* (SNPs), which account for about 90% of polymorphisms and represent a natural genetic variability at high density in the human genome, can also convey enhanced or reduced susceptibility toward specific diseases by impairing or improving physiological function as well as modulating drug response. New mutations arise in single individuals, in somatic cells, or in the germ line, where they can be transmitted to offspring. Mutations can

Guide to Mutation Detection, Edited by Graham R. Taylor and Ian N. M. Day.
ISBN 0-471-23444-3

be induced in our DNA by exposure to a variety of mutagens occurring in our external environment or to mutagens generated in the intracellular environment. However, the greater source of mutations is spontaneous errors in DNA replication and repair. Coding and noncoding DNA are equally susceptible to mutation; however, the consequences are restricted to the 3% of our 3×10^9-base pair human genome that is coding DNA. Mutations either do not change the sequence of the gene product (*silent* or *synonymous* mutations) or result in an altered sequence in a polypeptide or functional RNA (*nonsynonymous* mutations, which can have no effect, a beneficial effect, or a deleterious effect on gene expression resulting in disease or lethality). The great majority of new nonsynonymous mutations in coding DNA reduce the fitness of their carriers. They are therefore selected against and removed from the population. Occasionally a new mutation may be advantageous in heterozygotes and spread in the population (ΔF508 of the CFTR gene).

Large-scale chromosome abnormalities include loss or gain of chromosomes or breakage and rejoining of chromatids. They are rare as constitutional mutations but are often pathogenic. They are much more common as somatic mutations and often found in tumor cells. Such DNA rearrangements may be nonrandomly distributed in the genome as a consequence of selection for growth advantage of the mutant cell. The location of these lesions is also influenced by the inherent potential of some DNA sequences to be involved in recombination (either homologous unequal recombination or nonhomologous recombination). To facilitate the search for sequence characteristics that render certain DNA regions prone to rearrangement, a special gross rearrangement breakpoint database (GraDB) has been established (1).

Smaller-scale mutations mostly occur at three types of DNA sequence at a gene locus: (i) the coding sequence of the gene, where the great majority of recorded pathogenic mutations have been identified, (ii) intragenic noncoding sequences, especially those that are necessary for correct expression of the gene, and (iii) regulatory sequences outside exons, notably promoter elements. Three classes of small-scale mutations can be distinguished: base substitutions, deletions or insertions.

Base substitutions involve replacement usually of a single base; in rare cases several clustered bases may be replaced simultaneously as a result of gene conversion events, which mostly occur at certain repeated sequences. Base substitutions are the most common sequence changes both in coding and noncoding DNA. Surprisingly, *transitions* (pyrimidine C or T replaced by a pyrimidine, or purine A or G replaced by a purine) are more common than *transversions* (pyrimidine replaced by a purine or conversely), especially in mitochondrial DNA. The excess of transitions over transversions is at least partly due to the high frequency of C→T transitions resulting from cytosine methylation and subsequent spontaneous deamination in the CpG dinucleotide, which is often a hot spot for mutation in human genes: Its mutation rate is 8.5 times higher than the average dinucleotide mutation rate (2). Substitutions result into three different classes of effects on the genetic code: (i) *nonsense* mutations (a codon specifying an amino acid is replaced by a stop codon; almost always associated with a dramatic reduction in gene function), (ii) *missense* mutations (the altered codon specifies a different amino acid that is either chemically similar to it (conservative substitution) or chemically different (nonconservative substitution), and (iii) *synonymous* (*silent*) substitutions, which do not change the sequence of the gene product because the new codon specifies the same amino acid; they often occur at the third base position of a codon ("wobble"); however, in rare cases these substitutions may not be as neutral as expected, causing disease by activating a cryptic splice site or creating an unstable mRNA. Nucleotide substitutions occurring in noncoding DNA usually are neutral variations. Exceptions include certain changes in regulatory elements or in canonic intronic sequence positions, such as at splice junctions (donor or acceptor sites) or at splice branch sites. Moreover, a proportion of base substitutions in intronic or even in exonic regions will cause failure of splicing or the use of an alternative illegitimate (cryptic) or natural splice site, resulting in some exon skipping or intron retention in the mRNA.

Base insertions: One or a few nucleotides are inserted into a sequence; in rare cases, this involves transposition from another locus via an RNA intermediate. Large insertions and duplications are rare, as are insertions of transposable elements.

Base deletions: One or a few nucleotides are eliminated from a sequence.

Microinsertions and deletions are very common in noncoding DNA but rare in coding DNA, where they produce frameshifts. Larger deletions are rarer and occur mostly at regions containing tandem repeats or between interspersed repeats. Slipped strand mispairing of short tandem repeats during DNA replication predisposes to frameshifting deletions and insertions.

Nonsense mutations, frameshifting deletions and insertions, and splicing mutations skipping a single exon containing a number of nucleotides that cannot be divided by 3 introduce a premature termination codon in the mRNA. The chain-terminating mutations have three major possible consequences: (i) unstable mRNA, which may be degraded in vivo by a mechanism called nonsense-mediated decay, (ii) truncated polypeptide, and (iii) skipping of constitutive exons.

Dynamic mutations (unstable triplets): These mutations can be stably transmitted from carriers to their offspring without any change through generations. Over the past 10 years a new class of mutations has been discovered, which consists of rapid expansion of intragenic or extragenic triplet repeats that can affect gene expression and cause disease. The expanded triplet is unstable, expanding (most often) or contracting (rarely) the number of its repeats when transmitted from parent to child. There are 10 different trinucleotide repeats in the human genome. Most of these are known as useful polymorphic markers. However, close to or within a few genes, certain triplets show abnormal behavior. Usually, their length (their number of repeats) is stable in mitosis and meiosis in the normal population; however, above certain threshold lengths (in certain cases, these alleles are considered as a *premutation*), the repeats become extremely unstable, depending on the sex of the transmitting parent as well as the length and composition of the repeat.

Genes showing unstable expanding trinucleotide repeats can be divided into two major classes:

1. Modest expansions of (CAG)n repeats within the coding sequence, on the order of 30 repeats for the polymorphic alleles and 40–200 repeats for the pathological alleles (e.g., Huntington disease); the resulting protein product with an abnormal polyglutamine tract shows a gain of function
2. Genes with very large expansions of a noncoding repeat that can be found in the promoter, the untranslated regions, or intronic sequences; unstable pathologic alleles have several hundreds (or thousands) of copies that inhibit gene expression, causing loss of function (e.g., Fragile X syndrome). Recently, other classes of expanded triplets have been discovered to cause disease, including polyalanine tracts.

Epigenetic Modifications Can Cause Disease

A variety of mechanisms that affect the chromatin environment of a gene and hence its capacity for gene expression have been discovered in the last few years. They can be responsible for genetic changes that are heritable although they are not directly attributable to DNA sequence changes.

DNA methylation is an important mammalian factor of gene control, maintaining repression of transcription. Pathogenic silencing by methylation occurs in many tumors and can also be triggered by a local expansion of a trinucleotide repeat as in the case of fragile X syndrome.

In genomic imprinting of genes a small number of genes are subjected to allelic exclusion according to the parent of origin, which reflects a nonequivalence in expression of the maternal and paternal genomes at these loci. Disorders of imprinting may be caused by deletions or point mutations in imprinted genes themselves, if an individual inherits two copies of one gene from one parent and none from the other (uniparental disomy) or if the imprinting center is not functioning normally. There are now several examples in which alterations of certain chromosomal regions produce different phenotype depending on the parent of origin. The best example is deletion of 15q12, which on the paternal chromosome produces Prader–Willi syndrome and on the maternal chromosome produces Angelman syndrome. The parental imprint is erased and reset between each generation during gametogenesis. The mechanisms that are able to distinguish between maternally and paternally inherited alleles are not deciphered, but allele-specific DNA methylation has been shown to be a key component in maintaining the imprinted status.

Changes in *large-scale chromatin configuration* can also exacerbate or silence expression of intact genes. Examples include position effects in aniridia (silencing of PAX6 gene) or in facioscapulohumeral muscular dystrophy (probable gene silencing by a 3.2-kb deletion in 4q telomeric repetitive sequences).

The Mitochondrial DNA Is Prone to Mutations

Because of the very large size of the nuclear genome, most mutations occur in nuclear DNA sequences. However, the maternally inherited mitochondrial genome (16,000 bases, about 1/200,000 of the size of the nuclear genome) is a hot spot for pathogenic mutations because of two main factors: (i) 93% of mitochondrial DNA is coding and (ii) the mtDNA has a higher mutation rate than the nuclear genome, not being protected by histones, undergoing many rounds of replication, and not being repaired as efficiently as the nuclear genome. Thus the mitochondrial genome is a significant cause of human genetic disease. A characteristic feature of mitochondrial diseases is their *matrilineal inheritance* (because sperm mitochondria are lost soon after fertilization, sperm do not contribute mitochondria in the zygote), affecting both sexes but transmitted only by affected mothers. The zygote receives its total complement of mtDNA ($>10^5$ copies) entirely from the cytoplasm of the oocyte. If a portion of these mtDNA molecules contains a mutation, the random segregation of mutant and wild-type mitochondria during mitosis will result in daughter cells with different proportions of mitochondrial mutations (*heteroplasmy*). Only a few mitochondrial disorders are characterized by a mtDNA mutation that is present in virtually all cells (*homoplasmy*).

100,000 Mutations Already Identified as Causing Disease

The degree to which a pathogenic mutation results in an aberrant phenotype depends on several factors including (i) the mutation class and the way in which the expression of the mutant gene is altered (loss of function, gain of function, dominant-negative effect, overexpression or ectopic expression, etc.), (ii) the degree to which the aberrant phenotype is expressed in the heterozygote, and (iii) the proportion and nature of cells in which the mutant gene is present (inherited or somatic mutations) and, for a few number of genes, the parental origin of the mutation (a few genes are imprinted). Since Haldane first observed that most de novo mutations resulting in hemophilia were generated in the male germ line, direct observation of disease-causing mutations supports higher mutation rates in males, most likely to be related to the greater number of germ cell divisions in males.

Since 1979, where the first single-base pair disease-causing substitution was described for β-thalassemia, about 100,000 mutations have so far been reported in 2000 genes (data extrapolated from locus-specific mutation databases contents, see HUGO HGVS website). The spectrum of DNA disease-producing mutations ranges from large gene alterations (insertions, deletions, duplications, inversions, expansion of trinucleotide repeats)to microdeletions or -insertions and single-base substitutions. Nearly all common monogenic diseases (200–250) and a large number of rare ones (1000–1200) have been traced back to one or more defective genes (3000–4000). New genetic mechanisms have been uncovered, such as genomic imprinting, triplet-repeat expansion, deficiencies of DNA repair, and methylation defects.

The number and frequency of disease-causing or predisposing alleles varies enormously from locus to locus. Rare diseases are generally caused by a wide range of different mutations, with each mutation constituting only a small fraction of the spectrum. Some disorders, such as cystic fibrosis, are associated with a small number of common mutations and a background of more than 1200 very rare alleles already identified. The *common disease/common variant* (CD/CV) hypothesis, proposed several years ago (3, 4, 5) predicts that the genetic risk for common disorders will often be due to one or a few predominating disease alleles. Some prototypical examples include the apoE E4 allele in Alzheimer disease or factor V Leiden in deep venous thrombosis. Reich and Lander (6) provided a framework for understanding and predicting the allelic spectra of disease genes, assuming that the human population experienced a dramatic expansion from a small, ancestral population with a size on the order of 10,000 individuals to the modern population size of 6 milliards that began between 20,000 and 150,000 years ago. According to their hypothesis, the increase in allelic diversity will occur rapidly for the rare diseases but will require hundreds of thousands of years after the expansion for the common diseases, which is rather optimistic in the issue of mutation detection.

Genetic mosaicism can be defined as the presence in a single individual of two genetically distinct populations of cells that differ from each other at the level of DNA sequence but that derive from a single zygote. Germ-line and somatic mosaicism have now emerged as important factors that contribute to phenotypic variability. There are increasing reports of multiple different types of mosaicism detected in patients with inherited and noninherited disorders including chromosomal mosaicism, mitochondrial mosaicism (heteroplasmy), germ-line mosaicism (de novo mutation), somatic mosaicism (due either to de novo mutation during embryogenesis or to reversion to normal of inherited mutations), and mosaicism of neoplasia [where DNA changes occur in the tumor cells but not in the other cells of the body, such as loss of heterozygosity (LOH) by mitotic recombination or microsatellite instability in HNPCC syndromes] (7, 8). Germ-line mosaicism is suspected when several offspring of unaffected parent manifest the phenotype, which is otherwise transmitted as an autosomal dominant or X-linked disorder. Phenotypically normal individuals might transmit several gametes that are clonal descendants of a single progenitor cell in which a de novo mutation occurred during the early development of a parent. There is now a long list of monogenic disorders in which somatic mosaicism has been shown, including metabolic diseases, skeletal dysplasias, muscular dystrophy (DMD), clotting disorders, and tumor suppressor disorders. Mosaicism is underestimated, because genetic testing for mutations is made by analyzing DNA in blood cells, in which a low level of mosaicism will remain usually undetected.

Detecting Mutations: A Combination of Scanning and Screening

The first area benefiting directly from the discovery of causative mutations is that of molecular diagnostics in affected families. The ability to determine who is and who is not at risk for disease, sometimes before the onset of symptoms, is becoming essential for proper management of patients and their families. In individuals who inherit mutant genes, preventative measures might reduce morbidity and mortality and allow more thoughtful planning for the future. The benefits of genetic testing are equally important for those family members who are found not to carry the relevant mutation: These individuals are spared unnecessary medical procedures and tremendous anxiety.

In a few number of diseases there are sites within the gene that are mutated in virtually all cases. Examples include sickle cell disease, in which an A to T transversion at codon 6 of the β globin gene is ever-present, and Huntington disease, in which virtually all patients have an expanded tract of CAG trinucleotide repeats, creating a long polyglutamine stretch within the huntingtin protein. These mutations can be readily identified with assays designed to detect the specific alterations. However, the majority of inherited diseases are caused by diverse mutations scattered along the length of the affected gene. In non-genetically isolated populations without a history of severe bottlenecks, many highly penetrant disease genes have complex mutation spectra. The nature of mutations is diverse, and the ease of detecting them stretches across a continuum between "compliant mutations", which can be detected or identified by using routine PCR-based genetic tests, and "refractory mutations" (including deletions encompassing one or more exons, insertions, or translocations or a variety of alterations that affect the expression of the gene at the RNA or protein level), which will not be detected by routine testing (9).

This has led to the development of two broad categories of mutation detection technologies: The first group, designed to scan for unknown mutations within a gene, are highly informative but are tedious and are incompatible with high throughput and low cost. In the second group, direct methods focused on specific mutations have been developed; together with robotics, these methods for direct mutation analysis have helped in reducing cost and increasing throughput, but only a limited number of mutations can be analyzed. Because of the extreme diversity of DNA alterations, no single method of detection is applicable for all situations, and clinical laboratories often have to combine several techniques in order to screen for all possible changes in the homozygous and heterozygous states.

DETECTION OF KNOWN MUTATIONS

Testing for the presence or absence of a known sequence change is a much simpler problem than scanning a gene for the presence of any mutation. Consequently, a number of specific mutation detection assays are commercially available. In one set of methods, mutations are analyzed after the target sequence has been amplified by PCR: They are detected by restriction digest, by allele-specific hybridization, or by ligation or nonligation of adjacent probes. In a second set of methods, PCR is part of the detection system, and these methods rely on the selective extension of primers or on real-time quantitative analysis of the products. Non-gel-based detection systems have been developed for most of the assays described, making these methods favorable for application in routine laboratory settings. For each technique, reaction conditions must be standardized and appropriate internal controls must be included. One must keep in mind that misleading results may be obtained because of unknown polymorphisms within the target region affecting either restriction enzyme recognition sequences or hybridization of probes and binding of primers.

Restriction digestion analysis of PCR-amplified DNA has been the simplest way of testing for the presence or absence of a mutation when it creates or abolishes either a natural restriction site or one engineered by a form of PCR mutagenesis using a primer carefully designed to have a mismatched nucleotide that together with the mutant sequence creates a restriction site not present in the normal allele. Gains in throughput can be achieved using high-density gels such as the *microtiter array diagonal gel electrophoresis* (MADGE) format (10).

A number of methods rely on hybridization with specific oligonucleotide probes that can effectively discriminate between the wild-type and variant sequences.

Allele-specific oligonucleotide (ASO) *hybridization.* Using suitably stringent hybridization conditions, short allele-specific synthetic probes hybridize to the target sequence only when they match perfectly. Wild-type or mutant ASO probes are usually designed to have differences in the central segment of the sequence in order to maximize thermodynamic instability of mismatched duplexes. The original dot blot or slot blot formats have evolved into nonradioactive detection in reverse dot blots and multiple solid-phase or other multiplex allele-specific diagnostic assays. ASO hybridization is also the principle on which the design of DNA chips was based.

Allele-specific amplification (ASA). Complementary to either the wild-type or the mutant sequence, ASA has been assayed through several approaches, of which the most popular is the *amplification refractory mutation system* (ARMS) (11). In ARMS there is a lack of primer elongation due to a mismatch at the far 3′-end of the primer. Paired PCR reactions must be carried out with a common primer and either the primer for the normal sequence or the primer for the mutated sequence. The technique can be multiplexed to type up to 20 to 30 mutations simultaneously (e.g., cystic fibrosis mutations), and high-throughput formats can be achieved with closed-tube assays. Other methods rely on a lack of primer annealing due to a mismatch located within the primer.

Single nucleotide primer extension (PEX), or *"minisequencing."* Extension of the primer by one base occurs only when the labeled nucleotide is complementary to the nucleotide of the target DNA adjacent to the 3′-end of the primer. When incubated in the presence of dideoxynucleotide triphosphates labeled with different fluorophores, the allele-specific dye-labeled dNTP is linked to the primer in the presence of DNA polymerase and target DNA. Simply exciting the fluorescent dye in the reaction and determining the change in fluorescence can determine the genotype of the target DNA molecule. PEX is a very robust allelic discrimination method. It is highly flexible and requires the smallest number of primers/probes. Probe design and optimization of the assay are usually very straightforward. Minisequencing can be run on DNA sequencers, such as the "SnapShot" assay from Applied Biosystems. Mass spectrometry, in which molecules are ionized and volatilized in a vacuum and separated by their mass-to-charge ratios, offers an alternative highly automatable detection scheme using *matrix-assisted-laser desorption ionization–time-of-flight–mass spectrometry* (MALDI-TOF-MS). A powerful multiplex SNP analysis using solid-phase-capturable biotinylated dideoxynucleotide terminators was recently designed (12), in which measuring the mass of purified primer-extension products identifies the nucleotide at the polymorphic site. Both homozygous and heterozygous genotypes for human hereditary hemochromatosis C282Y and H63D were clearly determined.

Allele-specific ligation: Oligonucleotide ligation assay [OLA (13)]. Three oligonucleotides are used in OLA: two allele-specific probes (one specific for the wild-type allele and the other specific for the mutant allele) plus a fluorescent common probe. The forward primer has its 3′ base placed at the position of the variation and located immediately adjacent to the 5′-end of the common probe. Under appropriate conditions, ligation of both primers by DNA ligase will take place only when the 3′-end of the primer matches perfectly with the target sequence. The first step is PCR, the second ligation. The ligation products are then separated by electrophoresis. Solid-phase formats, single-tube multiplex amplification, and multicolor fluorescence detection on automated sequencers have been described. (By varying the combinations of color dyes and probe lengths, multiple mutations can be detected in a single reaction; see, e.g., the OLA test for 31 common CFTR mutations).

Real-Time or Homogeneous Solution Hybridization Mutation Typing Methods

Many of the techniques used routinely in a wide range of molecular genetics rely on PCR amplification of a target molecule followed by a variety of postamplification analyses to detect the specific analyte amplified. Development of thermocyclers that can take fluorescence measurements during PCR amplification has enabled genetic testing in single closed-tube reactions, with the ability to amplify, detect, and quantify DNA targets in the tube as PCR proceeds (reviewed in 14). Advantages over traditional methods include speed, improved reproducibility, reduced risk of contamination, possible automation and high-throughput processing in microplates, and the ability to more accurately quantify the amount of starting material present because the methods allow for "real-time" monitoring of the entire PCR reaction. The signal detected directly correlates with the amount of PCR accumulating, which in the log-linear phase of the amplification process is dependent on initial target copy number. Performing measurements in the log-linear phase of amplification enables more accurate quantification than was previously possible with gel-based techniques that examine end point reactions. Since its development in the 1990s, many different assay formats have been developed, and the number of real-time PCR machines of different design is continuously increasing. A recent review provides a survey of the instruments and assay formats available and discuses the pros and cons of each (15). The current chemistries used to monitor the amplification process include either nonspecific double-stranded DNA-binding dyes such as SYBRGreen, which produces enhanced fluorescence signals on binding to double-strand DNA duplexes as PCR proceeds and is mostly used for gene dosage, or specific probe hybridization-based systems that rely on the principle of *fluorescence resonance energy transfer* (FRET) for signal generation and are mostly used for genotyping. FRET is a process in which energy from an excited donor fluorophore is transferred to a nearby acceptor dye. When two fluorophores whose excitation and emission spectra overlap are in close physical proximity, excitation of one fluorophore will cause it to emit light at wavelengths that are absorbed by and that stimulate the second fluorophore, causing it to fluoresce.

The *LightCycler* system uses two adjacent probes that are labeled such that when both probes are hybridized to the target, the labels are brought close to each other; the donor fluorophore is excited by the light source of the instrument, and energy is transferred from the donor to the acceptor, producing an increase in fluorescence from the acceptor fluorophore. The system is programmed to monitor the melting curve analysis of allele-specific FRET probes after the PCR, allowing direct typing of sample without any further processing.

TaqMan probes, Molecular Beacons, and *Scorpion primers* rely on the close proximity of donor fluorophore and nonfluorophore acceptor molecules (quenchers). In the unhybridized probe conformation, little or no signal is generated (the fluorophore of the donor is quenched). On hybridization to the target, the fluorophore and quencher become separated through either conformational changes that occur (*Molecular Beacons* and *Scorpion primers*) or enzymatic cleavage of the fluorophore from the quencher as a result of the 5′ nuclease activity of *Taq* polymerase (*TaqMan probes*). The physical separation of the fluorophore and quencher moieties produces an increase in signal. Mutation typing is determined by measurement of the signal intensity of the two allele-specific reporter dyes after PCR amplification. Because of the presence of a hairpin-loop structure, *Molecular Beacons* and *Scorpion primers* might have a higher specificity than *TaqMan* probes.

Allele-specific nucleotide incorporation. Pyrosequencing is a real-time sequencing method in which DNA polymerase catalyzes the incorporation of a dNTP during the extension of the sequencing primer only if it is complementary to the base in the template strand (16). Unlike traditional sequencing, it does not require any electrophoresis. The generation of pyrophosphate when the dNTP is incorporated is coupled to a luciferase-catalyzed reaction, resulting in light emission and yielding a quantitative and distinctive pyrogram. Genotyping of 34 common mutations in the CFTR gene is commercially available, and current instrumentation sold by Pyrosequencing AB (Uppsala, Sweden) has been adapted for high-throughput detection of SNPs.

High-Density Oligonucleotide Arrays (DNA Chips) Promise High-Throughput Mutation Typing

Hybridization chips contain hundreds of oligonucleotides matching normal or mutant sequences in a gene or in several genes, thus offering simultaneous analysis of many polymorphisms or mutations. The DNA sample of interest is PCR amplified to incorporate fluorescently labeled nucleotides and then hybridized to the array. Each oligonucleotide in the high-density array acts as an allele-specific probe.

Perfectly matched sequences hybridize more efficiently to their corresponding oligomers on the array and, therefore, give stronger fluorescent signals over mismatched probe-target combinations. The hybridization signals are quantified by high-resolution fluorescent scanning and analyzed by computer software.

Minisequencing chips use arrayed oligonucleotide primers with free 3′-OH ends. Unlabeled PCR-amplified test DNA is hybridized to the array and acts as template for addition of a single labeled dNTP to each arrayed primer: Extension will occur only if the 3′ end of the primer perfectly matches the template.

Although DNA chip technology is in commercial use and has yielded vast amounts of genetic and cellular information, major improvements are needed to make this technology an accurate and inexpensive tool for the detection of multiple types of homozygous or heterozygous mutations in the diagnostic laboratories (17, 18, 19).

Mutation Detection Without PCR Amplification

Two methods are sensitive enough to work directly from genomic DNA.

The *Invader* assay (Third Wave Technology) is a FRET-based method in which two oligonucleotides (a wild-type or variant signal probe) plus an upstream Invader probe are used in each reaction. The signal probe contains a portion perfectly matched to the target and a fluorescently labeled flap. These probe hybridize in tandem to a specific region of genomic DNA. When the 3′-end of the Invader oligonucleotide overlaps the hybridization site of the 5′-end of the downstream signal probe by at least 1 bp (when the upstream probes "invades" at least one nucleotide into the downstream duplex), the unique structure will be recognized and cleaved by Cleavase enzymes. The cleaved signal probe dissociates, and a new signal probe anneals and is subsequently cleaved. The process repeats multiple times, resulting in an accumulation of reporter molecules and producing linear amplification of the fluorescent signal. A single nucleotide mismatch between the Invader probe and the template renders the conformation unrecognizable by Cleavase and there is no signal generated. Thus, instead of amplifying the target, the Invader procedure results in the accumulation of signal molecule only in case of a perfect match between the downstream signal probe and the target.

The *rolling circle amplification* (RCA) method is based on ligation-dependent circularization of allele specific DNA probes followed by exponential rolling circle amplification of the circularized probes (also called "padlock" probes). Mutations are interrogated with open circle probes that can be circularized by DNA ligase when the probe matches the genotype. An amplified detection signal is generated by *exponential rolling circle amplification* (ERCA) of the circularized probe. The ERCA products for normal and mutant alleles incorporate a different fluorescent primer.

PNA and LNA Probes

Sequence-specific base pairing between the strands of DNA according to the Watson–Crick model forms the basis of many detection systems. The crucial specificity of this hybridization reaction in discriminating between single base variations may be enhanced by using synthetic *peptide nucleic acids* (PNA) and *locked nucleic acid* (LNA) profluorescent probes (20, 21).

PROCEDURES FOR DETECTION OF UNKNOWN POINT MUTATIONS: PCR-BASED SCANNING METHODS

Direct detection of mutations is not simple, for many reasons that are related both to genes themselves and to technologies. It is the scanning of large genes for private mutations that is the rate-limiting step. For a number of diseases, complete sequencing of large genes cannot be a routine procedure in clinical laboratories, not only because it is time-consuming, costly, and tedious, but also because mutations can be missed. Although the introduction of capillary electrophoresis in sequencing technology has improved and simplified the process dramatically, DNA sequencing, like any other technique, does not detect 100% of point mutations. Moreover, depending on the size of the genes, it may not be practical to carry out sequencing of vast stretches of DNA without knowing whether or not mutations exist and without knowing in advance the sites of mutations in the target genes. The alternative approach is to amplify exons and introns of interest by PCR from genomic DNA, scan the PCR products for the presence of DNA variants by a scanning method, and then use limited sequencing to confirm and identify the mutation in the abnormal fragment (most scanning methods do not identify the precise nature of the mutant allele). The requirements for clinical routine diagnostic use and for research use are rather different in terms of mutation detection sensitivity and specificity, speed, and reproducibility of the assay. In the diagnostic lab, considerable time can be spent optimizing a test for a particular gene, provided the method once developed is quick and simple.

Methods that Detect Conformational Changes

Conformation-based methods of mutation detection rely on the fact that DNA fragments containing a sequence alteration have altered mobility under certain conditions of gel electrophoresis. A shift in the position of the band representing the mutant DNA relative to the position of the band for normal DNA connotes the presence of a mutation somewhere within the test fragment, but it does not indicate the precise position or identity of the mutation. A single-base mismatch can produce conformational changes in the double helix that cause the differential migration of homoduplexes and heteroduplexes containing mismatches. Several methods (DGGE, HA, DHPLC) rely on formation of heteroduplexes. Most mutations occur in heterozygous form; for homozygous mutations or X-linked mutations in males, it is necessary to add some reference wild-type DNA.

The first conformation-based procedure introduced for mutation detection was *denaturing gradient gel electrophoresis* (DGGE) (22, 23), which detects mutations as differences in partial melting behavior of double-stranded DNA and, hence, differences in migration distance when subjected to increasingly denaturing conditions, because regions of single-stranded DNA within double-stranded fragments greatly retard the migration of DNA through acrylamide gels. Visual inspection of the gel immediately detects heterozygote samples (presence of homo- and heteroduplexes). Primers are made with GC clamps or with modifiers that cross-link the ends of the PCR product; these clamps reduce the likelihood of complete denaturation of the PCR product. Theoretical melting profiles of each DNA fragment can be predicted by appropriate computer programs (MELT87 and SQHTX). However, running conditions must be defined for each PCR product and these laborious electrophoretic methods are evaluated by visual inspection, which makes standardization difficult. It is now well established that denaturing (DGGE) or *temperature gradient gel electrophoresis* (TGGE) (24) detect close to 98% of point mutations in fragments under 400 bp in size and under optimal conditions, so that it is appreciated by clinical laboratories. Some multiplexing of PCR products is possible, allowing the evaluation of several exons simultaneously. However, the principal disadvantage of DGGE is that it is time-consuming and of low throughput.

The DGGE method has seen interesting recent adaptations such as *constant denaturant capillary electrophoresis* (CDGE) (25) and *two-dimensional electrophoresis* (this latter combines the high resolution of 2-D electrophoresis with DGGE) (26).

Heteroduplex analysis (Het or *HA)* is a very simple method, performed in nondenaturing gels, that carries the advantage that the assay conditions do not have to be determined for each DNA fragment. However, it detects fewer mutations than DGGE. A variant of HA analysis is *conformation-sensitive gel electrophoresis (CSGE)*, in which an appropriate system of mildly denaturing solvents can amplify the tendency of single-base mismatch to produce conformational changes (27).

Denaturing high-performance chromatography (DHPLC) (reviewed in 28, 29; 30) exploits the differential retention of homoduplex and heteroduplex molecules under partially denaturing conditions during reverse-phase ion-exchange HPLC. It combines the nonionic size separation of DNA fragments with thermal denaturation. In addition to the fact that temperature conditions can be accurately reproduced, the main advantage over electrophoretic approaches is that the procedure can be automated for high-throughput mutation or SNP scanning. The wave DNA analysis system (commercialized by Transgenomic, Inc) uses *temperature-modulated heteroduplex analysis* (TMHA). The heteroduplexes can be separated from the homoduplexes by column chromatography at a temperature that partially denatures the mismatched DNA. Fully denaturing temperatures allow for separation of single-stranded molecules in applications such as oligonucleotide purification or primer extension assays, whereas partially denaturing conditions permit accurate separation of sequence variants. WaveMaker software provides instrument and analysis control to precisely optimize assay conditions. The system offers a powerful alternative to gel electrophoresis with rapid protocols and high-throughput capabilities, so that it has become the most popular method for mutation scanning in both research and diagnostic laboratories. The main disadvantage is the high cost of the system and the reagents.

Single-strand conformation polymorphism (SSCP) or *analysis* (SSCA) (31) has been the most widely used method for scanning genes, because of its technical simplicity. It relies on the principle that secondary structure contributes to the mobility of single-stranded DNA in electrophoretic gels under nondenaturing conditions. Single-stranded DNAs are generated by denaturation of the PCR products and separated on a nondenaturing polyacrylamide gel. A fragment with a single base modification generally forms a different conformer and migrates differently compared with wild-type DNA. However, it has a major limitation because of the unpredictable electrophoresis behavior of the mutant strands, so different sets of conditions of electrophoresis (e.g., gel temperature and/or presence of glycerol) must be used to achieve maximum sensitivity. Furthermore, the sensitivity decreases as the size of the fragment being analyzed increases. Multiple refinements and improvements of SSCA have been described, including methylcellulose as a molecular sieving agent and post-PCR fluorescent labeling dye. Reduction in migration times and better resolution and reproducibility may be achieved with capillary electrophoresis and fluorescence detection (18, 32). Variations of the method include *dideoxy fingerprinting* (ddF) (33) and *bi-directional dideoxy fingerprinting* (bi-ddF), in which dideoxy chain-terminated products are analyzed by SSCP.

Base Mismatch Recognition (Cleavage- or Binding-Based Methods)

Several techniques exploit the fact that mismatched bases are sensitive to binding or cleavage by enzymes or chemicals. After PCR amplification, the wild-type and variant products are subjected to denaturation/renaturation to create heteroduplex molecules. After incubation with resolvases or chemicals, the products are resolved electrophoretically side by side to score for the presence of mismatch-cleaved DNAs. Cleavage methods mostly work on heteroduplex molecules, cleaving the helix at the distorted region caused by the mismatch. This means that homozygous or hemizygous mutations would not be detected unless the sample is hybridized with a reference sample to form the mismatched heteroduplex. These methods are able to scan larger fragments than conformation-based techniques, and the size of the cleaved product roughly indicates the localization of the mutation; however, they require considerable post-PCR manipulation.

Ribonuclease mismatch cleavage. The single-strand specificity of RNase has been utilized to digest RNA:RNA or RNA:DNA heteroduplexes. The technique has been adapted to commercially available nonisotopic RNase cleavage assays (NIRCA) (34) with higher mutation rates than in the original versions. However, the major drawback of these methods is that they require in vitro synthesis of RNA.

The chemical cleavage method (CCM) (35) employs hydroxylamine and osmium tetroxide to modify DNA at the site of base mismatch. The modified DNA is then cleaved with piperidine, and the reaction products are separated by PAGE. CCM can detect mutations in DNA fragments up to 1.4 kb in length with almost 100% efficiency. CCM has evolved over the years from radiolabeled to nontoxic fluorescently labeled multiplex methods. However, CCM is not well suited to routine lab uses because of the number of analytic steps.

Enzyme mismatch cleavage (EMC) by resolvases (36) and other endonucleases including single-strand-specific nucleases (37) holds promise for the future, because standardization and automation with capillary electrophoresis and microdevices should be achieved with relative ease. However, more experience is needed with these methods, particularly for the analysis of heterozygous states.

Base excision sequence scanning (BESS-T and BESS-G) (38), also known as glycosylase-mediated polymorphism detection assay (GMPD), does not require the formation of heteroduplexes: It detects mutations by cleavage of PCR products at modified nucleotides (for instance, thymidine replaced by uridine), generating defined series of fragments similar to a sequencing ladder. The usage of uracylglycosylase may be interesting because in about 80% of all base substitutions, a thymidine is involved at the sense or the antisense DNA strand.

Mismatch binding proteins (MutS and homologs) have been used either in direct or in exonuclease protection assays. Given the normal role of bacterial mismatch-repair proteins to replace misincorporated bases in newly synthesized DNA, extensive efforts have been made to adapt these proteins for mutation detection assays. These proteins bind to the site of mismatch and do not cleave the DNA.

Mutation Detection at the Peptide Level

The *protein truncation test* (PTT) (39) selectively detects translation-terminating mutations, which are revealed on the protein level by SDS-PAGE. PCR is performed with RNA or DNA using forward primers containing a T7 promoter sequence and an eukaryotic translation initiation sequence. Subsequently, PCR products are used as templates in coupled transcription-translation reactions. The size of labeled translation products is analyzed by gel electrophoresis: Stop codons generated by frameshift or nonsense mutations lead to a reduced size of translated proteins. The assay is best suited for diseases in which translating termination mutations predominate, as it avoids the confusion with silent polymorphisms. The PTT assay is usually performed on cDNA copies of the mRNA of interest, where segments gathering together many exons can be scanned in only one reaction. For instance, RT-PCR-PTT has been successfully applied to the detection of point mutation in the large DMD (Duchenne muscular dystrophy) gene. In some cases, it can be done directly from genomic DNA to monitor large exons of several kbs. The translated products may be visually inspected or separated by chromatography or capillary electrophoresis, allowing an objective measurement of truncated proteins. A recent modification using *N*-terminally tagged test peptides and measuring peptide masses by MALDI-TOF should allow detection of not only truncations but also amino acid substitutions (40). Recently a *high-throughput solid-phase protein truncation test* (HTS-PTT) has been described in which capture and detection of translation products are accomplished in a single well with a standard 96-well microtiter plate *enzyme-linked immunosorbent assay* (ELISA) format and chemiluminescence readout (41).

Direct Sequencing

Direct DNA sequencing is, in theory, the most accurate technique for point mutation detection. However, even the present state-of-the-art capillary Sanger-DNA sequencing with laser-induced multicolor fluorescence detection faces difficulties in accurately sequencing GC-rich regions owing to the compression of the DNA fragments. In cases where the mutation affects only one of the two alleles of the gene, the peak heights corresponding to wild-type and mutant alleles are not always present at the predicted ratio of 1:1. Therefore, most genetic testing laboratories sequence both strands of DNA to achieve maximum sensitivities and to provide independent confirmation of any putative mutation identified. Direct sequencing of PCR products is often used as a final step of any mutation scanning procedure, both confirming and identifying the sequence alteration.

Pyrosequencing is an alternative sequencing approach, in which each nucleotide is identified as soon as it has been incorporated into a growing DNA strand by DNA polymerase (42). DNA polymerase catalyzes the incorporation of the dNTP into the DNA strand if it is complementary to the base in the template strand. Each incorporation event releases a pyrophosphate, which is converted to ATP by a sulfurylase. This ATP then drives the luciferase-mediated conversion of luciferin to oxyluciferin that generates visible light in amounts that are proportional to the amount of ATP. ATP and unincorporated dNTPs are continuously degraded by apyrase. The light is switched off, and the next dNTP is added. As the process continues, the complementary DNA strand is built up and the nucleotide sequence is determined from the signal peak in the pyrogram.

The next wave of sequencing could be mass spectrometry (MS). Because DNA sequences are determined by accurately measuring the mass of the DNA fragments, DNA must be stringently purified and be free from salts and other contaminants. In *solid-phase-capturable* (SPC) sequencing (biotinylated dideoxynucleotide terminators) (43), DNA sequencing fragments that are terminated by biotin-ddNTPs at the 3′-end in a cycle-sequencing reaction are bound to a streptavidin-coated solid phase. All excess primers, salts, and falsely terminated DNA fragments are washed away to provide pure DNA samples for MALDI-TOF MS detection. Segments up to 100 bp can be read in a very short time. Moreover, in case of frameshift mutation in the heterozygote, the sequences of both alleles can be simultaneously read and the mutation site can be unambiguously identified (44).

mRNA Analysis

Mutation detection in clinical diagnostic labs is applied predominantly to patients' genomic DNA. Although this approach is expected to detect the majority of mutations causing single gene disorders, an increased number of mutations are coming to light that will not be detected with standard mutation screening methods on genomic DNA. To offer comprehensive mutation screening in genes with large introns such as the dystrophin gene, an RNA-based approach is likely to be required. If a gene with many exons has to be scanned for unknown mutations, testing by RT-PCR offers several advantages: (i) the coding sequence can be scanned through a much smaller number of PCR reactions than genomic DNA, (ii) the procedure can detect aberrant splicing caused by sequence changes in introns or in exons, and (iii) some large deletions or duplications that would remain undetectable with DNA testing can be detected by aberrant sizes of RT-PCR products. However, RNA is more difficult to obtain and handle because of rapid degradation of mRNA and low expression in available tissues including illegitimate or ectopic expression from white blood cells. A potential limitation is unstable mRNA generated by some mutations (truncative mutations can result in nonsense-mediated mRNA decay) so that the RT-PCR product from an heterozygous individual may show only the normal allele. It is possible to confirm the expression of the two alleles through heterozygosity for a polymorphism.

A potential advantage of mRNA analysis is avoidance of the coamplification of a gene with pseudogenes (in the genomic DNA approach, it is necessary to design primers specific to the gene and not to its pseudogenes).

Sequencing RT-PCR products without previous scanning may be, for some genes, the method of choice to detect and identify the mutation in a single step. However, much time can be wasted investigating artifacts, especially if the sequencing template is not of the highest quality. Whatever the method, it is necessary to confirm RNA alterations in genomic DNA of the patients. The Estivill group reported the first description of an alternative splicing induced by cold shock conditions in a human pre-mRNA. The accuracy of the splicing mechanism was perturbed after samples were kept for a short period of time at room temperature, resulting in the insertion of a 31-bp cryptic exon in NF1 mRNA (45). Although rare, this phenomenon must be taken into account when using RNA technology as it can lead to misdiagnosis of a disease-causing mutation.

■ PROCEDURES FOR THE DETECTION OF REFRACTORY MUTATIONS

Many mutations are difficult or impossible to detect with the techniques described above. For example, if the genomic region examined is deleted from the mutant allele, the PCR product that is obtained from genomic DNA will be exclusively derived from the wild-type allele, leading to the false conclusion that this region is normal. The type of mutations that are refractory to detection include deletions encompassing one or more exons, insertions, translocations or a variety of alterations that affect the expression of the gene at the RNA or protein level. Intronic mutations that affect splicing can be revealed through the analysis of RNA. Polymorphisms, when present within an mRNA transcript, can be used to assess relative levels of expression of the two alleles of the gene. Quantitative hybridization, quantitative PCR, or Southern blotting can theoretically detect deletions of one or a few exons. Gross rearrangements are easily detected at the cytogenetic level by using FISH-based methods (deletions or duplications on the order of megabases), array-CGH-based methods (deletions or duplications on the order of 200–500 kb) or Southern blotting (deletions or duplications on the order of kilobases). The Southern blot method is on the way to being abandoned by most diagnostic laboratories because it requires large amounts of DNA of good quality, uses radioactive probes, and takes several days. Consequently, as midsize deletions or duplications are rarely screened, they are usually underestimated. The publication of the draft sequence of the human genome has facilitated the development of novel techniques for assessment of copy number at multiple loci (reviewed in 46).

Quantitative real-time PCR is now the method of choice for absolute and relative quantification of DNA and RNA template molecules in a variety of applications including allele imbalances, detection of deletions in genes, and gene dosage. Real-time PCR is also becoming increasingly important in the diagnosis of tumors, such as for the detection and monitoring of minimal residual diseases and the identification of micrometastases in colorectal cancer, neuroblastoma, and prostate cancer. It has been used to quantify amplifications of oncogenes or deletions of tumor suppressor genes in tumor samples. Also, the results of DNA chip experiments are validated by real-time PCR quantifications. However, these technologies pose many challenges in the clinical diagnostic setting, where the interpretation of data generated can be affected by many factors including the cutoffs to distinguish positive from negative samples, the assay controls and quantification calibrators used, and the ability to accurately distinguish specific from nonspecific products. There is a need for greater standardization of the methods used and interlaboratory validation of the technologies. Also, the number of alleles that can be analyzed at the same time is limited.

Large heterozygous deletions or duplications in genes can easily be detected by using various formats of fluorescent multiplex PCR assays designed to amplify multiple target sequences simultaneously under quantitative conditions (47, 48, 49); the sensitivity of the method has recently been improved in *quantitative multiplex PCR of short fluorescent fragment* (QMPSF) (50).

Two new methods rely on semiquantitative analysis of PCR products with specifically bound probes that are amplified with universal primers, which allows multiplexing of numerous targets.

Multiplex amplifiable probe hybridization (MAPH) relies on the ability of multiple short DNA probes (100–300 bp) to be recovered quantitatively by use of a common primer pair after hybridization to genomic DNA immobilized on a membrane (51). The probes are designed to be sufficiently different in size to be resolved by electrophoresis. Each probe is flanked with a pair of common synthetic primer binding sites. After washing to remove unbound probe, the remaining specifically bound probe fragments (that are present in amounts proportional to their target copy number) are stripped from the membrane and are further amplified by use of the common primer pair. PCR products are then separated by electrophoresis, and the relative quantitation of band intensities or peak heights (depending on the method of detection) will indicate deletions or duplications. The method has been applied to the detection of constitutional genomic deletions in genes (52) or in multiple subtelomeric regions of chromosomes (53). Recently it was demonstrated that MAPH can also be a reproducible high-throughput method suitable for the assessment of acquired genomic imbalances of multiple loci in tumor DNA samples with heterogeneous cell populations (54).

In *multiplex ligation-dependent probe amplification* (MLPA) (55), each target sequence in genomic DNA is hybridized in solution to two adjacent complementary probes. The probes are subsequently joined by a ligation reaction, and copy sequences are amplified in a multiplex PCR reaction using fluorescent-labeled universal primers attached to the probes. Only probes hybridized to the target sequence will be ligated and subsequently amplified in the PCR reaction. One of the two adjacent probes has a variable length random fragment to allow size-based electrophoretic separation of multiplex PCR products. The long probes are M13 derived, whereas the short probes are synthetic. The peak area of each amplified product will reflect the relative copy number of the target sequence. Up to 40–50 target sequences can be analyzed simultaneously (56). Commercially available kits have been designed for several tumor suppressor genes as well as subtelomeric regions of all chromosomes. Recently, the use of MLPA for rapid high-throughput prenatal detection of common aneuploidies has been assessed (57).

Several microarray-based comparative genomic hybridization assays have been tested recently for detecting gross duplications and deletions, especially in tumor cells, using DNA purified from BAC clones spaced along a chromosome segment (58), and there are now commercially available arrays (59).

In the approach called *conversion* (converting diploid to haploid mutants), patient cells are fused with a specially designed rodent cell line, creating hybrids that stably retain a subset of human chromosomes (60). About 25% of the derived hybrids contain a single copy of any human chromosome of interest. The diploid nature of the human genes has been converted to a haploid state, in which mutations are easier to detect. For example, deletion of a single exon will be easily detected from genomic analysis of hybrids, because no PCR product will be observed from the hybrid containing the mutant allele. Mutations affecting the expression of the gene can be revealed through analysis of the RT-PCR products of the haploid hybrids, whereas analyses of diploid genome are uninformative or difficult to interpret. However, because of the time and expense associated with the generation of hybrids, this approach cannot be used in diagnostic labs.

■ CONCLUSION: NEW CLINICAL PRACTICE

Disorders such as congenital disorders and hereditary diseases are now recognized as major health burdens. The birth incidence of infants with congenital disorders is generally estimated to be about 30–65 per 1000, and an estimated 6 of 10 people will, by the age of 60, develop a disease that is at least partially genetically determined. Because of recent progress in molecular genetics, it is now evident that inherited predisposition is important in a number of common diseases of later life, such as atherosclerosis, coronary heart disease, hypertension, diabetes mellitus, some rheumatic, oncological, and mental illnesses, and cancer, which develop into severe handicaps in predisposed people. It is probable that genetic approaches to the prevention of common diseases will emerge as one of the dominant strategies for the improvement of health. Moreover, pharmacogenomics is an emerging field examining the genetic basis for individual variations in response to therapeutics (drug efficacy and toxicity). Mutation and SNP detection is therefore becoming more and more important, and much effort will be invested in the development of cost-effective sequencing methods and high-throughput scanning or screening techniques.

- *Molecular diagnosis* is the detection of pathogenic mutations in DNA and RNA samples to aid in detection, diagnosis, subclassification, prognosis, and monitoring response to therapy (61). Molecular genetic testing includes several categories of testing within genetic services that can be divided into those services that target whole population with a view to identifying those at increased risk and those that focus on the needs of families who are affected by a genetic disorder (62). There are several categories of genetic tests :
- *Clinical diagnosis*. DNA test to confirm clinical diagnosis, for example, positive Friedreich ataxia testing is 100% sensitive and 100% specific.
- *Carrier testing or prenatal testing.* Detect heterozygotes among individuals at risk for reproductive planning purposes. In the US, in a 30-year period, 51,000 carriers for Tay–Sachs disease in Ashkenazi populations have been identified, resulting in the detection of 1400 two-carrier couples (63).
- *Prenatal diagnosis.* For couples known to be at high risk of transmitting disease alleles, PND permits informed decision making about continuing the pregnancy, for example, severe hemoglobinopathies.
- *Pre-symptomatic (or predictive) DNA testing.* Allows genetic disorder to be detected in advance of clinical presentation and includes at least three different indications: (i) predictive testing without treatment—for example, Huntington disease screening identifies people who will develop the disease with the aim of enabling personal decision making; (ii) predictive testing with treatment—for example, FAP testing identifies family members requiring early colectomy. Once the parental FAP mutation is known, the offspring can have the test to determine whether he has inherited the normal parental gene (he is no longer at risk) or the mutated one (he will develop FAP at some future date); and (iii) predispositional testing—for example, BRCA1 testing indicates the probability of developing breast cancer and related cancers.
- *Community or population screening.* An example is *newborn screening*—for example, cystic fibrosis testing provides improved nutrition and pulmonary management of affected children. Infants are first tested with the use of an immunoreactive trypsinogen assay; if the result is positive, the test is followed up with a DNA test (31 common mutations) of the original specimen of dried blood spot.

■ QUALITY CONTROL OF MOLECULAR GENETIC TESTING

The genetic test is usually ordered by a physician with a background in genetics and is not offered directly to the consumer, so that the physician can explain, in collaboration with the laboratory, the benefits, limitations, and possible adverse consequences of the test (64). Preanalytical information is essential for assessing the appropriateness of testing for a given patient and, depending on the disorder, may be critical for interpretation of test results. For example, interpretation of CF mutation analysis requires knowledge of the indication for testing, family history, and ethnic background of the patient because the mutation detection rates are less than 97% for almost all ethnic groups. The interpretation of a heterozygous genotype (detection of only one mutation) is fundamentally different for a diagnostic versus a carrier test. Moreover, for carrier tests, the prior and revised carrier risks are based on family history, as well as the population prevalence of carriers in the patient's ethnic group. Carrier risk revision for a patient with negative mutation analysis and a positive family history is enhanced by knowledge of the familial mutation (64).

As rapid technological advances produce systems that are capable of increased sensitivity, throughput, and quantitative potential, new sets of validation issues must be addressed. These include the reproducibility of the techniques, both intra- and interlaboratory, in terms of sensitivity, accuracy, precision, and the interpretation of data where arbitrary selection of cutoff points to separate positive and negative results are often used. The issue of quality assurance is also highlighted in considering that predictive and susceptibility genetic testing is often performed on asymptomatic persons and interpretation of results may or may not be supported by other findings, such as family history. Furthermore, molecular genetic tests are often performed only once in the life of an individual and may never be reported or confirmed. There is need also for better understanding of appropriate genetic testing practices among physicians. In Europe, QA standards for human molecular genetic testing are continually updated through guidelines and recommendations of the EMQN network. There is a major need for positive control samples containing well-defined mutations associated with disease. Very few laboratories have sufficient resources to themselves develop immortalized cell repositories representing most common human mutations (65).

Because large heterozygous deletions involving one or several exons most probably will be missed by routine PCR-based techniques, diagnostic DNA analyses should include a comprehensive examination of the whole relevant gene in the patient and confirmation of carrier status in both parents. This should allow one to distinguish homozygosity for a point mutation from a loss of heterozygosity due to a large deletion in one allele.

■ NEW INSIGHTS IN HEREDITY

The improved understanding of the molecular etiology of genetic disorders through the identification of thousands of causative mutations is now altering our perception of disease transmission. The classic model of individual alleles segregating into families according to Mendel's law, which allowed the discovery of many single-gene disorders, does not explain a growing number of disorders in which it appears that phenotypic effects can result from the combined action of alleles in many genes. Even for a disease like cystic fibrosis, considered to be a pure Mendelian recessive disease, mutational CFTR genotypes cannot predict the phenotype in each patient, which probably depends on a discrete number of alleles at different loci, some of them already identified. In reality, there is a continuum between Mendelian and complex traits (66), with *oligogenic disorders* emerging in the middle of this spectrum. Retinitis pigmentosa (RP) was the first example of proven *digenism*; Kajiwara et al. (67) showed that heterozygous mutations in both the ROM1 gene (retinal outer segment membrane protein 1) and the RDS gene (peripherin/retinal degeneration) were required in some seemingly dominant RP pedigrees to cause disease. The digenic RDS mutation prevents the formation of functional RDS-RDS homocomplexes, with the null-ROM1 mutation further decreasing the amount of functional complexes available, leading to photoreceptor degeneration.

Need for Functional Assays of Mutants and Variants

The methods described above will detect nucleic acids alterations but do not necessarily define their biological significance. Some sequence changes, such as those causing frameshift or chain terminations, are obviously functionally deleterious to the protein, but missense mutations leading to amino acid substitutions or base change in introns and untranslated regions are often problematic. Additional information including family and population studies may help to determine the significance of a sequence alteration. Extensive databases for genes and diseases can be extremely useful for interpreting the significance of missense mutations and other sequence variants. Equally important are functional studies of the mutant proteins after in vitro expression, which make invaluable the contribution of basic research to genetic testing.

In conclusion, much of the molecular biology in diagnostic use is still both tedious and time-consuming, involving endless pipetting, mixing, incubating, pouring gels, deciphering electropherograms, extracting nucleic acids, and so forth. Despite major improvements, scanning genes for unknown mutations in a service laboratory still remains a considerable problem.

REFERENCES

1. Abeysinghe SS, Stenson PD, Krawczak M, Cooper DN (2004) Gross rearrangement breakpoint database (GRaBD). Hum Mutat 23: 219–221.
2. Krawczak M , Cooper DN (1995) Core database. Nature 374: 402.
3. Lander ES (1996) The new genomics: Global views of biology. Science 274: 536–539.
4. Cargill M, Altshuler D, Ireland J, Sklar P, Ardlie K, Patil N, Shaw N, Lane CR, Lim EP, Kalyanaraman N, Nemesh J, Ziaugra L, Friedland L, Rolfe A, Warrington J, Lipshutz R, Daley GQ, Lander ES (1999) Characterization of single-nucleotide polymorphisms in coding regions of human genes. Nat Genet 3: 231–238.
5. Chakravarti A (1999) Population genetics—making sense out of sequence. Nat Genet Suppl: 56–60.
6. Reich DE, Lander ES (2001) On the allelic spectrum of human disease. Trends Genet 17: 502–510.
7. Hirschhorn R (2003) In vivo reversion to normal of inherited mutations in humans. J Med Genet. 40: 721–728.
8. Youssoufian H, Pyeritz RE (2002) Mechanisms and consequences of somatic mosaicism in humans. Nat Rev Genet 3: 748–758.
9. Yan H, Kinzler KW, Vogelstein B (2000) Tech. sight. Genetic testing—present and future. Science 289: 1890–1892.
10. O'Dell SD, Chen X, Day IN (2000) Higher resolution microplate array diagonal gel electrophoresis: Application to a multiallelic minisatellite. Hum Mutat 15: 565–576.
11. Newton CR, Graham A, Heptinstall LE, Powell SJ, Summers C, Kalsheker N, Smith JC, Markham AF (1989) Analysis of any point mutation in DNA. The amplification refractory mutation system (ARMS). Nucleic Acids Res 17: 2503–2516.
12. Kim S, Shi S, Bonome T, Ulz ME, Edwards JR, Fodstad H, Russo JJ, Ju J (2003) Multiplex genotyping of the human beta 2-adrenergic receptor gene using solid-phase capturable dideoxynucleotides and mass spectrometry. Anal Biochem 316: 251–258.

13. Nickerson DA, Kaiser R, Lappin S, Stewart J, Hood L, Landegren U (1990) Automated DNA diagnostics using an ELISA-based oligonucleotide ligation assay. Proc Natl Acad Sci USA 87: 8923–8927.
14. Foy CA, Parkes HC (2001) Emerging homogeneous DNA-based technologies in the clinical laboratory. Clin Chem 47: 990–1000.
15. Wilhelm J, Pingoud A (2003) Real-time polymerase chain reaction. Chembiochem 4: 1120–1128.
16. Ronaghi M, Nygren M, Lundeberg J, Nyren P (1999) Analyses of secondary structures in DNA by pyrosequencing. Anal Biochem 267: 65–71.
17. Chen L, Ren J (2004) High-throughput DNA analysis by microchip electrophoresis. Comb Chem High Throughput Screen 7: 29–43.
18. Larsen LA, Christiansen M, Vuust J, Andersen PS (2001) Recent developments in high-throughput mutation screening. Pharmacogenomics 2: 387–399.
19. Mantripragada KK, Buckley PG, de Stahl TD, Dumanski JP (2004) Genomic microarrays in the spotlight. Trends Genet 20: 87–94.
20. Latorra D, Campbell K, Wolter A, Hurley JM (2003) Enhanced allele-specific PCR discrimination in SNP genotyping using 3′ locked nucleic acid (LNA) primers. Hum Mutat 22: 79–85.
21. Igloi GL (2003) Single-nucleotide polymorphism detection using peptide nucleic acids. Expert Rev Mol Diagn 3: 17–26.
22. Myers RM, Larin Z, Maniatis T (1985) Detection of single base substitutions by ribonuclease cleavage at mismatches in RNA-DNA duplexes. Science 230: 1242–1246.
23. Myers RM, Maniatis T, Lerman LS (1987) Detection and localization of single base changes by denaturing gradient gel-electrophoresis. Methods Enzymol 155: 501–527.
24. Henco K, Harders J, Wiese U, Riesner D (1994) Temperature gradient gel electrophoresis (TGGE) for the detection of polymorphic DNA and RNA. Methods Mol Biol 31: 211–228.
25. Hovig E, Smith-Sorensen B, Brogger A, Borresen AL (1991) Constant denaturant gel electrophoresis, a modification of denaturing gradient gel electrophoresis, in mutation detection. Mutat Res 262: 63–71.
26. Vijg J, van Orsouw NJ (1999) Two-dimensional gene scanning: Exploring human genetic variability. Electrophoresis 20: 1239–1249.
27. Ganguly A (2002) An update on conformation sensitive gel electrophoresis. Hum Mutat 19: 334–342.
28. Xiao W, Oefner PJ (2001) Denaturing high-performance liquid chromatography: A review. Hum Mutat 17: 439–474.
29. Frueh FW, Noyer-Weidner M (2003) The use of denaturing high-performance liquid chromatography (DHPLC) for the analysis of genetic variations: Impact for diagnostics and pharmacogenetics. Clin Chem Lab Med 41: 452–461.
30. Kwok PY, Chen X (2003) Detection of single nucleotide polymorphisms. Curr Issues Mol Biol 5: 43–60.
31. Orita M et al. (1989) Detection of polymorphisms of human DNA by gel electrophoresis as single-strand conformation polymorphisms. Proc Natl Acad Sci USA 86: 2766–2770.
32. Doi K, Doi H, Noiri E, Nakao A, Fujita T, Tokunaga K (2004) High-throughput single nucleotide polymorphism typing by fluorescent single-strand conformation polymorphism analysis with capillary electrophoresis. Electrophoresis 25: 833–838.
33. Liu Q, Feng J, Sommer SS (1996) Bi-directional dideoxy fingerprinting (Bi-ddF): A rapid method for quantitative detection of mutations in genomic regions of 300–600 bp. Hum Mol Genet 5: 107–114.
34. Goldrick MM (2001) RNase cleavage-based methods for mutation/SNP detection, past and present. Hum Mutat 18: 190–204.
35. Cotton RGH, Rodrigues NR, Campbell RD (1988) Reactivity of cytosine and thymine in single base-pair mismatches with hydroxylamine and osmium tetroxide and its application to the study of mutations. Proc Natl Acad Sci USA 85: 4397–4401.

36. Solaro PC, Birkenkamp K, Pfeiffer P, Kemper B (1993) Endonuclease VII of phage T4 triggers mismatch correction in vitro. J Mol Biol 230: 868–877.
37. Till BJ, Burtner C, Comai L, Henikoff S (2004) Mismatch cleavage by single-strand specific nucleases. Nucleic Acids Res 32: 2632–2641.
38. Hawkins GA, Hoffman LM (1999) Rapid DNA mutation identification and fingerprinting using base excision sequence scanning. Electrophoresis 20: 1171–1176.
39. Roest PA, Roberts RG, Sugino S, van Ommen GJ, den Dunnen JT (1993) Protein truncation test (PTT) for rapid detection of translation-terminating mutations. Hum Mol Genet 2: 1719–1721.
40. Garwin AM et al (2000) MALDI-TOF based mutation detection using tagged in vitro synthesized peptides. Nat Biotechnol 18: 95–97.
41. Gite S, Lim M, Carlson R, Olejnik J, Zehnbauer B, Rothschild K (2003) A high-throughput nonisotopic protein truncation test. Nat Biotechnol 21: 194–197.
42. Garcia CA, Ahmadian A, Gharizadeh B, Lundeberg J, Ronaghi M, Nyren P (2000) Mutation detection by pyrosequencing: sequencing of exons 5–8 of the p53 tumor suppressor gene. Gene 253: 249–257.
43. Ju J (2002) DNA sequencing with solid-phase-capturable dideoxynucleotides and energy transfer primers. Anal Biochem 309: 35–39.
44. Ruparel H, Ulz ME, Kim S, Ju J (2004) Digital detection of genetic mutations using SPC-sequencing. Genome Res 14: 296–300.
45. Ars E, Serra E, de la Luna S, Estivill X, Lazaro C (2000) Cold shock induces the insertion of a cryptic exon in the neurofibromatosis type 1 (NF1) mRNA. Nucleic Acids Res 28: 1307–1312.
46. Sellner LN, Taylor GR (2004) MLPA and MAPH: New techniques for detection of gene deletions. Hum Mutat 23: 413–419.
47. Yau SC, Bobrow M, Mathew CG, Abbs SJ (1996) Accurate diagnosis of carriers of deletions and duplications in Duchenne/Becker muscular dystrophy by fluorescent dosage analysis. J Med Genet 33: 550–558.
48. Robinson MD, Chu CE, Turner G, Bishop DT, Taylor GR (2000) Exon deletions and duplications in BRCA1 detected by semiquantitative PCR. Genet Test 4: 49–54.
49. Saugier-Veber P, Drouot N, Lefebvre S, Charbonnier F, Vial E, Munnich A, Frebourg T (2001) Detection of heterozygous SMN1 deletions in SMA families using a simple fluorescent multiplex PCR method. J Med Genet 38: 240–243.
50. Casilli F, Di Rocco ZC, Gad S, Tournier I, Stoppa-Lyonnet D, Frebourg T, Tosi M (2002) Rapid detection of novel BRCA1 rearrangements in high-risk breast-ovarian cancer families using multiplex PCR of short fluorescent fragments. Hum Mutat 20: 218–226.
51. Armour JA, Barton DE, Cockburn DJ, Taylor GR (2002) The detection of large deletions or duplications in genomic DNA. Hum Mutat 20: 325–337.
52. White S, Kalf M, Liu Q, Villerius M, Engelsma D, Kriek M, Vollebregt E, Bakker B, van Ommen GJ, Breuning MH, den Dunnen JT (2002) Comprehensive detection of genomic duplications and deletions in the DMD gene, by use of multiplex amplifiable probe hybridization. Am J Hum Genet 71: 365–374.
53. Hollox EJ, Atia T, Cross G, Parkin T, Armour JA (2002) High throughput screening of human subtelomeric DNA for copy number changes using multiplex amplifiable probe hybridisation (MAPH). J Med Genet 39: 790–795.
54. Reid AG, Tarpey PS, Nacheva EP (2003) High-resolution analysis of acquired genomic imbalances in bone marrow samples from chronic myeloid leukemia patients by use of multiple short DNA probes. Genes Chromosomes Cancer 37: 282–290.
55. Schouten JP, McElgunn CJ, Waaijer R, Zwijnenburg D, Diepvens F, Pals G (2002) Relative quantification of 40 nucleic acid sequences by multiplex ligation-dependent probe amplification. Nucleic Acids Res 30: e57.
56. Hogervorst FB, Nederlof PM, Gille JJ, McElgunn CJ, Grippeling M, Pruntel R, Regnerus R, van Welsem T, van Spaendonk R, Menko FH, Kluijt I, Dommering C, Verhoef S, Schouten JP, van't Veer LJ, Pals G (2003) Large genomic deletions and duplications in the BRCA1 gene identified by a novel quantitative method. Cancer Res 63: 1449–1453.

57. Slater HR, Bruno DL, Ren H, Pertile M, Schouten JP, Choo KH (2003) Rapid, high throughput prenatal detection of aneuploidy using a novel quantitative method (MLPA). J Med Genet 40: 907–912.
58. Cai WW, Mao JH, Chow CW, Damani S, Balmain A, Bradley A (2002) Genome-wide detection of chromosomal imbalances in tumors using BAC microarrays. Nat Biotechnol 20: 393–396.
59. Hui AB, Lo KW, Teo PM, To KF, Huang DP (2002) Genome wide detection of oncogene amplifications in nasopharyngeal carcinoma by array based comparative genomic hybridization. Int J Oncol 20: 467–473.
60. Yan H, Papadopoulos N, Marra G, Perrera C, Jiricny J, Boland CR, Lynch HT, Chadwick RB, de la Chapelle A, Berg K, Eshleman JR, Yuan W, Markowitz S, Laken SJ, Lengauer C, Kinzler KW, Vogelstein B (2000) Conversion of diploidy to haploidy. Nature 403: 723–724.
61. Trent RJ, Yu B, Caramins M (2004) Challenges for clinical genetic DNA testing. Expert Rev Mol Diagn 4: 201–208.
62. Donnai D (2002) Genetic services. Clin Genet 61: 1–6.
63. Khoury MJ (2003) Genetics and genomics in practice: The continuum from genetic disease to genetic information in health and disease. Genet Med 5: 261–268.
64. Amos J, Patnaik M (2002) Commercial molecular diagnostics in the U.S.: The Human Genome Project to the clinical laboratory. Hum Mutat 9: 324–333.
65. Williams LO, Cole EC, Lubin IM, Iglesias NI, Jordan RL, Elliott LE (2003) Quality assurance in human molecular genetics testing: status and recommendations. Arch Pathol Lab Med 127: 1353–1358.
66. Badano JL, Katsanis N (2002) Beyond Mendel: An evolving view of human genetic disease transmission. Nat Rev Genet 3: 779–789.
67. Kajiwara K, Berson EL, Dryja TP (1994) Digenic retinitis pigmentosa due to mutations at the unlinked peripherin/RDS and ROM1 loci. Science 264: 1604–1648.

3

A FULL-COVERAGE, HIGH-RESOLUTION HUMAN CHROMOSOME 22 MICROARRAY FOR CLINICAL AND RESEARCH APPLICATIONS

Jan P. Dumanski and Patrick G. Buckley
Department of Genetics and Pathology, Uppsala University, Sweden

■ ABSTRACT

We have constructed the first comprehensive microarray representing a human chromosome for analysis of DNA copy number variation. This chromosome 22 array covers 34.7 Mb, representing 1.1% of the genome, with an average resolution of 75 kb. To demonstrate the utility of the array, we have applied it to profile glioblastoma, acral melanoma, dermatofibrosarcoma, DiGeorge syndrome, and neurofibromatosis 2 (NF2). We accurately diagnosed homozygous/heterozygous deletions, amplifications/gains, translocation breakpoints and *IGLV/IGLC* locus instability. Two significant methodological advances in array construction were also developed and validated. We further identified a known chromosome 22-located gene as a candidate tumor suppressor in glioblastoma. A broad spectrum of issues in medical research and diagnostics can be approached with our array. This well-annotated and gene-rich autosome contains numerous uncharacterized disease genes. It is therefore crucial to associate these genes to specific 22q-related conditions, and this array will be instrumental toward this goal.

Guide to Mutation Detection, Edited by Graham R. Taylor and Ian N. M. Day.
ISBN 0-471-23444-3

■ INTRODUCTION

Chromosome 22 is the second smallest human chromosome and has been a testing ground for recent genomic advances allowing fundamental biological questions to be approached from a global perspective. It contains ~600 genes, which represent 2.4% of all known genes. The first comprehensive linkage map over chromosome 22 was constructed in 1991, the complete YAC-based physical map was established in 1995, and in 1999 it was the first sequenced chromosome (1–3). Microarray-based comparative genomic hybridization (array-CGH) is rapidly emerging as a powerful tool for a variety of genetic applications (4, 5), particularly as it provides significant advantages over conventional techniques for detection of gene dosage alterations. Figure 3.1 outlines the main methodological differences between array-CGH and the commonly used metaphase-CGH. A genome-wide scanning array, with an average resolution of 1.3 Mb, was recently been published (6). This array consists of BAC clones evenly spaced across the genome printed on a glass slide. This approach for the generation of genomic microarrays is summarized in Figure 3.2. We have constructed and tested the first high-resolution chromosomal genomic microarray, which will allow the study of disease genes and other research questions related to this autosome. Use of the genomic DNA clones across the entire long arm of chromosome 22 provides both long-range coverage of 34.7 Mb and an exceptional average resolution of 75 kb.

■ RESULTS

Preparation of the Array

We aimed to fully cover known segments of chromosome 22q with genomic clones from the minimal tiling path (3) without gaps larger than 100 kb. Therefore, we selected a set of 460 loci derived from chromosome 22, together with nine X chromosome loci and 31 loci from other chromosomes, as controls. There are 167 cosmids/fosmids and 329 BACs/PACs on the array. Four loci were covered with a novel, repeat-free, nonredundant PCR-based strategy. In addition, 20 clones were amplified in parallel with phi29 DNA polymerase and represented as independent measurement points. Each locus has an ID number, with the chromosome 22 loci assigned numbers 1–480 (including phi29-amplified clones), ordered from centromere to telomere. The chromosome 22 part of the array spans 34.7 Mb, which results in an average resolution of 75 kb. The array contains a 161-kb gap (0.46% of 34.7 Mb), which was caused by the failure in confirming the correct identity of one clone with our quality control criteria. Other regions not represented on the array were excluded based on bioinformatic analysis (see METHODS), which clearly indicated that these sequences are unsuitable for CGH analysis. All clones within these intervals were found to contain high repeat content (up to 99%) and/or stretches of strong sequence similarity (spanning ≥15 kb with overall sequence identity of ≥98%) to other human chromosomes and/or other regions on 22q. For instance, the pericentromeric region showed strong similarities to chromosome 14, which is in agreement with recent analysis of segmental duplications in the genome (7). Thus these chromosome 22 loci are practically unapproachable by current methods.

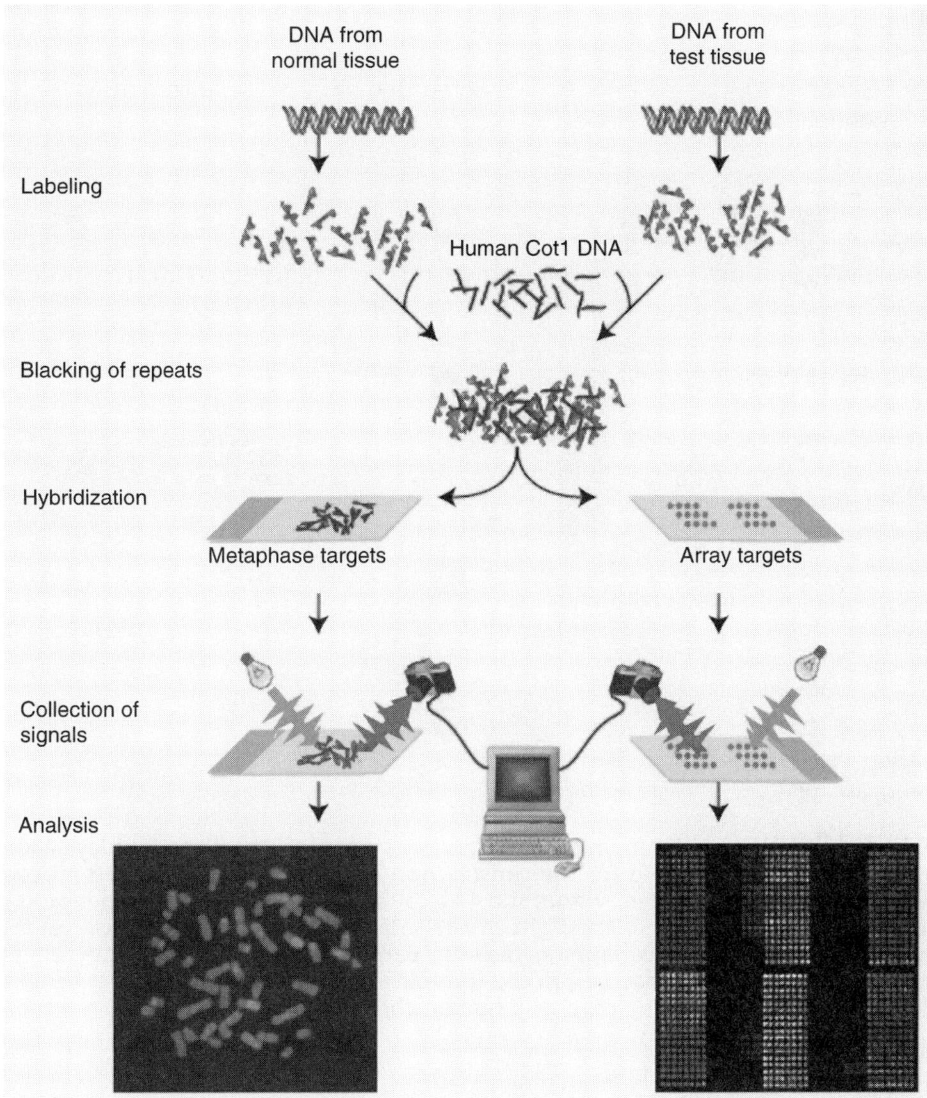

FIGURE 3.1. Comparison of metaphase- and array-based comparative genomic hybridization. Comparative genomic hybridization is based on the hybridization of differentially labeled test and reference DNA cohybridized to an immobilized DNA support. Both metaphase- and array-CGH use human Cot-1 DNA, which is rich in common repeats and is used to block repetitive sequences in both the test and reference probes. After an overnight hybridization the fluorescent signals are detected and converted into digital data. Ratios are then calculated to determine whether there is a gene copy number imbalance present in the test sample compared with the reference. Metaphase- and array-CGH mainly differ with regard to the nature of the immobilized DNA. In metaphase-CGH the DNA is complete chromosomes fixed on a glass slide, whereas array-CGH uses genomic clones to cover large genomic regions. Therefore, array-CGH has a much higher resolution of analysis as a result of the increased number of measurement points across a single locus.

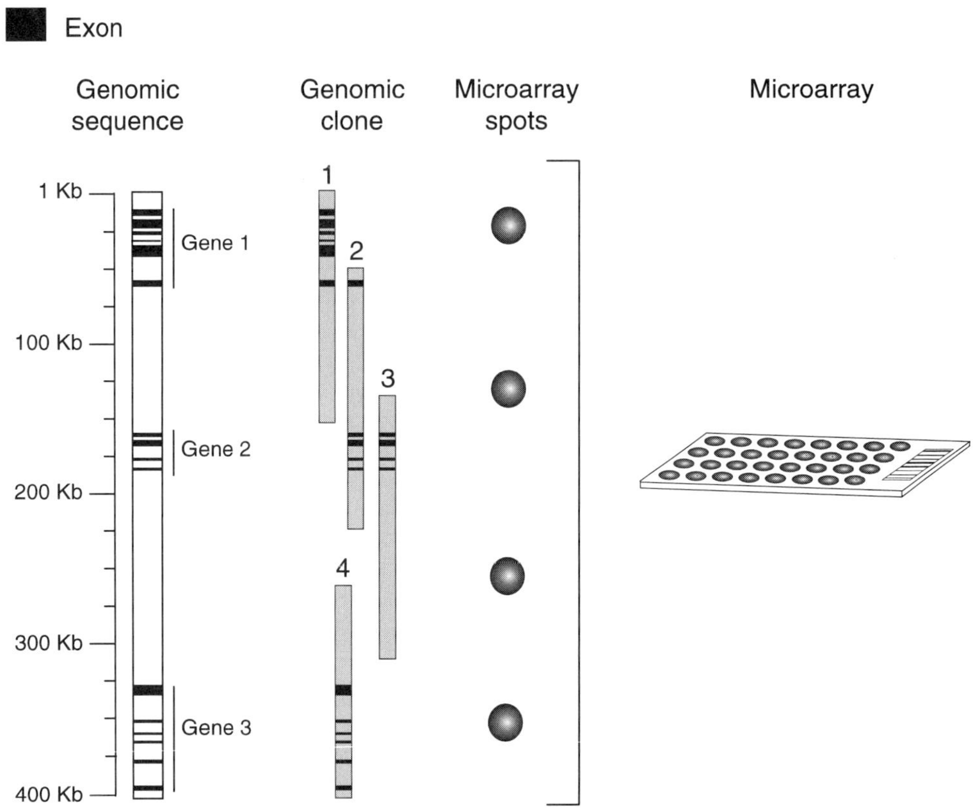

FIGURE 3.2. Scheme of genomic microarray construction. This figure displays 400 kb of sequence that is targeted for the construction of a genomic clone-based microarray. There are 3 genes present within the 400-kb sequence, and exons are denoted by the horizontal black bars. To the right of the sequence are the genomic clones that cover this region in an overlapping fashion. Each clone constitutes a spot on the genomic microarray and may be printed in replicates. With this approach a genomic microarray may contiguously cover full human chromosomes.

Validation of the Array

To evaluate the chromosome 22 array for detection of copy number variation we carried out a series of experiments. In the normal XY/ normal XX hybridization, all chromosome 22 loci and control loci from other autosomes showed two copies and chromosome X controls provided a sensitive means to detect one copy. Meningioma T24 is a previously reported tumor with homozygous and heterozygous deletions (Fig. 3.3(a)) (8, 9). Hence, it allows detection of zero, one, and two copies of loci across chromosome 22. As expected, in addition to the homozygous deletion, we also observed the large terminal heterozygous deletion.

The neurofibromatosis type 2 (NF2) case JP [constitutional DNA derived from lymphoblastoid cell line (LCL)] is another well-characterized patient with a 6-Mb heterozygous interstitial deletion (Fig. 3.3(b)) (9, 10). We detected this deletion between ID140 and 275. We also observed a region of genomic instability within the immunoglobulin lambda variable and constant chain loci (*IGLV*/*IGLC*). This region displayed a single homozygous deletion in ID78 surrounded by at least three pockets of heterozygously deleted loci. To verify this finding, we examined schwannoma DNA from the same patient and confirmed the presence of the 6-Mb heterozygous deletion (Fig. 3.3(c)). In contrast, the *IGLV/IGLC* region exhibited a shift toward gain in copy number. Color switch experiments were performed for all the above-described samples confirming these results.

RESEARCH APPLICATIONS OF CHROMOSOME 22 ARRAY

Chromosome 22 contains several uncharacterized disease-causing genes. For instance, previous low-resolution loss-of-heterozygosity studies of glioblastoma indicated that chromosome 22 harbors a tumor suppressor gene(s) implicated in development of this aggressive and incurable tumor (11, 12). We applied the array to profile two previously studied glioblastomas (unpublished). Array-CGH confirmed the presence of large heterozygous deletions in both cases. Unexpectedly, we also detected a homozygous deletion of ID186. This observation was confirmed by Southern blot analysis using probes derived from this locus. The nonzero value for the ratio in the homozygously deleted region may be due to a normal tissue component in the tumor, inefficient blocking of repetitive sequences, autofluorescence of the target clones, or that cosmid is larger than the deletion. This cosmid contains a single gene, implicating it as a candidate tumor suppressor.

Another type of tumor studied was acral melanoma, which is a rare form of cutaneous melanoma that does not seem to be related to UV exposure (13). Previous metaphase-CGH showed that acral melanomas differed from other melanomas by amplifications/gains involving regions on 11q13, 22q11–13, and 5p15 (13). Fifteen acral melanomas were hybridized, and eight samples showed different patterns of amplifications/gains, with respect to their regional distribution along 22q and level of copy number variation. Three profiles of acral melanomas on the chromosome 22q array will be shown later. We also confirmed the previous report of 5p15 gains (13), because a number of control loci from chromosome 5p15 revealed gains in several tumors. Thus analysis of a larger series of acral melanomas at the same resolution across the whole genome is likely to reveal new information and might elucidate mechanisms underlying tumorigenesis of these tumors.

THE CHROMOSOME 22 ARRAY AS A DIAGNOSTIC TOOL

We also employed the array as a diagnostic tool in several well-studied disorders, such as NF2, DiGeorge syndrome, and dermatofibrosarcoma protuberans (DFSP). Using peripheral blood DNA from a female NF2 patient p41, with a moderate disease phenotype, we detected and sized a large heterozygous deletion. The deletion covered ID132–177 (~3.3 Mb), which included the *NF2* gene. This is a rare example of an NF2 patient with an interstitial deletion extending far toward the centromere from the *NF2* gene (9, 10) (Fig. 3.3).

We further applied this array in the analysis of previously characterized DiGeorge syndrome patients with deletions in the DiGeorge critical region (DGCR). Two of these subjects will be shown later. As expected, the deletion spans the region between low copy repeats A–D (14), which are located in ID16–47 (2.4 Mb). This demonstrates the usefulness of the array in the diagnosis of DiGeorge syndrome. However, array-CGH results are difficult to interpret for some loci within DGCR.

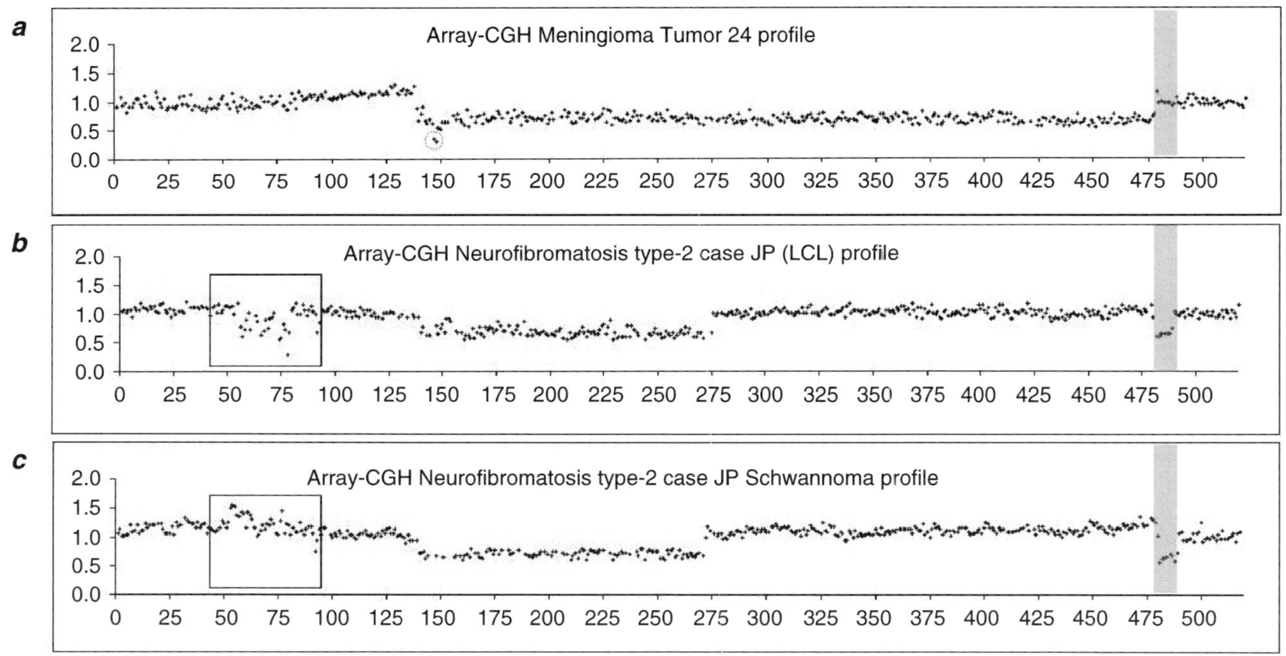

FIGURE 3.3. Analysis of patient samples with the chromosome 22 array. *a*, *b*, and *c* display 3 array-CGH experiments of the entire array, which contains a total of 520 loci. The *X*-axis displays 480 chromosome 22 loci, ordered from centromere (*left*) to telomere (*right*). Dots within the shaded box indicate control loci from chromosome X (9 clones). Dots plotted after the chromosome X controls are derived from chromosomes other than X and 22 (31 clones). Each dot on a chart represents the normalized average ratio between fluorescent signals for each locus (*Y*-axis), which are derived from 3–4 independent replica spots on the array, in a single hybridization experiment. Panel ***a*** shows a normalized average ratio from an experiment using meningioma T24 tumor DNA derived from a female patient vs. normal female DNA (9). Circle outlines 2 consecutive loci (ID147 and 148, average ratio 0.37 and 0.32, respectively), which are deleted homozygously in this tumor. Panels ***b*** and ***c*** show the results from two experiments using DNA derived from LCL of NF2 case JP and the schwannoma tumor DNA, derived from the same patient (9, 10). Both profiles display a large 22q interstitial deletion; however, there is a distinct difference in the array-CGH profile within the IGLV/IGLC region (rectangle) in both DNAs derived from LCLs compared with tumor DNA from the same patient.

Bioinformatic analysis of these loci provides clues to their unexpected behavior. For instance, a majority of clones are cosmids with a high content of common repeats (>50%). We therefore assume that the deletion is contiguous. The difficulties in analysis of DGCR are also in agreement with reports of segmental duplications on 22q (3, 7). These results demonstrate the need for an array specifically tailored for analysis of DGCR. In summary, in all experiments with DiGeorge patients at least 14 clones within the DGCR are reliable for diagnosis of disease-causing deletions.

The array has further been applied in analysis of DFSP. DFSP is a cutaneous tumor the cytogenetic features of which include the translocation t(17;22)(q22;q13) or, more commonly, supernumerary ring chromosomes containing material from 17q and 22q (15, 16). Previous studies revealed that these rearrangements result in a fusion between collagen type I, alpha 1 gene (*COL1A1*) from chromosome 17 and the platelet-derived growth factor, B-chain gene (*PDGFB*) from chromosome 22. Array-CGH was carried out on two DFSPs. These tumors contain the *COL1A1/PDGFB* fusion gene (17, 18), and in both cases we detected the breakpoint at ID325. This clone contains the entire *PDGFB* gene, which is truncated as a consequence of the translocation. Strikingly, the 22q loci that follow after ID325 toward the telomere (ID326–480) behave as though they were affected by a terminal, heterozygous deletion. The fluorescence ratios for these clones in both experiments are at the level of chromosome X controls. We also observed a low-level gain in both tumors, involving the 22q segment centromeric to the translocation breakpoint, which is in agreement with previous cytogenetic reports (19). Both tumors were studied cytogenetically and were found to have a diploid nature, except for chromosomes 17 and 22 (18). Thus non-22 autosomal controls in array-CGH confirm the diploid nature of these tumors and allow proper normalization of CGH results. Another finding, seen in both tumors, was the presence of several independent peaks of amplification at an even higher level, which is compatible with the genetic instability of ring chromosomes (20).

DEVELOPMENTS OF ARRAY-CGH METHODOLOGY

We experimented with a novel, strictly sequence-defined, and PCR-based strategy for preparation of gene dosage arrays (patent pending) (21), using pools of nonredundant DNA fragments in four 22q loci. The rationale was to test an approach that would overcome the limitations of array-CGH for regions covered by genomic clones with a high repeat content or containing duplications of sequence in the human genome. Results from multiple experiments fully confirm that the pools perform as reliable replacements for genomic clones. Two of these experiments will be shown in detail later. Furthermore, there was no detectable difference in the results obtained from unmodified primers compared with DNA amplified using 5′-amino-labeled oligonucleotides.

The other successful development was the application of phi29 DNA polymerase in the preparation of DNA before printing of the array. One of the main constraints of standard array-CGH protocols is a low yield of DNA obtained from PACs and BACs. This often requires costly repetitions of large-scale preparations. The goal was to test the performance of DNA amplified with phi29 polymerase (22, 23). Twenty clones were processed in two ways: (i) using standard Qiagen-based protocols and (ii) using 10 ng of DNA derived from a Qiagen preparation as template. Approximately 50 μg of DNA was produced in a single overnight phi29 reaction, which will suffice for production of >6000 microarrays. The phi29-amplified DNA was quality controlled as described in METHODS. The results of phi29-amplified DNA from two array-CGH experiments will be presented in detail later. In summary, phi29 polymerase-based protocol performs well and is a reliable way to produce DNA for array-CGH.

DISCUSSION

Reliable and high-resolution measurement of copy number variation is fundamental in many areas of diagnostics and medical research. This array is the first in a new generation of high-resolution genetic tools. We provide 99.5% coverage across 34.7 Mb (1.1% of the human genome) with an unprecedented average resolution of ~75 kb. Several examples shown in this study highlight the power and potential of this array.

CHROMOSOME 22 ARRAY IN THE RESEARCH AND MOLECULAR DIAGNOSIS OF DISEASE

DFSP is a locally aggressive tumor that is sometimes misdiagnosed as dermatofibroma, a common and benign lesion (24, 25). This leads to inadequate primary treatment and high recurrence rate (24, 25). All previously studied cases of DFSP displayed a presence of the *COL1A1/PDGFB* fusion gene, which offers the possibility to develop diagnostic tools (18). Our array offers a quick molecular diagnostic kit based on detection of the translocation breakpoint on 22q in combination with loss of genetic material from 22q distal to the breakpoint. This is the first report of array-CGH in detection of translocation breakpoints, which has a diagnostic value.

Detection of a ~40-kb homozygous deletion while scanning 34 Mb of 22q in glioblastoma is an illustrative example of the sensitivity and comprehensiveness of our analysis. It is also a strong indication that investigation of a larger tumor series with this degree of resolution will reveal a highly complex picture of aberrations. It should also be stressed here that the segment encompassing the homozygously deleted cosmid is not located within the primary glioblastoma candidate region on 22q. It was previously positioned closer to the telomere of 22q with low-resolution genotyping (11, 12).

We observed genetic instability in the *IGLV/IGLC* region, which was most pronounced in DNA from lymphoblastoid cell lines (LCLs) derived from Epstein–Barr virus (EBV)-transformed B-lymphocytes. EBV-carrying LCLs are polyclonal after transformation, oligoclonal on prolonged in vitro culturing, eventually becoming monoclonal (26). The LCL from NF2 case JP displayed heterozygous and homozygous deletions within *IGLV/IGLC*. The other studied LCLs displayed heterozygous deletions or no detectable aberrations in this locus. Deletions within *IGLV/IGLC* can be explained by rearrangements of immunoglobulin loci during B-cell maturation, and a more detailed analysis of LCLs at different stages of in vitro culturing should be carried out. Results from the schwannoma of NF2 case JP, when compared with the analysis of LCL from the same patient, suggest that the *IGLV/IGLC* locus displays a general genetic instability that is not restricted to B-lymphocytes. An alternative explanation to this instability could be that there is a copy number variation of regions with very high sequence similarity to the *IGLV/IGLC* loci. Another even more important implication of the above results is that this array, at the current level of resolution across the entire 22q, provides an opportunity to study the possible presence of deletion/amplification polymorphisms in human DNA samples derived from normal, nonrelated, multiethnic populations.

FUTURE STRATEGIES FOR ARRAY CONSTRUCTION AND APPLICATIONS OF THE ARRAY

Chromosome 22 is a suitable experimental platform for developing new technologies in array-CGH. It is a challenging subject because the tiling path of 22q is rich in inter- and intrachromosomal segmental duplications (also denoted as low copy repeats), common repeats, and pseudogenes (3, 7), which restrict its analysis. Current protocols for array-CGH rely on human Cot-1 DNA for suppression of background derived from common repeats. However, Cot-1 DNA is expensive and the quality of different commercially available batches is extremely variable. Thus a decrease in the amount of Cot-1 DNA in, or its full elimination from, array-CGH is desirable. Additional drawbacks associated with genomic clones include their limited availability, contamination with bacteriophages, inserts that are prone to rearrangements, and purchasing of expensive clones from private distributors. Our repeat-free strategy eliminates a majority of the above limitations and would permit analysis of any sequenced genome. The above reasoning and difficulties encountered in diagnosing DiGeorge syndrome-associated deletions highlight a need for a second-generation, higher-resolution array that is entirely based on the repeat-free strategy. It should be further stressed that the repeat-free strategy also allows for a significant increase in array-CGH resolution, currently down to a 20-kb genomic segment (Kiran Mantripragada et al., in preparation). A further considerable improvement of the methodology is the application of phi29 polymerase in preparation of ready-to-print DNA that entails a considerable advantage with regard to work input.

A wide range of questions in basic research and diagnostics may now be approached with this array. These include comprehensive epigenetic profiling of 22q-located genes and high-resolution analysis of replication timing across 22q. Moreover, chromosome 22 contains a considerable number of uncharacterized disease genes, for example, familial schizophrenia susceptibility, glioblastoma, and other types of astrocytoma, ependymoma, meningioma, schwannomatosis, pheochromocytoma, and breast and colon cancer (27). As chromosome 22 is well characterized with regard to active gene content, it is therefore critical to link the known genes to the above-mentioned conditions. Our high-resolution array will be instrumental toward this goal.

METHODS

Selection and Preparation of Genomic Target Clones

We used the tiling path of clones derived from the effort toward sequencing of chromosome 22 (3) (www.ensembl.org). The identity of each clone was verified and quality controlled by agarose gel electrophoresis of native/*Eco*RI -digested DNA and STS-PCR using at least one marker for each clone. DNA from Qiagen preparations was quantified with the picoGreen assay (Molecular Probes, Inc., Eugene, OR).

Preparation of the PCR Pools

The genomic sequences were downloaded from NCBI (www.ncbi.nlm.nih.gov). The common repeats were masked with *RepeatMasker* (ftp.genome.washington.edu/RM/), and the ".masked" output file was analyzed by *blastn* against the *nr*-section of the EMBL database. Further, we used the Oligo6 program to design primers in the nonredundant regions. Standard protocols were employed in amplification of DNA using total human DNA as template. The size of amplified fragments was controlled, and quantification was carried out by picoGreen assay. All fragments from a single locus were pooled before dissolving in printing solution. Detailed information can be found at http://puffer.genpat.uu.se/chrom_22_array/chrom22.htm.

Amplification of DNA by phi29 Polymerase

DNA from genomic clones (10 ng) was placed into a 96-well plate in a total volume of 100 µl containing 37 mM Tris_HCl, pH 7.5, 50 mM KCl, 10 mM $MgCl_2$, 5 mM $(NH_4)_2SO_4$, 1.0 mM dNTPs, 50 µM exonuclease-resistant hexamer, 1 unit/ml of yeast pyrophosphatase (Roche Molecular Biochemicals, Indianapolis IN), and 800 units/ml phi29 DNA polymerase (Amersham Biosciences, Piscataway NJ). Reactions were incubated for 16 ho at 30°C and terminated by heating to 65°C for 3 min in a PCR System Thermocycler (Applied Biosystems, Foster City, CA) (23).

Printing of the Array, Hybridizations, Scanning, and Image Analysis

Methods used for array production, DNA labeling, hybridization, and posthybridization processing are shown at http://puffer.genpat.uu.se/chrom_22_ array/chrom22.htm. DNA was sonicated, precipitated, and dissolved in 80% DMSO. DNA was printed on aminosilane-coated slides with the ProSys robot (Cartesian Technologies, Irvine, CA) using 12 SMP3 pins (TeleChem International, Sunnyvale, CA) and a center-to-center distance of 250 µm between adjacent spots. Each clone was printed in quadruplicate. One microgram of test and reference DNA was labeled by random priming with Cy3 dCTP (PA 53021, Amersham Biosciences) and CY5 dCTP (NEN Life Science Products, Boston, MA), respectively, followed by hybridization (9, 28). Image acquisition was performed with the GenePix 4000B scanner (Axon Instruments Inc, Union City, CA). Analysis of hybridization intensity was carried out with the GenePixPro image analysis software (Axon Instruments). Ratio of means for the intensities between test DNA and reference DNA was calculated, as well as the average and standard deviation of the four replicas for each clone. Clones displaying standard deviation > 5% between a minimum number of three spots were discarded from further analysis. The ratios of the chromosome 22 target clones were normalized against the average ratio of the target clones from other autosomal controls.

ACKNOWLEDGMENTS

This work was supported by grants from the U.S. Army Medical Research and Materiel Command, award no. DAMD17-00-1-0536, the Swedish Cancer Foundation, the Swedish Research Council, and Uppsala University.

REFERENCES

1. Dumanski JP et al. (1991) A map of 22 loci on human chromosome 22. Genomics 11: 709–719
2. Collins JE et al. (1995) A high resolution integrated yeast artificial chromosome clone map of human chromosome 22. Nature 377: 367–379.
3. Dunham I et al. The DNA sequence of human chromosome 22. Nature 402: 489–495.
4. Solinas-Toldo S et al. (1997) Matrix-based comparative genomic hybridization: biochips to screen for genomic imbalances. Genes Chromosomes Cancer 20: 399–407.
5. Pinkel D et al. (1998) High resolution analysis of DNA copy number variation using comparative genomic hybridization to microarrays. Nat Genet 20: 207–211.
6. Snijders AM et al. (2001) Assembly of microarrays for genome-wide measurement of DNA copy number. Nat Genet 29: 263–264.
7. Bailey JA et al. (2002) Human-specific duplication and mosaic transcripts: the recent paralogous structure of chromosome 22. Am J Hum Genet 70: 83–100.
8. Peyrard M et al. (1994) Characterization of a new member of the human b-adaptin gene family from chromosome 22q12, a candidate meningioma gene. Hum Mol Genet 3: 1393–1399.
9. Bruder CE et al. (2001)High resolution deletion analysis of constitutional DNA from neurofibromatosis type 2 (NF2) patients using microarray-CGH. Hum Mol Genet 10: 271–282.
10. Bruder CE et al. (1999) Severe phenotype of neurofibromatosis type 2 in a patient with a 7.4-MB constitutional deletion on chromosome 22: Possible localization of a neurofibromatosis type 2 modifier gene? Genes Chromosomes Cancer 25: 184–190.
11. Ino Y et al. (1999) Common regions of deletion on chromosome 22q12.3–q13.1 and 22q13.2 in human astrocytomas appear related to malignancy grade. J Neuropathol Exp Neurol 58: 881–885.
12. Oskam NT, Bijleveld EH, Hulsebos TJ (2000) A region of common deletion in 22q13.3 in human glioma associated with astrocytoma progression. Int J Cancer 85: 336–339.
13. Bastian BC et al. (2000) Gene amplifications characterize acral melanoma and permit the detection of occult tumor cells in the surrounding skin. Cancer Res 60: 1968–1973.
14. Shaikh TH et al. (2000) Chromosome 22-specific low copy repeats and the 22q11.2 deletion syndrome: Genomic organization and deletion endpoint analysis. Hum Mol Genet 9: 489–501.
15. Pedeutour F et al. (1996) Translocation t(17;22) (q22;q13) in dermatofibrosarcoma protuberans. A new tumor-associated chromosome rearrangement. Cytogenet Cell Genet 72: 171–174.
16. Nishio J et al. (2001) Supernumerary ring chromosome in a Bednar tumor (pigmented dermatofibrosarcoma protuberans) is composed of interspersed sequences from chromosomes 17 and 22: A fluorescence in situ hybridization and comparative genomic hybridization analysis. Genes Chromosomes Cancer 30: 305–309.
17. Simon MP et al. (1997)Deregulation of the platelet-derived growth factor B-chain gene via fusion with collagen gene COL1A1 in dermatofibrosarcoma protuberans and giant-cell fibroblastoma. Nat Genet 15: 95–98.

18. O'Brien KP et al. (1998) Various regions within the alpha-helical domain of the COL1A1 gene are fused to the second exon of the PDGFB gene in dermatofibrosarcomas and giant-cell fibroblastomas. Genes Chromosomes Cancer 23: 187–193.
19. Pedeutour F et al. (1995) Ring chromosomes in dermatofibrosarcoma protuberans are low-level amplifiers of chromosome 17 and 22 sequences. Cancer Res 55: 2400–2403.
20. Gisselsson D et al. (1999) The structure and dynamics of ring chromosomes in human neoplastic and non-neoplastic cells. Hum Genet 104: 315–325.
21. Dumanski JP (2002) Gene dosage arrays, their production and use thereof, Swedish patent application, no. SE0200865-4.
22. Dean FB, Nelson JR, Giesler TL, Lasken RSR (2001) Rapid amplification of plasmid and phage DNA using Phi 29 DNA polymerase and multiply-primed rolling circle amplification. Genome Res 11: 1095–1099.
23. Dean FB et al. (2002) Comprehensive human genome amplification using multiple displacement amplification. Proc Natl Acad Sci USA 99: 5261–5266.
24. Rutgers EJ, Kroon BB, Albus-Lutter CE, Gortzak E (1992) Dermatofibrosarcoma protuberans: treatment and prognosis. Eur J Surg Oncol 18: 241–248.
25. Fish FS (1996) Soft tissue sarcomas in dermatology. Dermatol Surg 22: 268–273.
26. Nilsson K (1979) The nature of lymphoid cell lines and their relationship to the virus. In *The Epstein-Barr Virus*, Epstein M, Achong B (eds.), Springer-Verlag, New York.
27. Dumanski JP (1996) The human chromosome 22-located genes and malignancies of the central nervous system. Neuropathol Appl Neurobiol 22: 412–417.
28. Feinberg AP, Vogelstein B (1984) A technique for radiolabelling DNA restriction endonuclease fragments to high specific activity. Addendum. Anal Biochem 137: 266–267.

4

DETECTION OF LARGE HETEROZYGOUS DELETIONS AND DUPLICATIONS

Mario Tosi
Institute National de la Sante et de la Recherche Médicale, Faculté de Médecine et Pharmacie, Rouen, France

■ INTRODUCTION

Recognition of the important consequences of gene deletions or duplications and of other rearrangements of genetic material has stimulated the development of sensitive methods to detect them. The combination of restriction enzyme analysis, gel electrophoresis, and molecular hybridization has opened the way to major discoveries, such as the demonstration and characterization of immunoglobulin gene rearrangements, the characterization of transposable elements and of retroviral insertions, and the discovery of gene fusions in cancer and of instabilities due to the presence of repetitive sequences in many genes involved in genetic diseases. The Southern blot method, which has made many of these breakthroughs possible, still represents one of the major tools of molecular biology and genetics. However, it suffers from some limitations that have become more and more apparent with the spread of PCR-based techniques in all areas of biology. Southern blot analysis, in fact, requires genomic DNA of good quality in amounts that are in the order of tens of micrograms, that is, at least ten times more than the amounts that are used in most PCR-based applications. Moreover, even using multiple restriction enzyme digests, the interpretation of the data is not always straightforward, and in any case Southern blot hybridization can be time consuming, particularly for large genes. On the other hand it should be noted that all PCR-based mutation-screening

Guide to Mutation Detection, Edited by Graham R. Taylor and Ian N. M. Day.
ISBN 0-471-23444-3

methods that are commonly used amplify only the normal allele when an heterozygous deletion is present and, in the case of duplications, a larger PCR product is often not seen, because of competition with the normal product.

■ DETECTION OF GENE REARRANGEMENTS

Screening not only for point mutations or microdeletions/insertions but also for large rearrangements is now considered a necessary component of the molecular diagnosis of many genetic diseases and particularly those due to defects in genes that contain a large number of repetitive elements (e.g., *Alu* repeats). In several cases such gene rearrangements were detected in parallel studies at the mRNA level, which sometimes reveal the consequence of the gene rearrangement in the form of a size change of an RT-PCR product. However, in addition to the limitation to genes that are expressed in peripheral blood cells and to the difficulty of establishing high-quality collections of RNA samples, only a fraction of the gene rearrangements can be detected with this approach, because of the instability of mRNAs carrying out-of-frame deletions or insertions.

Another approach that has been applied to particular genes was based on long-range PCR. Assuming that the target gene is affected by deletions/duplications that do not extend beyond the known gene boundaries, one can in principle find such rearrangements by covering the gene region with combinations of long-range PCR (1). For large genes, a large number of combinations are however necessary to ensure that at least one primer pair for long-PCR encompasses the rearrangement. Moreover, gene duplications, even those that duplicate relatively short stretches of DNA, are often missed by long-range PCR, because the normal allele, which yields a shorter product, competes with the amplification of the affected allele.

The following recent methods will be reviewed in more detail: (i) real-time quantitative PCR, (ii) DNA combing, (iii) quantitative multiplex PCR of short fluorescent fragments, and (iv) multiple amplifiable probe hybridization.

Quantitative real-time PCR (using the TaqMan™ technology from Applied Biosystems or the LightCycler™ technology from Roche) (Fig. 4.1) is a high-precision approach to quantitative analyses at the mRNA or DNA level, including gene dosage. The advantage of real-time PCR is that for each target DNA segment that is amplified in each of the samples to be tested a large number of time points are measured. The value attributed to each DNA segment within each sample is the number of cycles (or fractions of cycles) at which the increase of fluorescence, which is specifically related to the amplification level, reaches a suitably defined threshold. A number of internal controls are included as reference for constant copy number and as a basis for corrections of variations in PCR efficiency. This method has been applied in many laboratories to the precise definition of the copy number in genes affected by partial or total deletions or duplications (2–4). For genes consisting of large numbers of exons the main disadvantage of quantitative real-time PCR is the cost of reagents. Using SYBR-Green dye for labeling may reduce costs (5).

DNA rearrangements can be visualized under the optical microscope using *fluorescent hybridization to stretched*, or *"combed"* DNA (see, e.g., 6). As illustrated by, for example, recent applications to the detection of large rearrangements of the *BRCA1* gene involved in predisposition to breast and ovarian cancer (7), this approach can be sensitive, down to a resolution of ~2 kb. However, this method requires special equipment and software and relies on the availability of intact

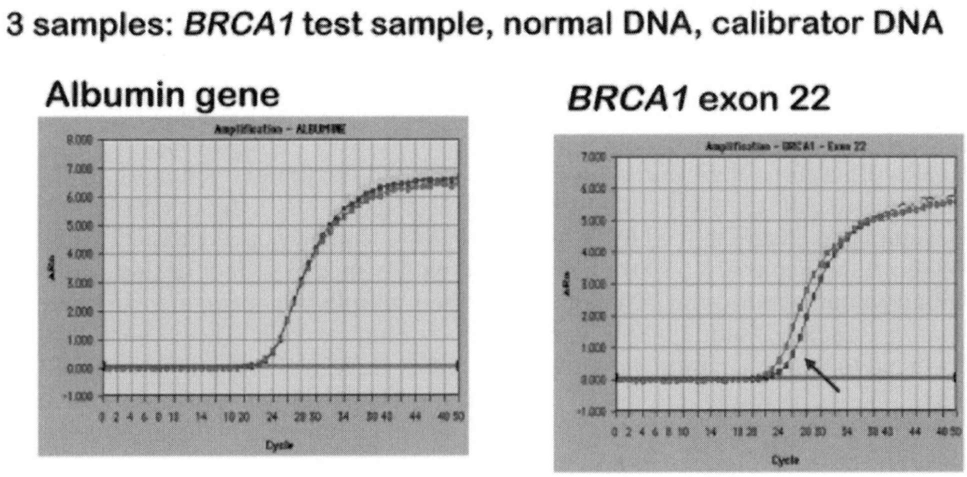

	Sample	$C_{t\,ALB}$	$C_{t\,Ex22}$	NEx22
■	Calibrator	21.72	21.77	1.00
□	Ex22 normal	21.81	21.75	1.03
■	Ex22 deletion	21.93	23.01	0.47

FIGURE 4.1. Real-time PCR using TaqMan to detect a deletion of *BRCA1* exon 22. (From M. Barrois, IGR, Villejuif, France)

TABLE 4.1. PCR-Based Quantitative Methods: Multiplex PCR of Fluorescent Fragments and Real-Time PCR

Multiplex PCR of Fluorescent Fragments	Real-time PCR
One-step method	Specific fluorescence measures the yield of PCR at each cycle
End point measurement	Real-time measurement
Hybridization kinetics of multiple primers can differ	Allows absolute measurement of copy number
Interference between primers is possible	High-precision method
	Closed format
	Several quantitative labeling formats are possible

DNA. Therefore, it is best suited to the analysis of DNA from established cell lines. Nevertheless, it is important to note that this is presently the only method that allows one to detect with high resolution cryptic rearrangements that do not result in loss or acquisition of DNA material, like inversions of a gene or of a short segment of a chromosome.

Various formats of *fluorescent multiplex PCR* have been used to amplify simultaneously multiple target sequences under quantitative conditions, thus allowing, in several genes, the detection of the 2-fold reduction resulting from heterozygous exon deletions and the detection of the 1.5-fold increase resulting from heterozygous exon duplications (8–10). In contrast to real-time PCR (Table 4.1), in which the slope of the amplification is measured for each target, these are end point methods, that is, amplification yields are measured at the end of the PCR (usually in the exponential phase). It is therefore essential in quantitative multiplex PCR experiments to minimize the variability of PCR efficiency between samples, and this condition can be fulfilled only if the DNA samples to be tested are of the same quality. Therefore, the same protocol must be used for the extraction of all DNAs to be tested. Alternatively, rather simple postextraction DNA treatments have been developed to ensure that all DNA samples behave similarly during PCR. The choice of short amplicons (150–300 bp), combined with a special primer design, has been found to homogenize the PCR kinetics of the multiple target fragments (up to 10) that are amplified simultaneously in the same multiplex PCR. These characteristics have led to the designation *quantitative multiplex PCR of short fragments* (QMPSF) (Ref. 11) for a recent version of this method (Fig. 4.2). Its sensitivity has been documented by, for example, the detection of exon deletions and duplications in the *MLH1* and *MSH2* genes in hereditary nonpolyposis colorectal cancer (HNPCC) families (12), by the demonstration of a high prevalence of *MSH2* rearrangements in this genetic predisposition (13) and by the detection of novel deletions and duplications in the *BRCA1* gene (11). A disadvantage of QMPSF and of all methods based on quantitative PCR is their liability to false positives resulting from allele dropout, that is, lack of amplification of one allele, if the sequence of one of the primers is affected by a polymorphism. However, known polymorphic positions are usually avoided at the stage of primer selection. Nevertheless, one must be aware of this potential problem, and, if the intensity of a single amplicon is reduced, the suggestive evidence of a deletion must be confirmed by using a different primer pair.

Multiple amplifiable probe hybridization (MAPH) (Refs. 14, 15) (Fig. 4.3) has the advantage of being less affected by variations in the quality of the DNA samples to

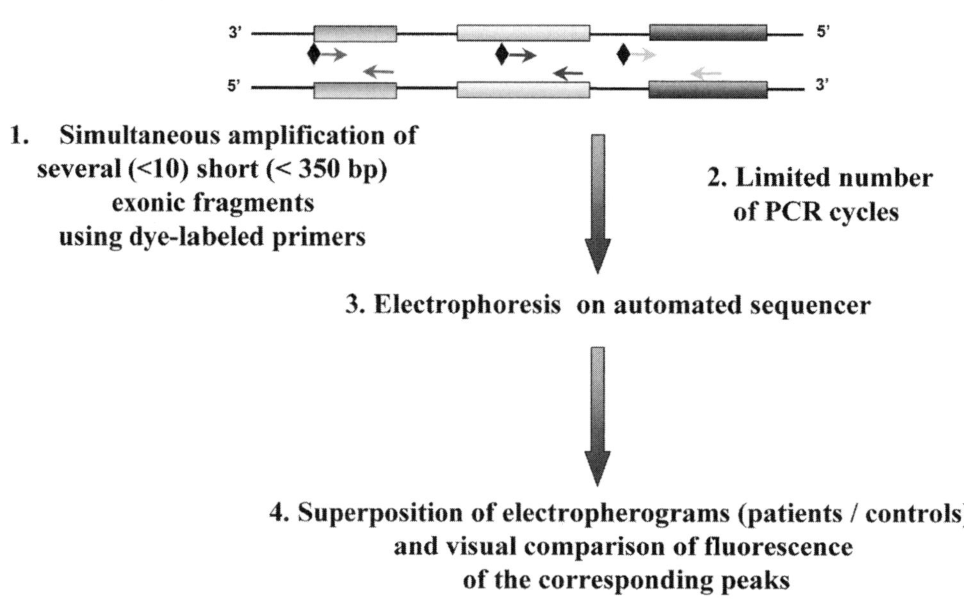

FIGURE 4.2. Quantitative multiplex PCR of short fluorescent fragments (QMPSF).

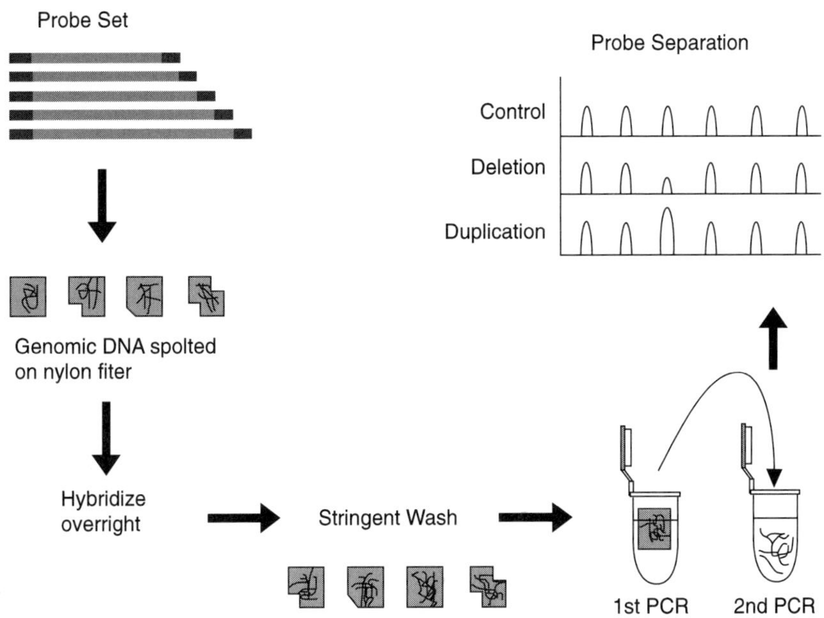

FIGURE 4.3. MAPH: Multiple hybridizable probe hybridization (16).

be tested. Moreover, MAPH is not affected by the presence of SNPs and, once established for a particular application, it can be used with high throughput. In this method, the test DNA is spotted and immobilized on a membrane and is washed before hybridization of multiple probes. For each target gene, the set of probes is constructed in such a way that they are all flanked by the same pair of short sequences, thus allowing their PCR amplification. One of these primers can be radioactively labeled, as in the initial MAPH protocol, but a fluorescent label is presently used in most cases. Thus the quantitative aspect of MAPH resides in the hybridization step, which is similar to that of Southern blot hybridization, and PCR is used to amplify the probes that have quantitatively hybridized to their targets and to measure their amounts. Up to 40 probes can be hybridized simultaneously to many immobilized DNA samples, if the size of these probes is chosen in a way that they are all distinguishable by electrophoresis (14). MAPH, like QMPSF, has the advantage of not requiring special materials or equipments, but the development of good sets of probes for each MAPH application is not trivial and should probably be realized by specialized groups.

Multiplex ligation-dependent probe amplification (MPLA) is discussed in a separate chapter.

MAPH and MLPA have been reviewed recently (17).

■ DETECTION OF CHROMOSOMAL REARRANGEMENTS

Standard karyotyping methods do not have sufficient resolution to detect certain types of rearrangements that are associated with serious and rather frequent genetic diseases, for example, the subtelomeric rearrangements that are quite often found associated with mental retardation.

Fluorescent in situ hybridization (FISH), which is the method most commonly used in this field, depends on the availability of good-quality samples (good mitotic index, good metaphases). Additional limitations are the cost of reagents and the time and expertise needed to analyze each sample.

Array CGH, that is, comparative genomic hybridization combined with the use of microarrays (18, 19) is reviewed elsewhere in this volume. The use of large numbers of large-insert genomic clones spotted on glass slides allows one to reach a resolution on the order of 20–100 kb. Array CGH requires special equipment and has recently undergone several refinements. It has recently been applied to a number of deletion syndromes, including the detection of subtelomeric deletions (20).

Two of the methods that have already been described above in the section on detection of gene deletions have also been used to detect cryptic chromosomal rearrangements. MAPH has been used to screen for subtelomeric chromosomal abnormalities (21), and QMPSF has been used to reveal novel deletions in the DiGeorge region on chromosome 22q (22).

As is always the case when comparing methods, one has to find the best balance between advantages and disadvantages, depending on the nature of the project and the environment in which the method is being used.

REFERENCES

1. Payne SR, Newman B, King M-C (2000) Complex germline rearrangement of BRCA1 associated with breast and ovarian cancer. Genes Chromosomes Cancer 29: 58–62.
2. Bièche I, Olivi M, Champème MH, Vidaud D, Lidereau R, Vidaud M (1998) A novel approach to quantitative PCR using real-time detection: Application to the detection of gene amplification in breast cancer. Int J Cancer 78: 661–666.
3. Boulay JL, Reuter J, Ritschard R, Terracciano L, Herrmann R, Rochlitz C (1999) Gene dosage by quantitative real-time PCR. Biotechniques 27: 228–230.
4. Laurendeau I, Bahuau M, Vodovar N, Larramendy C, Olivi M, Bieche I, Vidaud D, Vidaud M (1999) TaqMan PCR-based gene dosage assay for predictive testing in individuals from a cancer family with INK4 locus haploinsufficiency. Clin Chem 45: 982–986.
5. Boehm D, Herold S, Kuechler A, Liehr T, Laccone F (2004). Rapid detection of subtelomeric deletion/duplication by novel real-time quantitative PCR using SYBR-Green dye. Hum Mutat. 23: 368–378.
6. Herrick J, Michalet X, Conti C, Schurra C, Bensimon A (2000) Quantifying single gene copy number by measuring fluorescent probe lengths on combed genomic DNA. Proc Natl Acad Sci USA 97: 222–227.
7. Gad S, Aurias A, Puget N, Mairal A, Schurra C, Montagna M, Pages S, Caux V, Mazoyer S, Bensimon A, Stoppa-Lyonnet D (2001) Color bar coding the BRCA1 gene on combed DNA: a useful strategy for detecting large gene rearrangements. Genes Chromosomes Cancer 31: 75–84.
8. Yau SC, Bobrow M, Mathew G, Abbs SJ (1996) Accurate diagnosis of carriers of deletions and duplications in Duchenne/Becker muscular dystrophy by fluorescent dosage analysis. J Med Genet 33: 550–558.
9. Robinson MD, Chu CE, Turner G, Bishop DT, Taylor GR (2000) Exon deletions and duplications in BRCA1 detected by semiquantitative PCR. Genet Test 4: 49–54
10. Saugier-Veber P, Drouot N, Lefebvre S, Charbonnier F, Vial E, Munnich A, Frebourg T (2001). Detection of heterozygous SMN1 deletions in SMA families using a simple fluorescent multiplex PCR method. J Med Genet 38: 240–243
11. Casilli F, Di Rocco ZC, Gad S, Tournier I, Stoppa-Lyonnet D, Frebourg T, Tosi M (2002) Rapid detection of novel BRCA1 rearrangements in high-risk breast-ovarian

cancer families using multiplex PCR of short fluorescent fragments. Hum Mutat. 20: 218–226.

12. Charbonnier F, Raux G, Wang Q, Drouot N, Cordier F, Limacher JM, Saurin JC, Puisieux A, Olschwang S, Frebourg T (2000) Detection of exon deletions and duplications of the mismatch repair genes in hereditary nonpolyposis colorectal cancer families using multiplex polymerase chain reaction of short fluorescent fragments. Cancer Res. 60: 2760–2763.
13. Charbonnier F, Olschwang S, Wang Q, Boisson C, Martin C, Buisine MP, Puisieux P, Frebourg T (2002) MSH2 in contrast to MLH1 and MSH6 is frequently inactivated by genomic rearrangements in hereditary non polyposis colorectal cancer. Cancer Res 62: 848–853.
14. Armour JAL, Sismani C, Patsalis PC, Cross G (2000) Measurement of locus copy number by hybridization with amplifiable probes. Nucleic Acids Res 28: 605–609.
15. Armour JAL, Barton DE, Cockburn DJ, Taylor GR (2002) The detection of large deletions or duplications in genomic DNA. Hum Mutat 20: 325–337.
16. White S, Kalf M, Liu Q, Villerius M, Engelsma D et al. (2002) Comprehensive detection of genomic duplications and deletions in the DMD gene, by use of multiplex amplifiable probe hybridization. Am J Hum Genet 71: 365–374.
17. Sellner LN, Taylor GR (2004) MLPA and MAPH: New techniques for detection of gene deletions. Hum Mutat. 23: 413–419.
18. Pinkel D, Segraves R, Sudar D, Clark S, Poole I et al. (1998) High resolution analysis of DNA copy number variation using comparative genomic hybridization to microarrays. Nat Genet 20: 207–211.
19. Snijders AM, Nowak N, Segraves R, Blackwood S, Brown N et al. (2001) Assembly of Microarrays for genome-wide measurement of DNA copy number. Nat Genet 29: 263–264.
20. Veltman JA, Schoenmakers EF, Eussen BH, Janssen I, Merkx G et al. (2002) High-throughput analysis of subtelomeric chromosome rearrangements by use of array-based comparative genomic hybridization. Am J Hum Genet 70: 1269–1276.
21. Sismani C, Armour JA, Flint J, Girgalli C, Regan R, Patsalis PC (2001) Screening for subtelomeric chromosome abnormalities in children with idiopathic mental retardation using multiprobe telomeric FISH and the new MAPH telomeric assay. Eur J Hum Genet 9: 527–532.
22. Jacquet H, Roax G, Thibault, Hecketsweiler B, Houy E et al. (2002) PRODH mutations and hyperprolinemia in a subset of schizophrenic patients. Hum Molecular Genetics 11: 2243–2249.

5

PRINCIPLES OF MUTATION SCANNING BY DHPLC

John F. Harvey
National Genetics Reference Laboratory, Salisbury District Hospital, Salisbury, Wiltshire, UK

■ INTRODUCTION

This chapter describes the Transgenomic WAVE® DNA Fragment Analysis system with DNA Sep® Technology as used in molecular genetics diagnostic laboratories for the detection of unknown mutations. Four software packages are currently available: WAVEmaker 3.4, 4, 4.1, and Navigator. Denaturing high-performance liquid chromatography (DHPLC), also known as temperature-modulated heteroduplex analysis (TMHA), provides a rapid automated scanning method for mutations, known or unknown. Mutations are visualized as a characteristic pattern of peaks corresponding to the mixture of heteroduplexes and homoduplexes formed when wild-type and mutant DNA are hybridized (1). DHPLC has been shown to clearly resolve mutations in various genes with detection rates ranging from 94% to 100% (2–9). These high rates of detection rely on optimization of the run conditions, including careful primer design, the choice of fragment sequence and length, and array of melt temperatures necessary to accommodate melt domains of each fragment. It is reported in the literature that the sensitivity of detection by DHPLC is higher than for alternative gel-based analysis techniques; that is, mutations not detected by gel-based techniques could be detected by DHPLC (10–12).

Guide to Mutation Detection, Edited by Graham R. Taylor and Ian N. M. Day.
ISBN 0-471-23444-3

■ PRINCIPLES OF MUTATION DETECTION BY DHPLC

Individuals who are heterozygous for a mutation or polymorphism have a 1-to-1 ratio of wild-type and mutant DNA. A mixture of hetero- and homoduplexes is formed when the PCR product is hybridized by heating to 95°C and cooling slowly (Fig. 5.1). The DNA from individuals who have two identical mutant alleles (homozygous mutation) must be mixed with wild-type DNA and hybridized. After this treatment, a sample will contain a mixture of hetero- and homoduplexes. However, DNA from homozygous wild-type individuals will form only one species, the homoduplex wild type.

The WAVE nucleic acid analysis platform is based on ion-pair reversed-phase high-performance liquid chromatography (IP-RP-HPLC) and can be operated in three modes as determined by the temperature at which separation is performed (Fig. 5.2).

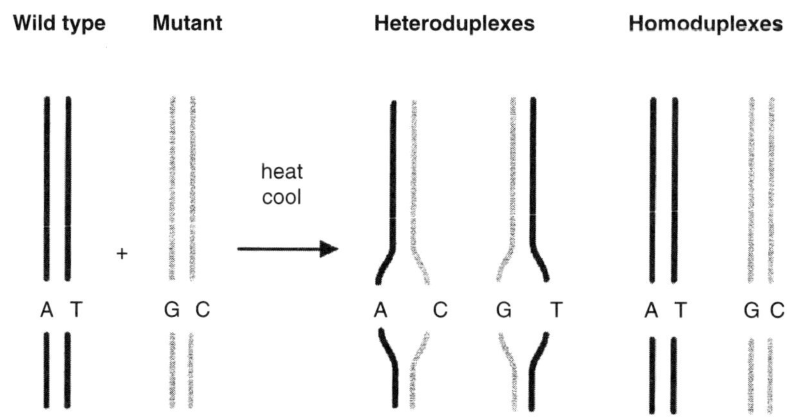

FIGURE 5.1. Formation of a 1 : 1 mixture of heteroduplexes and homoduplexes through hybridization.

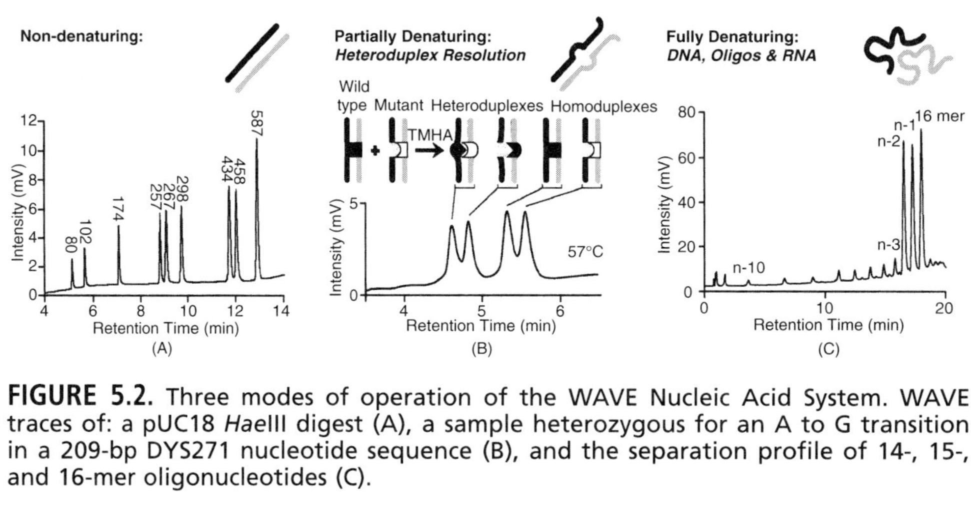

FIGURE 5.2. Three modes of operation of the WAVE Nucleic Acid System. WAVE traces of: a pUC18 *Hae*III digest (A), a sample heterozygous for an A to G transition in a 209-bp DYS271 nucleotide sequence (B), and the separation profile of 14-, 15-, and 16-mer oligonucleotides (C).

The process uses two phases: a stationary phase consisting of a DNASep® cartridge and a mobile phase of chemicals including acetonitrile (ACN, an organic cosolvent used to selectively elute samples) and the ion pairing agent triethylammonium acetate (TEAA), which ensures that the DNA interacts with the stationary phase. Under nondenaturing conditions (<50°C), double-stranded DNA (dsDNA) will be eluted according to size (not sequence) with the WAVE buffer system (Fig. 5.2A). Essentially, the anionic nucleic acid phosphate backbone of DNA interacts electrostatically (noncovalently) with the ammonium ion of the TEAA on the basis of size. The larger the DNA fragment, the more interactions it makes with TEAA. Once it is associated with TEAA, it effectively has a hydrophobic outer coating of TEAA alkyl groups that allows it to interact with the hydrophobic DNASep matrix, with the number of bonds formed being determined by the DNA fragment size. Elution of DNA from the DNASep matrix is brought about by the use of the hydrophobic eluant ACN. As a gradient (a constantly increasing concentration) of ACN is run through the column, DNAs of increasing size are eluted (Fig. 5.2A). This nondenaturing mode works well for quantitative work comparing peak height or area ratios for duplication/deletion detection (13).

At fully denaturing temperatures, single-stranded molecules (RNA or DNA) can be separated (Fig. 5.2C). This mode can be used for oligonucleotide quality control and purification. Oligos can be effectively separated from failure peaks with a fragment collector. There is a range of columns and buffers available (Transgenomic Ltd., Crewe, UK) for high-purification work (Transgenomic application Note 103).

Separating fragments under partially denaturing conditions is the focus of this chapter. Denaturing and hybridizing a sample containing two products that differ by as little as one nucleotide results in the formation of two homoduplexes and two heteroduplexes. These may be visualized as four distinct peaks (Fig. 5.2B), but any difference in peak profile compared with a homozygote wild-type control is indicative of a mutation. In practice, a four-peak pattern is seldom seen, as the homo- and heteroduplex elution profiles associated with many mutations, overlap. A distinct mutant pattern is nevertheless usually seen when the fragment-specific T_m (temperature at which the fragment is ~75% helical) is selected for analysis and at temperatures corresponding to ~70–90% melt.

At a column temperature determined with the melting algorithm WAVEmaker software (Transgenomic Ltd., Crewe, UK) DHPLC separates the DNA molecules with the heteroduplexes eluting first (Fig. 5.2B). This is the result of bubble formation in the dsDNA duplex that reduces the number of charges accessible for electrostatic interactions between the DNA bases and the ion-pairing layer of the column stationary phase. The charge density of the DNA molecule is reduced within the single-stranded region.

■ MATERIALS

Hardware

The current DHPLC WAVE systems include the 2100A, 3500A, or HT and the MD equipped with DNASep® Cartridges (Transgenomic, San Jose, CA; see Table 5.1). Each integrated system comprises an autosampler, pump, column oven, DNASep® column, in-line degasser, UV detector, WAVE utility software 3.4, 4.0, 4.1, or

TABLE 5.1. The WAVE Nucleic Acid Fragment Analysis Systems

Feature	2100/3500	2100 A/3500 A	3500HT	MD
Column	4.6 mm DNASep Cartridge	4.6 mm DNASep Cartridge	6.5 mm DNASep Cartridge	4.6 mm DNASep Cartridge
Accelerator	None (external can be applied)	Internal	Internal	External
Oven calibration	Single, dual, or multipoint	Single, dual, or multipoint	Multipoint	Multipoint
Accuracy	0.1°C	0.1°C	0.1°C	0.1°C
Mixer	2 ml	2 ml	0.250 ml	—
Flow rate	0.9 ml/min	0.9 ml/min	0.9 ml/min (normal) 1.5 ml/min (rapid)	0.9 ml/min
Throughput	±200 injections/day	±200 injections/day	±400 injections/day	±200 injections/day

WAVEMAKER software or WAVE Navigator software can be run with all system configurations.
The UV detector is standard for all systems. Fluorescence detector and fragment collector are optional.

Navigator computer system, monitor, and color printer. Optional extras include a fragment collector, for isolating individual peaks for further analysis, sequencing, etc. (Transgenomic application note 104), and, an external accelerator for the 2100 A (internal accelerators are an integral part of the 3500 A and 3500HT systems, reducing the time of sample throughput, e.g. from ~6.8 to 5.6 min or ~3.2 min for the HT system). The addition of an accelerator to the original 2100 A for high-throughput work is useful, but it needs careful optimizing and current methods must be modified to ensure peaks are captured in the smaller window provided. To increase sensitivity (100×) a fluorescence detector may be added. This is used with a dye (Cy5 or Cy3) intercollator system that obviates the need for fluorescently tagged primers and allows potential pooling strategies.

Chromatography is performed with buffers A: aqueous solution of 0.1 M TEAA and B: 0.1 M TEAA in 25% ACN. The DNASep Cartridge is charged with a nucleic acids sample at a software-predicted ratio of buffers A and B, which ensures binding of DNA to the column matrix. Elution of the DNA, in a manner dependent on its lengths and state of naturation/denaturation, is then achieved by a systematic, software-predicted increase in elution buffer B relative to A. The elution of the nucleic acids is monitored on-line by UV detection. Eluent flow rates are 0.9 ml/min (standard analysis mode) or 1.5 ml/min (rapid analysis mode). Chromatograms are recorded at 260 nm; all analyses under nondenaturing conditions are run at 50°C. Mutation detection under partially denaturing conditions is performed at or near the software-predicted temperature indicated for each chromatogram.

DNASep Cartridge Technology

The DNASep cartridge contains a robust, nonporous, alkylated polystyrene-divinylbenzene C18 separation matrix (14, 15), which exhibits a higher affinity for dsDNA than for single-stranded DNA (ssDNA). Mutation detection is performed at or slightly below the melting temperature of the homoduplex. At this temperature heteroduplexes, if present, are partially melted at the mismatch site and therefore exhibit partial ssDNA character. This results in a decreased retention time of partially denatured heteroduplexes compared with the corresponding homoduplexes.

Software

WAVEMAKER and WAVE Navigator Software

WAVEMAKER and WAVE Navigator software packages (Transgenomic, San Jose, CA) control the operation of the HPLC hardware. Melt profiles, analysis temperatures, and HPLC gradient conditions are predicted by the software. The analysis temperatures for mutation detection are derived from software-predicted melt profiles of percent helical fraction versus base position. Both software packages employ a modified Fixman–Friere algorithm to predict the sequence-dependent melt behavior of DNA fragments (16) and control the acquisition and handling of data and support data analysis. WAVE Navigator includes enhanced features such as a relational database management system, automated mutation calling, across project data analysis, remote operation capabilities, and robust data storage capabilities.

■ METHODS

Primer Design

The major procedures involved in DHPLC mutation detection include PCR primer design; PCR (oil free); heteroduplex formation; melting profile determination of the fragment; and method preparation, which includes setting of the separation gradient, making a new project that includes sample sheet preparation, sample running, and sample analysis. Many operators will choose to use primer sets originally designed for gel SSCP/heteroduplex analysis. However, consideration should be given to the predicted fragment melting profiles. For the detection of unknown mutations it is probably best to select fragments of 150–450 bp. (Although not recommended, single base mutations can be detected in >1 kb fragments). If the sequence is known, construct the melting profile using appropriate software (see below). Where possible, locate primers near the high points on this profile so that GC-rich regions are at the ends of fragments rather than in the middle. Avoid inclusion of regions that differ by more than 15°C in the same fragment. If this is unavoidable, it may be necessary to substitute dGTP with N^7-dGTP (7-deaza-2′-deoxyguanosine 5′-triphosphate) to lower the melting temperature of GC-rich regions or, exceptionally, to include a short GC clamp of 3 or 4 bases. Once the approximate locations of primers have been decided, more detailed primer design can begin.

The use of primers with nontemplate tails such as universal sequencing primers or T7 promoters should be avoided, although it may be possible to use these with *Pfu* polymerase (Stratagene, La Jolla, CA) or Optimase (Transgenomic Ltd., Crewe, UK). The recommended primer melting temperature T_m is 56°C. The difference in T_m between primer and template should be about 25°C. Any possible primer dimers should be less stable than the 3′-pentomer by at least 5 kcal/mol. Any primer self-annealed loops should have a T_m of less than 12°C. Primers should be of high purity without failure sequences. To avoid degradation, it is preferable to store primers in Tris·HCl (pH 8.0) buffer rather than in pure water. Many websites are now available for effective primer design, for example Primer 3 software program (http://www.genome.wi.mit.adm/genome_software/other/primer3/html). Also, http://www.mutationdiscovery.com is proving a useful site, and a new module for amplicon design has recently been added for premium users. Regions of interest are indicated on a locus, and a simple user interface provides access to Primer 3-based algorithm for generating flanking primers. The resulting amplicon can be annotated back onto the Mutation Discovery locus sequence. A particularly useful feature is the inclusion of published SNPs on the displayed sequence, which allows careful choice of primers to avoid these polymorphisms.

Melt Profiling

Claims regarding the sensitivity of DHPLC vary but largely reflect the committed expertise of the users and the care taken in designing the analysis and the sequence examined, for example, the position of CG-rich regions, the size of fragment, and the number of injection temperatures used. Increasing the fragment length screened, for example to >500 bp, invariably results in the need to accommodate more melt domains and hence more injections per sample at a range of oven temperatures.

To predict the melting profile of DNA fragments, use the most up-to-date WAVEmaker or WAVE Navigator software. This should ensure the most accurate estimation of melting. The modified Melting Curves Report in Figure 5.3 was generated (using WAVEmaker software) for detection of SNPs in a 492-bp fragment and resulted in the discovery of the T to G transversion represented in Figure 5.4.

It is important to ensure that the whole sequence (including clamps if used) is included in the melt analysis; otherwise, gradient mispredictions will occur. Select amplicon for analysis on the "DNA" page according to the manual. In Navigator four windows are viewed: %helical fraction/tm; base position/tm; %helical fraction/base position; and tm/base position. Aim for melt temperatures giving about 75% helical within a range from 50% to 95% helical. Two to three temperatures may be necessary to accommodate all melt domains. Five melt option temperatures can be entered and viewed simultaneously. The predicted melt temperature/profile is seldom sufficient or totally reliable; a trial and error approach is advised to select the best temperatures. The melts predicted by other software are not always transferable. Small GC-rich regions within fragments can be troublesome, giving a "tight ceiling effect" on the %helical fraction/base position chart. These must be accommodated by careful melt temperature selection with 0.1°C temperature increments, repositioning primer sets with these regions at the end of fragments, or including GC clamps (17). The software allows for automatic addition of clamps to the amplicon, allowing the immediate affect on melt profiles to be determined. Sometimes direct sequencing may be the best option for analyzing "difficult" fragments.

Identifying the Presence of a Genetic Variation—Mutation Calling

Single nucleotide changes as well as short insertions and deletions can be detected by DHPLC. However, although elution profiles are consistent for a specific mutation, different mutations may share indistinguishable profiles, therefore sequencing

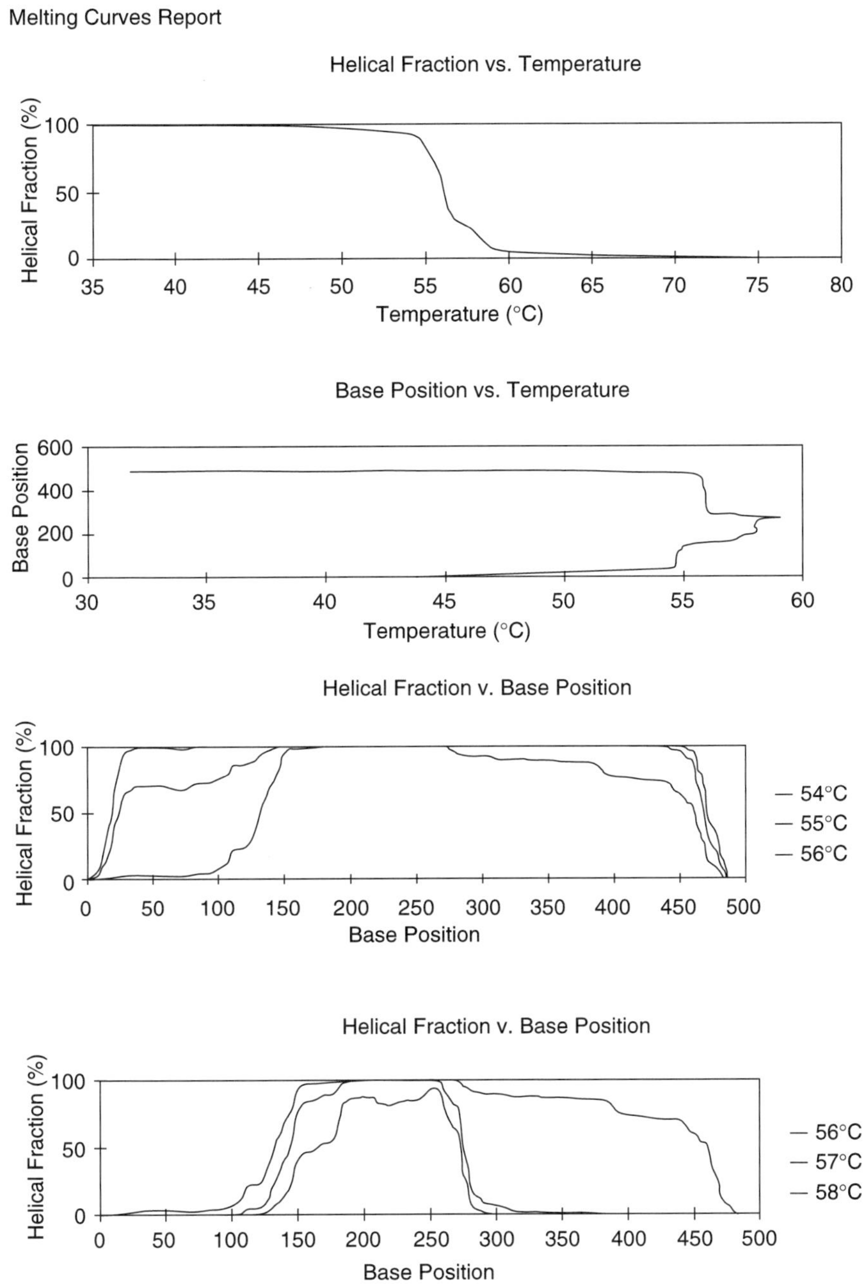

GTGAAATAAATGAATGATAACTGAAACAGGGAATTGGCCAAGAAAGACAAAGAGAAGGAATGAAACAATAAATAATAACACAGAAAGTGAAATTAGAATCAAACATAAA
CGGACCTATAGAATAACTATCTGACAATGGAAAGTATGACACTGCCATCAATCAAGAGACTCTAGATATGAAGCCAGAGGGAACTTCGTGAAGGTGAGCTAATCAAC
ATAAATCAGGAAAGCCATTGTAAGAAAAGGATGGGGCGGTCCCTGAAGAAATGGCACTAGTGAAAATTTTCACATTAGAGAAACTCTTGGAGATATTTCACAATATTGA
CAGTGCAAATGATAAGATACTGAAGCTGATTCAAACTTGGAAAAGAGTATGACAACTCACCAAGGTATAGAAAAGATTCTTGATCCATACTGAAAGTTACTGTAAGACA
AGGTGAACACTATTCACACTATTCCCGATACAACAATTTTTCACAAAGATATAAA

FIGURE 5.3. Melting curves report for a 492-bp DNA fragment showing two melt domains and predicted T_m of 55°C and 58°C to accommodate a GC-rich domain.

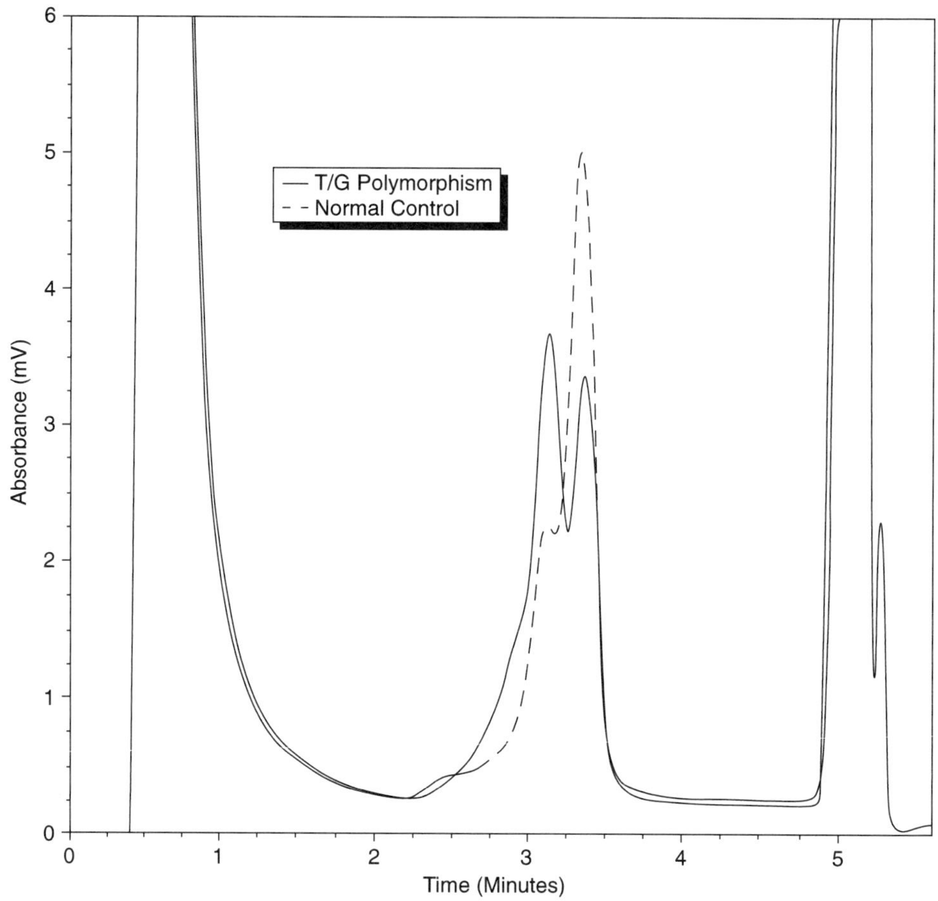

FIGURE 5.4. WAVE trace of a T/G transversion at base 123 in a 492-bp fragment, found at T_m 55°C as predicted by WAVEmaker software (Fig. 5.3). The normal wild-type control peak is superimposed on the mutant, double-peak trace, a feature now provided by Navigator software. Data courtesy of Rachel Wycherley, National Genetics Reference Laboratory (Wessex), UK.

is advised for the characterization of variants. DHPLC and sequencing analyses provide an efficient two-tier approach to mutation analysis. However, specific mutation analysis can be achieved by DHPLC using primer extension methods (18).

Mutations are identified based on a comparison between the chromatograms of wild-type amplicons that are devoid of sequence variations, and chromatograms of amplicons that carry mutations that have formed heteroduplexes with wild-type amplicons, recorded under identical chromatographic conditions. DHPLC has been shown to detect mosaic mutations that account for only a few percent of alleles in a PCR product and that remain undetected by direct sequence analysis (19). Mutations can be detected at very low levels with DHPLC (20, 21). This is useful for mitochondrial analysis, tumor samples, and somatic mosaics. The difficulty occurs when a pattern change has been seen resulting from a low-level mutation that is not able to be verified by sequencing because sequencing does not generally call mutations when present at less than 30%. If there is a definite and reproducible change that cannot be verified by direct sequencing of the DNA, enriching the DNA for mutant peak with a fragment collector followed directly by either sequencing or cloning and sequencing should enable localization of the mutation. (NB: collection can be done without a fragment collector by manual collection and careful timing.) Figure 5.4 shows a double-peak DHPLC trace of a polymorphic T/G transversion at base 123 in a 492-bp fragment at T_m 55°C as predicted by WAVEmaker software. The normal wild-type control peak is superimposed for comparison.

Characteristic examples of SNP mutation calling are presented in Figures 5.4 and 5.5. In Figure 5.5 the wild-type (a fragment of exon 11, BRCA1) chromatogram exhibits a single peak with a shoulder pattern. With increasing analysis temperature the PCR product progressively melts, resulting in a steady increase in its single-stranded character and causing a reduction in retention time. The single-peak chromatogram of the wild-type amplicon serves as a reference for mutation calling.

The presence of sequence variations/mutations leads to the formation of two duplexes (see above). In the case of the wild-type/mutant chromatogram (Fig. 5.5, 56.0–56.5°C), the presence of a sequence variation becomes apparent by the appearance of an additional peak in the chromatograms. In contrast to the

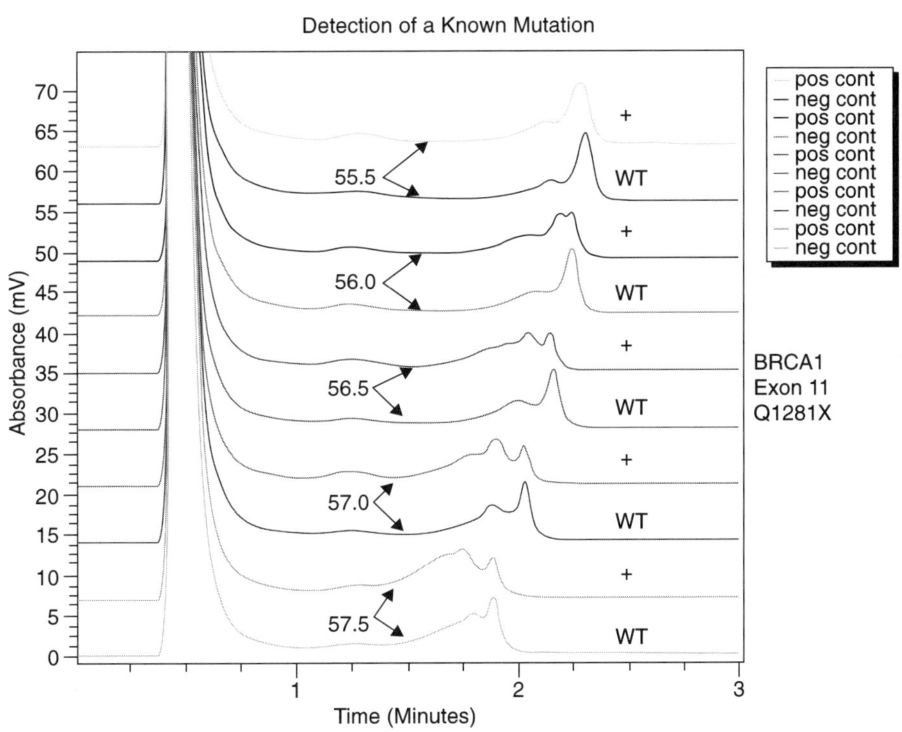

FIGURE 5.5. Mutation analysis by DHPLC of a point mutation Q1281X (Glutamine to a Stop) in exon 11 of *BRCA1*. PCR products were denatured and slowly reannealed to form mixtures of heteroduplexes and homoduplexes. DHPLC was performed over a temperature range from 55.5 to 57.5°C; the effect of oven temperature on mutant (+) and wild-type sample (WT) is demonstrated. All chromatograms were recorded under identical gradient conditions. Figure kindly provided by Gert Matthijs and Bruno Vankeirsbilck, Center for Human Genetics, Leuven.

chromatograms for the wild-type amplicon (Fig. 5.5, top), the 1:1 wild-type-mutant mixture contains two distinct duplexes, one peak of two overlapping homoduplexes that correspond to one wild-type and one mutant homoduplex and two heteroduplexes, which are formed by the (+/–) and the (–/+) strands of the wild-type and mutant allele PCR products. Because homoduplexes have higher melt temperatures than the corresponding mismatch-containing heteroduplexes, the retention times of peaks representing homoduplexes are always longer than the retention times of their heteroduplex counterparts under identical gradient conditions and at analysis temperatures at or near the melt temperature of the homoduplex.

In the wild-type/mutant chromatograms presented here, the two heteroduplex species and two homoduplex species are not resolved from each other at any of the temperatures tested. This is a typical result; the separation of four distinct peaks is seldom seen except for the mutation standard provided, DYS271 (Transgenomic Ltd, Crewe, UK).

It should be noted that many mutations do not give clearly defined heteroduplex peaks but may be simply represented as a subtle shoulder on the homoduplex peak. Failure to detect mutations may arise for a number of reasons. Careful maintenance (according to manufacturers' instructions) and instrument checks are essential (22).

Instrument Controls/Mutation Standards

The fragment size standard (pUC18/*Hae*III) and mutation detection control (DYS271) function as instrument controls. These controls should be run through the system at the recommended oven temperatures as a matter of course whenever the buffers are changed or ideally at the beginning of each run. In addition, controls should also be run after any system change, for example, changing the DNASep cartridge or in-line filter; otherwise, an instrument problem or a loss of heteroduplex resolution on the cartridge may not be obvious during the analysis. Results from the controls as well as system performance should be recorded in a log book on a regular basis. It is crucial that optimal resolution is obtained for DYS271. This is when the distance between the homoduplex peaks equals the distance between the heteroduplex peaks. If the two peaks are not equal or the peak sizes are disproportionate, oven calibration is suspect (Fig. 5.6).

Provided the mutation standards are working correctly the most likely problems are due to:

1. *Poor-quality PCR*: Very low yields can affect the ability to see mutations while very high yields can lead to large proportions of misincorporations and hence increase difficulty in calling mutations. These problems can also be due to variation in the fidelity of PCR enzymes.
2. *Mutations that occur in a high-melt GC-rich pocket* in a fragment with otherwise normal nucleotide ratios. To resolve these, it may be necessary to redesign fragments, preferably with the GC-rich regions toward the ends of the fragments. GC clamps may be used (17).
3. *The small size of the fragment.* Mutations in fragments less than 110–120 base pairs, which may melt too rapidly, can remain undetected unless some method of fragment extension is used.

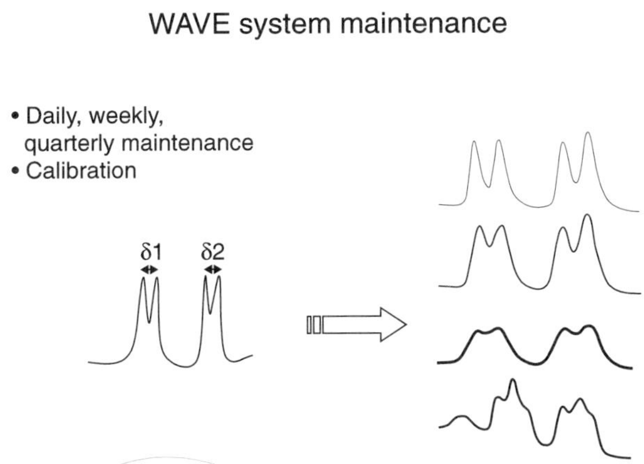

FIGURE 5.6. Diagram of a typical WAVE trace of the mutation standard DYS271 at optimal resolution (left). Abnormal traces represented (right) indicate system problems. Standards act as useful instrument controls and should be run frequently to ensure that conditions are optimal. Figure kindly provided by Gert Matthijs and Bruno Vankeirsbilck. Center for Human Genetics, Leuven.

Mutation Detection in the Rapid Analysis Mode

With WAVE System Models 2100 A and 3500 A approximately 200 samples can be analyzed per day assuming one injection per sample. If, as is common, three injections are used for a range of oven temperatures to accommodate the different melt domains in a fragment, the sample throughput will be reduced accordingly to ~60–70 samples per day. This corresponds to an average run time per sample of 7.2 min. The number of samples analyzed can be doubled with WAVE System Model 3500HT. The increased sample throughput is made possible, in part, by system modifications such as (i) change of the DNASep cartridge geometry, which permits an increase in flow rate from 0.9 ml/min to 1.5 ml/min, (ii) use of the smaller eluent mixer (0.250 ml vs. 2.00 ml), which supports faster and more precise changes in eluent composition, and (iii) the use of an accelerator, which allows for an accelerated cartridge-cleaning step as well as reduced equilibration times between injections (see Table 5.1).

Sample and Eluent Compatibility with the DNASep Cartridge

It is crucial for consistent, high-quality mutation analyses on the WAVE System that samples and buffer preparations are devoid of contaminants and additives that diminish the performance of the DNASep cartridge. Any compound present at concentrations high enough to interfere with the interaction between the separation matrix and the nucleic acids will diminish cartridge performance. The damage done by some components is reversible, whereas others cause permanent damage.

Importance of Polymerase Fidelity for Mutation Detection

DNA polymerases inherently incorporate errors randomly during amplification (23–25). The extent of error incorporation depends in part on a number of factors such as dNTP and Mg^{2+} concentration. However, the primary factor determining the error incorporation rate is the polymerase itself. Polymerases with $3' \rightarrow 5'$ exonuclease activity, generally referred to as proofreading activity, incorporate the least number of errors and exhibit highest replication fidelity (26–29). Errors incorporated at random during amplification have the potential to interfere with identification of mutations present in the amplified template. Thus the use of a high-fidelity polymerase such as Transgenomic's Optimase™ polymerase is recommended for the amplification of PCR products to be used for mutation detection.

All WAVE Nucleic Acid Fragment Analysis System configurations provide rapid and reasonably high-throughput DNA fragment separations without time-consuming post-PCR processing, using DNASep polymeric cartridges for the purpose of detecting genetic variations. Sensitivity and specificity of genetic variance detection with the WAVE System are superior to other methods and equivalent to sequencing at about one-fifth of the cost, reducing the volume of labor-intensive sequence data analysis (30), although this in turn may be reduced with the development and introduction of new sequence analysis packages. One current limitation of WAVE System analysis arises from the use of a single cartridge that restricts the sample throughput potential as compared with multicolumn/multicapillary analysis systems. However, the WAVE System still provides a sensitive, efficient technology platform for medium to low throughput applications.

REFERENCES

1. Oefner PJ, Underhill PA (1998) DNA mutation detection using denaturing high-performance liquid chromatography (DHPLC). In *Current Protocols in Human Genetics* Dracopoli NC, Haines JL, Korf BR, Moir DT, Morton CC, Seidman DE et al. (eds), Wiley, New York, pp. 7.10.1–7.10.12.
2. Haynes S, Eccles D, Harvey J (2000) DHPLC analysis of HNPCC: A rapid sensitive exon screen of hMLH1 and hMSH2. J Med Genet 37 (Suppl.1): S42.
3. Wagner T, Stoppa-Lyonnet D, Fleischmann E, Muhr DI, Pages S, Sandberg T et al. (1999) Denaturing high-performance liquid chromatography detects reliably BRCA1 and BRCA2 mutations. Genomics 62: 369–376.
4. Gross E, Arnold N, Goette J, Schwartz-Boeger U, Kiechle M (1999) A comparison of BRCA1 mutation analysis by direct sequencing. Hum Genet 105: 72–78.
5. Giordano M, Oefner PJ, Underhill PA, Cavilli-Sforza LL, Tose R, Richiardi PM (1999) Identification by denaturing high-performance liquid chromatography of numerous polymorphisms in a candidate region for multiple sclerosis susceptibility. Genomics 56: 247–253.
6. Liu WL, Oefner PN, Qian C, Odom RS, Francke U (1998) Denaturing HPLC-identified novel FBN1 mutations, polymorphisms, and sequence variants in Marfan syndrome and related connective tissue disorders. Genet Test 1: 237–242.
7. Jones AC, Sampson JR, Hoogendoorn B, Cohen D, Cheadle JP (2000) Application and evaluation of denaturing HPLC for molecular genetic analysis in tuberous sclerosis. Hum Genet 106: 663–668.
8. Klein B, Weirich G, Brauch H (2001) DHPLC-based germline mutation screening in the analysis of the *VHL* tumor suppressor gene: usefulness and limitations. Hum Genet 108: 376–384.
9. Holinksi-Feder E, Mller-Koch Y, Friedl W, Moeslain G, Keller G, Plaschke J et al. (2001) DHPLC mutation analysis of the hereditary nonpolyposis colon cancer (HNPCC) genes hMLH1 and hMSH2. J Biochem Biophys Methods 47: 21–32.

10. Choy YS, Dabora SL, Hall F, Ramesh V, Niida Y, Franz D, Kasprzyk-Obara J, Reeve MP, Kwiatkowski DJ (1999) Superiority of denaturing high performance liquid chromatography over single-stranded conformation and conformation-sensitive gel electrophoresis for mutation detection in TSC2. Ann Hum Genet 63: 383–391.
11. Gross D, Arnold N, Pfeifer K, Bandick K, Kiechle M (2000) Identification of specific BRCA1 and BRCA2 variants by DHPLC. Hum Mutat 16: 345–353.
12. Wagner T, Stoppa-Lyonnet D, Fleischmann E, Muhr D, Pages S, Sandberg T, Caux V, Moeslinger R, Langbauer G, Borg A, Oefner P (1999) Denaturing high-performance liquid chromatography detects reliably BRCA1 and BRCA2 mutations. Genomics 15: 369–376
13. Donohue E, Baty DU (2000) The use of DHPLC technology to detect PLP gene mutations in Pelizaeus-Merzbacher patients. J Med Genet 37 (Suppl. 1): S77.
14. Huber CG, Oefner PJ, Bonn GK (1993) High-resolution liquid chromatography of oligonucleotides on nonporous alkylated styrene-divinylbenzene copolymers. Anal Biochem 212: 351–358.
15. Huber CG, Oefner PJ, Preuss E, Bonn GK (1993) High-resolution liquid chromatography of DNA fragments on non-porous poly(styrene-divinylbenzene) particles. Nucleic Acids Res 21: 1061–1066.
16. Hecker KH, Asea A, Kobayashi K, Green S, Tang D, Calderwood SK (2000) Mutation detection in the human HSP7OB gene by denaturing high-performance liquid chromatography. Cell Stress Chaperones 5: 415–424.
17. Narayanawami G, Taylor PD (2001) Improved efficiency of mutation detection by denaturing high-performance liquid chromatography using modified primers and hybridization procedure. Genet Test 5: 9–16
18. Hoogendoorn B, Norton N, Kirov G, Williams N, Hamshere ML, Spurlock G et al. (2000) Cheap, accurate and rapid allele frequency estimation of single nucleotide polymorphisms by primer extension and DHPLC in DNA pools. Hum Genet 107: 488–493.
19. Jones AC, Sampson JR, Cheadle JP (2001) Low level mosaicism detectable by DHPLC but not by direct sequencing. Hum Mutat 17: 233–234.
20. Leonard D, Travis L, Addya K et al. (2002) p53 mutations in leukemia and myelodysplastic syndrome after ovarian cancer. Clin Cancer Res 8: 973–985.
21. Jones A, Sampsom J, Cheadle J (2001) Low level mosaicism detectable by DHPLC but not by current sequencing. Hum Mutat 17: 233–234.
22. CMGS Best Practice Guidelines for Use of the WAVE System in Diagnostic Service. http://www.cmgs.org/BPG/Guidelines/2002/dhplc.htm
23. Cline J, Braman JC, Hogrefe HH (1996) PCR fidelity of Pfu DNA polymerase and other thermostable DNA polymerases. Nucleic Acids Res 24: 3546–3551.
24. Mattila P, Korpela J, Tenkanen T, Pitkanen K (1991) Fidelity of DNA synthesis by the *Thermococcus litoralis* DNA polymerase—an extremely heat stable enzyme with proof reading activity. Nucleic Acids Res 19: 4967–4973.
25. Cha RS, Thilly WG (1993) Specificity, efficiency and fidelity of PCR. PCR Methods Applications 3: S18–S29.
26. Clayton LK, Goodman MF, Branscomb EW, Galas DJ (1979) Error induction and correction by mutant and wild type T4 DNA polymerases: Kinetic error discrimination mechanisms. J Biol Chem 254: 1902–1912.
27. Goodman MF, Fygenson DK (1998) DNA polymerase fidelity: From genetics to a biochemical understanding. Genetics 148: 1475–1482.
28. Bessman MJ, Muzycyka N, Goodman MF, Schnaar RL (1974) Studies of the biochemical basis of spontaneous mutation. II. The incorporation of a base and its analogue into DNA by wild type, mutator and antimutator polymerases. J Biol Chem 88: 409–421.
29. Reha-Krantz L (1998) Regulation of DNA polymerase exonucleolytic proof-reading activity: Studies of T4 "antimutator" DNA polymerases. Genetics 148: 1551–1557.
30. Xiao W, Oefner PJ (2001) Denaturing high-performance liquid chromatography: A review. Hum Mutat 17: 439–474.

6

ENZYMATIC AND CHEMICAL CLEAVAGE METHODS

Richard G. H. Cotton
Genomic Disorders Research Centre, St. Vincent's Hospital, The University of Melbourne, Victoria, Australia

■ INTRODUCTION

Mutation detection methods are divided into two major categories depending on what they are trying to detect. Methods for detection of mutations that are known are called genotyping or diagnostic methods. Methods that are used to find mutations for the first time are called scanning or discovery methods. This group of methods are much more complex and expensive than genotyping methods. Thus they are in need of novel approaches to make them simpler.

This group of scanning methods, of which there are (besides sequencing) 8 to 10, can be divided into two major groups. There are methods that depend on the physical properties of the DNA to indicate whether a base change is present in the DNA, and those that depend on cleavage of DNA at or near a mismatch, allowing both an indication that a mutation is present *and* its position.

Mismatches are formed when DNA duplexes, which are otherwise identical, differ by a single base pair and are melted, mixed, and hybridized to form heteroduplexes. For example:

```
AGGTC          AGATC          Boil          AGGTC          AGATC
         +                     →                      +
TCCAG          TCTAG          Cool          TCTAG          TCCAG
```

Guide to Mutation Detection, Edited by Graham R. Taylor and Ian N. M. Day.
ISBN 0-471-23444-3

Note that the original duplexes will be reformed and particularly that two different but reciprocal mismatches are formed. This then gives the methods targeting mismatches two chances of detecting the mismatch and hence the mutation.

When one considers all 12 possible base changes, this leads to 8 possible mismatches: AA, AG, AC, CC, CT, TT, TG, and GG. Thus an ideal method should target all these mismatches, or at least one mismatch in each of the reciprocal pairs of mismatches, for example, AA, TT; CC, GG; CT, GA; and CA, GT

Deletion or insertion mutations will give rise to unmatched bases in heteroduplexes, and generally methods targeting mismatches are much more effective on unmatched bases.

Mismatched base pairs have been found to have a different structure from matched or Watson–Crick base pairs (Fig. 6.1). Generally such base pairs result in the bases being much more accessible to chemical reagents and the mismatched base pairs being much more recognizable by enzymes. Hence the basis of chemical and enzyme cleavage methods.

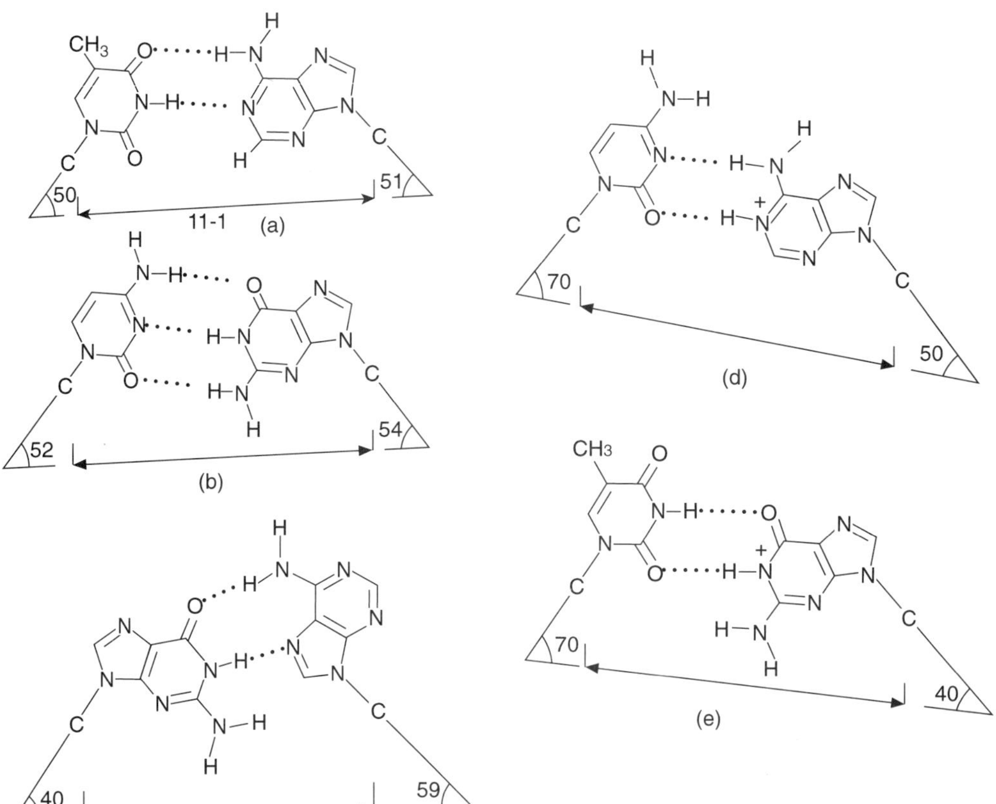

FIGURE 6.1. Mismatch structure. (a) A-T (b) G-C (c) G-A (d) C-A (e) T-G.

■ CHEMICAL CLEAVAGE METHODS

Principle

Empirical studies found that mismatched T bases were more reactive with osmium tetroxide than matched T bases. Similarly, mismatched C bases were more reactive with hydroxylamine than matched C bases (1). The reaction of the two reciprocal heteroduplexes with these two chemicals allows two reactive bases in each of the four sets of reciprocal pairs/mismatched base pairs given above. This allows two chances of detecting each pair and hence each mutation. This presumes that each mismatch possible (TT TG TC) is reactive with osmium tetroxide and each C mismatch (CC CA and CT) is reactive with hydroxylamine. In fact, this was found to be true, except that 1/3 of TG mismatches were found to have low reactivity with osmium tetroxide. In such cases the reactivity of CA is relied on to detect the mutation.

In practice, the reactivity of C and T bases is detected by cleaving the DNA chain at the reacted base with alkali, usually piperidine. The cleavage is then detected by separation of the products on a sequencing gel. The size of the fragments indicates the position of the mismatch/mutation (Figs. 6.2 and 6.3).

Because of the two definite chances of detecting a mutation as well as positioning it. this method is regarded as a precision method with 100% detection. This is underlined by experience in a UK laboratory, where in 400 successive cases of Factor IX deficiency, a mutation was found in each.

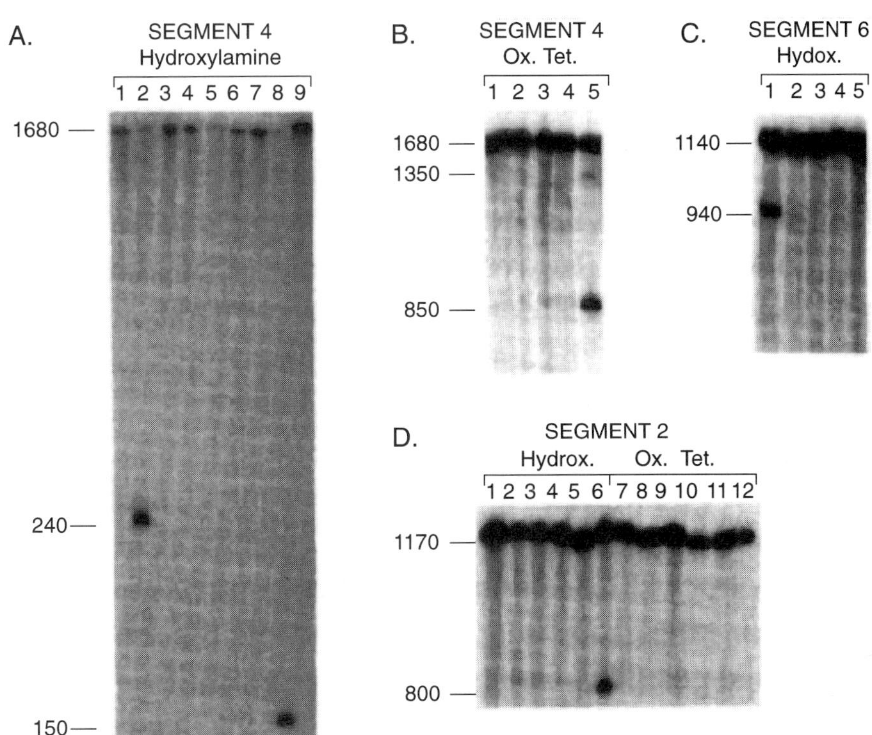

FIGURE 6.2. CCM of factor VIII cDNA 1140–1680 bp.

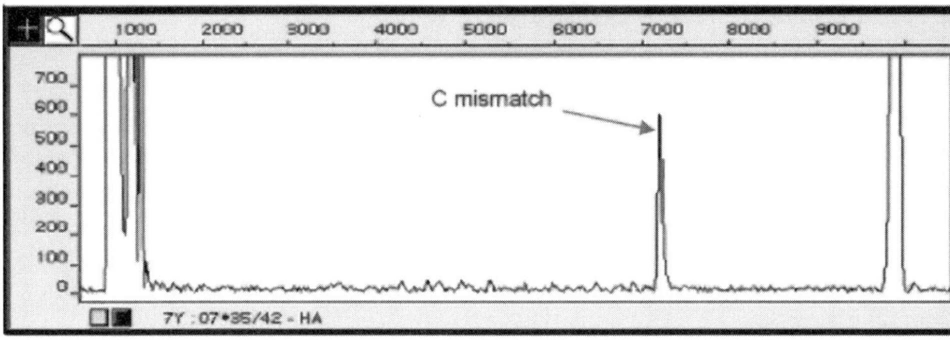

FIGURE 6.3. 14809delC cytochrome b.

Recent Improvements

The multiple steps, including tedious ethanol precipitates and washing, and the toxic chemicals have inhibited some users, even though such individuals use more dangerous chemicals on a daily basis. However, some of these problems have been alleviated in recent developments that allow users to take advantage of its considerable advantages.

Replacement of Osmium Tetroxide

Extensive studies and prolonged use have proven that potassium permanganate is able to function the same as osmium tetroxide, if not better (2). Potassium permanganate has probably been used for centuries to treat cuts and thus must be innocuous if handled appropriately, as with any chemical.

Reduction of Steps and Elimination of Ethanol Precipitation

We have found (3) that the method works just as well if the sample is treated with potassium permanganate and reacted and then hydroxylamine is immediately added and reacted. To avoid ethanol precipitation, these reactions are performed on silica beads so that before piperidine treatment, reagents are removed, the beads are reused, then piperidine/loading dye is added and reacted, and then the products that are now released from the beads (because of the alkali) are loaded to a sequencing gel (4).

Thus the steps now are:

PCR → Heteroduplex Formation → Addition of Beads → Addition of $KMnO_4$/Heat → Addition of Hydroxylamine/Heat → Wash Beads → Add Piperidine/Loading Dye/Heat → Load

■ ENZYME CLEAVAGE METHODS

Principle

Some enzymes are more reactive with specific structural distortions of duplex DNA than the bulk of the DNA sequence. These structural distortions include mismatches as they produce bubbles or bends in duplex DNA.

The dream of workers developing methods that cleave at mismatches is that an enzyme can be found that cleaves with little or no background cleavage and cleaves both strands. Thus, in the ideal world, we need a "restriction enzyme for mismatches." Although there are elements of the requirements in current methods, we have not obtained this ideal.

Clearly, the advantage of enzymes that cleave both strands is that there is no toxicity and products could simply be analyzed on agarose gels instead of sequencing gels.

Intuitively one would expect repair enzymes to be the best candidates for enzymes for this task. However, it turns out that this is not true because those that recognize all mismatches cut up to kilobases away and those that cut at the mismatches do not cut all mismatches, only A mismatches, for example.

Ribonuclease

The use of this enzyme was one of the first methods to be used as a scanning method to decide whether a piece of nucleic acid contains a mutation or not, to avoid excess sequencing that would otherwise have to be done (Figs. 6.4 and 6.5).

This method has an immediate disadvantage because RNA has to be produced from the DNA to be scanned, as the enzyme will only work on RNA. (This is not a disadvantage for those working with RNA viruses). However, it does have major advantages in that all mismatches are recognized and both strands are cut, allowing agarose analysis (5). A kit is available from Ambion for this test.

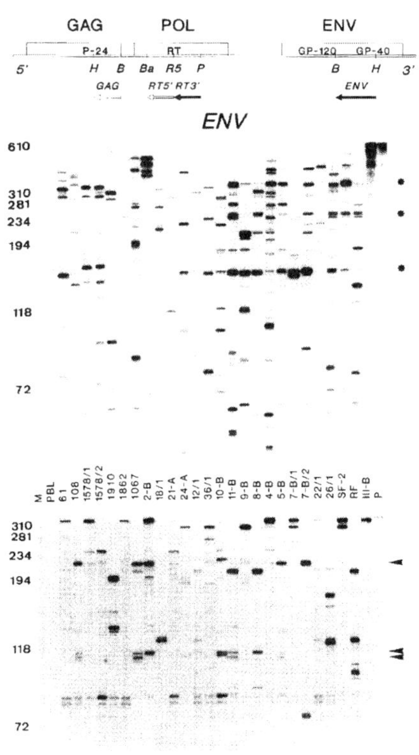

FIGURE 6.4. RNAse mismatch cleavage of RNA–RNA heteroduplices of HIV genes.

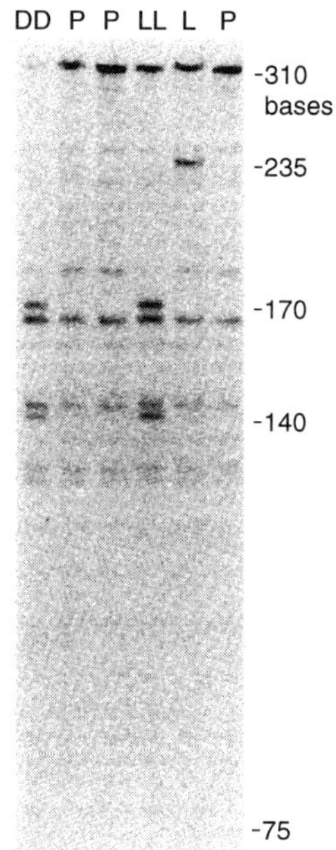

FIGURE 6.5. RNAse of cardiac myosin heavy chain mRNA.

In practice, PCR primers have to be used that contain promoters to allow RNA production for the experiments. The method is low tech and "cheap" and found by one hospital in London to be very suitable for a low budget.

Resolvases

Resolvases "resolve" cruciform "Holliday junctions," which are produced as a result of recombination, and in phages they are thought to cleave junctions that result from tangling when DNA is being packed into their heads. These enzymes cleave a range of complex substrates including single base pair mismatches. This finding prompted the examination of their ability to cleave all mismatches and hence their ability to detect all mutations.

We examined the ability of phage T4 endonuclease 7 to cleave several examples of each of the possible mismatch sets and found that one or both mismatches in each set was cleaved (6). However, it was quite clear that normal duplex DNA was also cleaved, presumably in areas of local particular physical structure induced by the specific sequence. This means the traces are not simple; however, an extra band in a consistent background is in fact easy to see, especially when a large number of the same DNA sample is run together on the same gel. This enzyme cleaves within 3 to 5 base pairs on the 5 side of the mismatch, but it appears that it is not a double-stranded cleavage, so agarose cannot be used.

This method was licensed by Amersham Pharmacia to be sold as a kit ("Passport"), but it has not been a big seller, possibly because the price was so high. It is estimated that it detects 95% of mutations (Fig. 6.6).

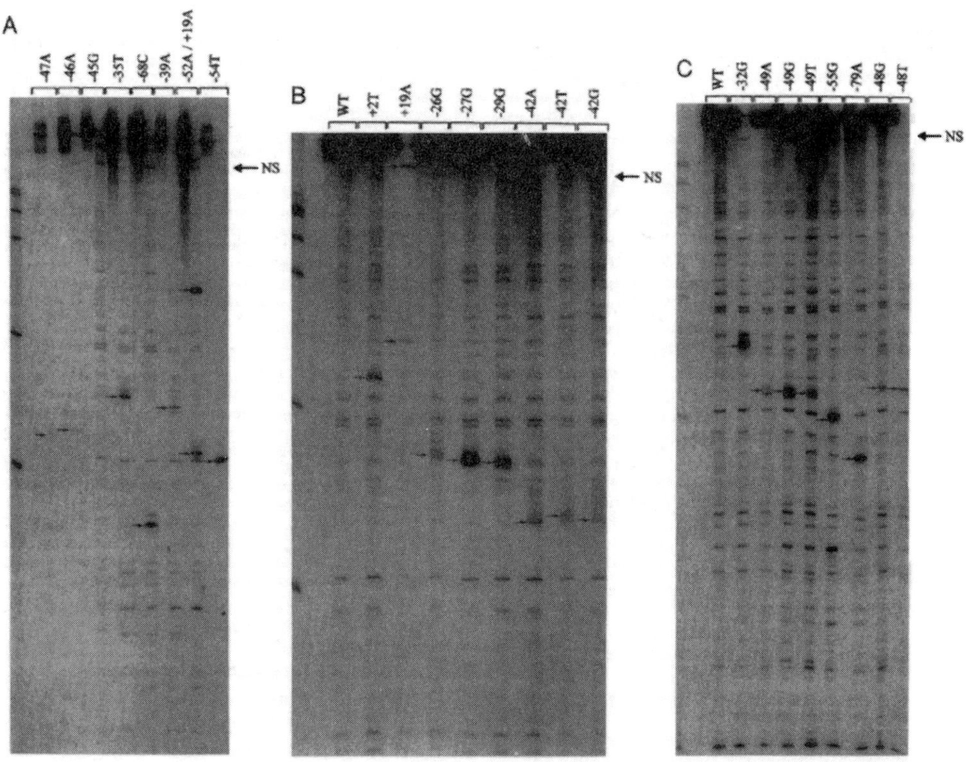

FIGURE 6.6. T4 endonuclease 7 cleavage of mismatches. Arrows mark cleavage.

■ OTHER ENZYMES

Mut-Y

This is a repair enzyme, and it cleaves only AC and AG mismatches on the A-containing strand. The AG mismatch is much more susceptible to cleavage than the AC mismatch (twenty times) (7).

CEL1

This is a mismatch cleaving activity from celery (8). It is clear that mismatches are cleaved; however, detailed characterization, such as whether all mismatch classes can be cleaved or whether the background is better than T4 endonuclease 7, has yet to be performed.

■ CONCLUSION

Cleavage methods can provide a huge advantage for mutation detection. This is because they are 95–100% effective and particularly because they position the mutation. As it is well known that sequencing can miss heterozygous mutations or such mutations can be difficult to see, it is a great advantage when one is directed to the position on the complex sequencing gel to find a subtle difference in the trace that may in fact be the mutation. However, we are still to be presented with the ideal cleavage method.

REFERENCES

1. Cotton RGH, Rodrigues NR, Campbell RD (1988) Reactivity of cytosine and thymine in single-base-pair mismatches with hydroxylamine and osmium tetroxide and its application to the study of mutations. Proc Natl Acad Sci USA 85: 4397–4401.
2. Lambrinakos A, Humphrey KE, Babon JJ, Ellis TP, Cotton RGH (1999) Reactivity of potassium permanganate and tetraethylammonium chloride with mismatched bases and a simple mutation detection protocol. Nucleic Acids Res 27: 1866–1874.
3. Ramus SJ, Cotton RGH (1996) Single-tube chemical cleavage of mismatch: successive treatment with hydroxylamine and osmium tetroxide. BioTechniques 21: 216–291.
4. Bui CT, Babon JJ, Lambrinakos A, Cotton RGH (2001) Detection of mutations in DNA by solid-phase chemical cleavage method: a simplified assay. In *Single Nucleotide Polymorphism: Methods and Protocols*, Pui-Yan Kwok (ed.), Humana Press, in press.
5. Goldrick MM (2001) RNase cleavage-based methods for mutation/SNP detection, past and present. Hum Mutat 18: 190–204.
6. Youil R, Kemper BW, Cotton RGH (1995) Screening for mutations by enzyme cleavage of mismatch using T4 endonuclease VII. Proc Natl Acad Sci USA 92: 87–91.
7. Lu AL, Hsu IC (1992) Detection of single DNA base mutations with mismatch repair enzyme. Genomics 14: 249–255.
8. Qiu P, Shandilaya H, D'Alessio JM, O'Connor K, Durocher J, Gerard GF (2004) Mutation detection using Surveyor™ nuclease. BioTechniques 36: 702–707.

7

QUALITY IN MUTATION DETECTION

David E. Barton
National Centre for Medical Genetics, Department of Medical Genetics, University College of Dublin, Dublin, Ireland

Most focus in mutation detection has been on assay sensitivity: "If a mutation is there, will I detect it?", but assay sensitivity is only a small part of the answer to the total quality question "How could my data be wrong, and what can I do to make sure they are correct?"

There are many reasons why your mutation detection data could be wrong. The reasons can be grouped into several subcategories, some of which are generic (i.e., they apply to all mutation detection technologies) and some specific to individual mutation screening methods. It is only by having systems in place to address all of the possible pitfalls that you can be confident that the result you produce is the correct one.

■ GENERIC PITFALLS

You Are Testing the Wrong Sample

Clearly, if you test the wrong sample, you are likely to get the wrong result. Sample mix-up occurs because of poor internal control systems. Guidelines for internal laboratory control have been published by EMQN (www.emqn.org). Although the increasing use of microtiter plates has reduced the number of pipetting steps, problems of plate identification and well identification have been introduced. Robotics and bar coding can reduce sample handling errors but are beyond the means of many laboratories.

Guide to Mutation Detection, Edited by Graham R. Taylor and Ian N. M. Day.
ISBN 0-471-23444-3

Testing the wrong tissue can also cause problems, especially when searching for somatic mutations or sequence changes. This is particularly true of tumor biopsies, which may contain a mixture of normal and tumor tissue. These problems can be overcome by the use of appropriately stained slides from the same tissue section.

PCR Has Altered Your Sample

Most sequencing technologies currently in use rely on PCR as the first step. It is important to be aware of the potential consequences of this enzymatic amplification of the target sequence. When you screen PCR products for mutations, you are asking the question "Does this PCR product contain a mutation?" and *not* "Does the original DNA sample contain a mutation?"

PCR Replication Errors

The PCR itself may have introduced a sequence change into your sample that does not exist in the template DNA. If the polymerase introduces an error in one of the early cycles of amplification, this change can be the predominant product when the reaction is complete. Such errors are random and should be revealed by repeat testing of the original sample. They are unlikely to convert a mutant allele to the wild-type sequence, so they are more likely to introduce false positive errors than false negatives.

PCR Amplification Errors

PCR primers only work reliably if they find a perfect match with the template, especially at their 3′ ends. Variations (SNPs) in the primer binding site can cause one allele not to amplify from genomic DNA samples. If only the normal allele amplifies from a sample containing a heterozygous mutation, the mutation will be missed. PCR primers for mutation scanning are usually based on intronic sequences, which are more likely than exons to contain polymorphic sequences.

The Mutant Sequence Is Not There

Clearly, if the sample being tested does not contain the sequence being screened, the mutation will not be detected. This can arise in several contexts.

Genomic Deletion

Many of the mutation screening methods currently in use will not detect large genomic deletions. PCR-based methods will amplify only the normal allele. The solution is either to use Southern blotting (of conventional and/or pulsed-field gels) or to use dosage methods to investigate copy number.

Absence of mRNA

Mutation screening methods based on cDNA are particularly prone to missing mutations because the mRNA is absent. Nonsense-mediated degradation removes mRNA species containing truncating mutations, leaving the normal message intact.

Promoter mutations can result in the absence of message from one allele and do not show up on exon-based mutation scans. Alternative splicing can be misinterpreted as deletions or insertions in cDNA-based assays. Genomic DNA contamination of cDNA preparations can cause similar artifacts.

Your Technology Is Deficient

Most techniques for detecting mutations are less than 100% sensitive (i.e., they can give a "wild-type" result when a mutation is present). Some are also less than 100% specific (false positives). It is important that users are aware of the limitations of their chosen method to avoid errors of interpretation. Some methods are particularly prone to problems with reproducibility (they do not always give the same result). SSCP and heteroduplex analysis, in particular, are sensitive to the running temperature, the buffer concentration, and the gel type and formulation.

SPECIFIC PITFALLS

As well as the generic problems outlined above, each method will have its own particular pitfalls that need to be avoided if possible. DNA sequencing is often taken to be the "gold standard" of mutation detection techniques because it is the only method that gives all the information required to describe the mutation. Although it is true that sequencing gives the most information, this should not be taken to mean that it is the most accurate method for detecting sequence changes, particularly heterozygous sequence changes. Experience with sequencing-based external quality assessment (EQA) schemes run by EMQN tells us that sequencing errors are quite common, even in the highly controlled setting of the diagnostic laboratory. There are numerous reasons why sequencing results can be incorrect.

You Are Sequencing from Plasmids

Sequencing from plasmid-cloned sequences can cause problems similar to those in sequencing from PCR products. Cloning can introduce a sequence change or amplify a change that is a minor species in the original sample. Unless a large number of clones are sequenced, you cannot be sure you have tested the mutant allele. In a small number of cases, the mutant allele may be toxic to the cloning host.

You Cannot Detect Heterozygous Sequence Changes

Much mutation screening in human genetics is carried out on samples expected to be heterozygous for any mutation present. This poses problems for direct sequencing, as heterozygotes are not always easy to detect, particularly if the background in your sequencing traces is too high. Differential incorporation of different dideoxy dye terminators can cause the mutant sequence peak to be lost in the background, especially if your software is not configured correctly for heterozygotes. Software programs such as PHRED will give you an objective assessment of the quality of your sequence data, and programs like Staden's TRACEDIFF can be used to highlight differences between reference sequence and test sequence traces. Other software for assessing sequence quality is also available.

You Have Not Designed Your Assay Properly

Assay design is fundamental to getting the results you want. If, for example, your primers are too close to the sequence to be tested, you may not be able to get good-quality sequence at the site of interest. Sequencing only one strand may save you time in the first instance, but it is not sufficient to produce reliable sequence.

You Are Not Interpreting Your Results Correctly

The EMQN data show us that most errors in reports of sequence data are due to misinterpretation of data that are correct. Such errors fall into several categories, the most common being simple typographical errors, incorrect numbering of bases, and using out-of-date sequence from the databases. It is essential to check the sequence databases regularly for updates of your sequences of interest. The databases are full of errors of both sequence and annotation; they change regularly as these errors are reported and corrected. Whenever possible, the Reference Sequence should be used for comparison with your test data (see http://www.ncbi.nlm.mh.gov/locuslink/refseq.html).

Of course, once you have found a sequence change, you are often faced with the challenge of deciding whether it is significant. This process has become increasingly difficult with the continuing discovery of elements like exonic splice enhancers (ESEs) and intronic binding sites for new splicing factors (see Refs. 1 and 2). No sequence change, intronic or exonic, can be dismissed as a "neutral polymorphism" without serious study, which can take much more time and effort than generating the sequence data in the first place.

■ REFERENCES

1. Cartegni L, Chew SL, Krainer AR (2002) Listening to silence and understanding nonsense: Exonic mutations that affect splicing. Nat Rev Genet 3: 285–298.
2. Pagani F, Buratti E, Stuani C, Bendix R, Dork T, Baralle FE (2002) A new type of mutation causes a splicing defect in ATM. Nat Genet 30: 426–429.

8

COMPARISON OF SCANNING METHODS FOR *BRCA1* MUTATION ANALYSIS

Norbert Arnold
Clinic for Gynecology and Obstetrics, University of Kiel, Kiel Germany

Mutation detection has found important applications in functional genomics and clinical diagnostics. To realize the full benefit of decoding the human genome, rapid and sensitive methods for screening genomic DNA for unknown sequence variation will be required. Technologies used for this task should be capable of fully automated high-throughput analysis that ideally do not require modified PCR primers, customized specific reagent arrays, detection labels, or any sample pretreatment other than PCR. Numerous techniques are available for mutation detection, such as direct sequencing (DS), protein truncation test (PTT) (1), single-strand conformation polymorphism (SSCP) (2), dideoxy fingerprinting assay (DDF) (3), denaturing gradient gel electrophoresis (DGGE) (4), two-dimensional gene scanning (TDGS) (5), conformation-sensitive gel electrophoresis (CSGE) (6), enzymatic mutation detection (EMD) (7), allele-specific oligonucleotide hybridization (ASO) (8), and immobilized DNA hybridization assays (9). Up to now, none of these methods meets this challenge. The widely used SSCP analysis is economical and simple but has low sensitivity, ranging from 60% to 90%. However, the major disadvantage of SSCP is the lack of its objective predictability. Thus, because there are no theoretical assurances, the nonappearance of a band shift does not prove the absence of a mutation. In contrast, automated direct sequencing is highly sensitive but costly and labor intensive. DDF is a combination of a Sanger sequencing reaction with multiple-fragment SSCP and is more sensitive than SSCP alone, but still labor intensive. With PTT, only sequence alterations leading to a truncated protein

Guide to Mutation Detection, Edited by Graham R. Taylor and Ian N. M. Day.
ISBN 0-471-23444-3

can be detected. Other alterations (missense, in-frame deletions, and insertions) escape the detection of the PTT assay. Other methods, including DGGE and CSGE, have been used with some success but, despite their improved sensitivity over SSCP, these methods are technically challenging and formatted for manual use only. TDGS is a method for analysing multiple DNA fragments in parallel for all possible sequence variations, using extensive multiplex PCR and two-dimensional electrophoretic separation on the basis of size and melting temperature. One source of error is the interpretation of the complex spot patterns produced by this method. Also, the sensitivity of TDGS is impaired by frequent preferential amplification of the nonmutant allele, resulting in very light heteroduplexes that obscure accurate reading of the gels (10). EMD methods for mutation scanning still lack the sensitivity and specificity of the chemical cleavage of the mismatch method. Other approaches such as immobilized DNA hybridization arrays still have significant false positive signal, as well as high cost per assay.

Therefore, denaturing high-performance liquid chromatography (DHPLC) has received increasing interest in clinical genetics, because the DHPLC method promises to fulfill these requirements. Figure 8.1 shows the system with its different components. The technique, which is based on heteroduplex detection, allows for automated identification of single-nucleotide substitutions and small deletions or insertions. Heteroduplex profiles are easily distinguished from homoduplex peaks (11). Furthermore, DHPLC has been shown to clearly resolve mutations in various genes with detection rates ranging from 92.5% to 100% (12).

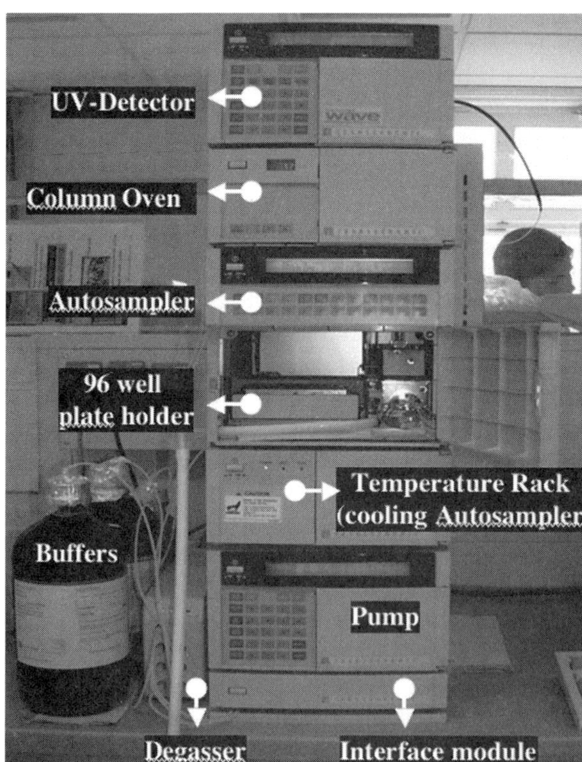

FIGURE 8.1. DHPLC system components.

In this review it will be shown that DHPLC fulfils the technical requirements to investigate several alterations in bigger sample sizes not only cost-effectively but also in a reasonable time. This will be demonstrated in the analysis of the two breast cancer susceptibility genes *BRCA1* and *BRCA2*. Together, the two genes comprise approximately 15,700 nucleotides of open reading frame. More than 1000 mutations of deduced or established clinical significance have been identified, and these are distributed throughout the 48 coding exons and respective splice junctions of the two genes [see Breast Cancer Information Core database (13)]. Yet even in families with both breast and ovarian cancer the detection rate for *BRCA1* and *BRCA2* mutations is 43% and 10%, respectively (14). Therefore, a selective prescreening to avoid labor- and cost-intensive sequencing of wild-type exons is an indispensable necessity.

In addition, amplimers containing known polymorphism with multiple alleles can markedly increase the sequencing cost for screening a large population. The known variants investigated in the study produced specific chromatogram profiles different from the known polymorphisms also occurring in the fragments in a specific number. Therefore, sequencing was unnecessary for DHPLC chromatograms characteristic for the specific polymorphisms. This possibility becomes an interesting point concerning genes such as *BRCA1*, where some fragments are highly polymorphic. If these harmless variations cannot be identified unambiguously, the advantage of a prescreening method turns out to be of minor impact, because 50% or more of the samples ultimately have to be sequenced. However, the reliability of a qualitative DHPLC analysis has to be tested for each gene individually because some sequence regions may produce very similar or even identical DHPLC peak patterns for different variants. The importance of multiple-temperature analysis as well as frequent column cleaning was emphasized. Successful separation of heteroduplexes on the DNASep® cartridge was influenced by several factors. We observed that peak resolution could decrease with increasing age of the column. Consequently, an objective assessment of column performance becomes necessary. This is especially the case in exons for which few control samples are available that can indicate optimal differentiation of peak profiles. Several factors can influence column resolution, for example, column batch, number of injections, additives in PCR reactions, or inadequate treatment of the column. Daily cleaning of the column with 75% acetonitrile at 60°C prolongs column performance tremendously. We have obtained up to 7000 injections on a single column using the resolution of the Mutation Standard (Transgenomic) as a determinant for column performance and resolution. Nevertheless, it seems very difficult to judge whether column performance is sufficient to predict specific sequence variations reliably. We suggest performing column testing at least once per week, which could consist of running a couple of samples with similar elution profiles. In this way, the resolution of the DHPLC peak patterns can be observed consecutively. When test samples for column performance indicate altered heteroduplex resolution and cleaning does not improve the performance, the column should be changed. Although analysis costs will increase if the column lifetime is not exhausted, the sensitivity for detecting alterations, which are difficult to resolve in the first place, will be increased if proper column performance is monitored. If homozygous alterations are expected in the analysis, especially in tumor samples, the analysis will require mixing with a reference DNA to obtain a heterozygous profile. Also, because the homozygous alteration could very often be recognized by a shift of the single peak in comparison to the wild-type peak, this strategy will ensure the results (Fig. 8.2).

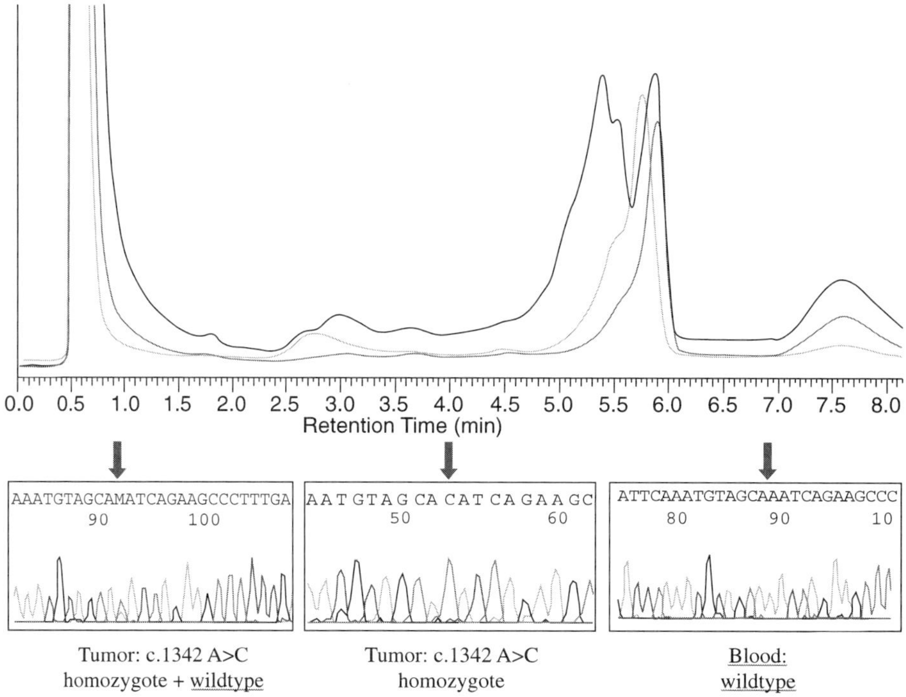

FIGURE 8.2. Homozygote and heterozygote HPLC traces.

The efficacy of DHPLC for mutational screening of the *BRCA1* and *2* genes is generally accepted. In a recent published study comparing SSCP, CSGE, TDGS, and DHPLC in *BRCA1* mutation analysis, only the laboratory using DHPLC correctly identified each of the mutations (10). Furthermore, the technique is now applied worldwide to mutation analysis in more than 50 genes.

■ REFERENCES

1. Hogervorst FB, Cornelis RS, Bout M, van Vliet M, Oosterwijk JC, Olmer R et al. (1995) Rapid detection of *BRCA1* mutations by the protein truncation test. Nat Genet 10: 208–212.
2. Markoff A, Savov A, Vladimirov V, Bogdanova N, Kremensky I, Ganev V (1997) Optimization of single-strand conformation polymorphism analysis in the presence of polyethylene glycol. Clin Chem 43: 30–33.
3. Lancaster JM, Berchuck A, Futreal PA, Wiseman RW (1997) Dideoxy fingerprinting assay for BRCA1 mutation analysis. Mol Carcinog 19: 176–179.
4. Fodde R, Losekoot M (1994) Mutation detection by denaturing gradient gel electrophoresis (DGGE). Hum Mutat 3: 83–94.
5. van Orsouw NJ, Dhanda RK, Elhaji Y, Narod SA, Li FP, Eng C, Vijg J (1999) A highly accurate, low cost test for BRCA1 mutations. J Med Genet 36: 747–753.
6. Markoff A, Sormbroen H, Bogdanova N, Preisler-Adams S, Ganev V, Dworniczak B, Horst J (1998) Comparison of conformation-sensitive gel electrophoresis and single-strand conformation polymorphism analysis for detection of mutations in the *BRCA1* gene using optimized conformation analysis protocols. Eur J Hum Genet 6: 145–150.

7. Taylor GR, Deeble J (1999) Enzymatic methods for mutation scanning. Genet Anal 14: 181–186.
8. Richter S, Seth A (1998) One step direct detection of recurrent mutations in the breast cancer susceptibility gene *BRCA1*. Int J Oncol 12: 1263–1267.
9. Tonisson N, Kurg A, Kaasik K, Lohmussaar E, Metspalu A (2000). Unravelling genetic data by arrayed primer extension. Clin Chem Lab Med 38: 165–170.
10. Eng C, Brody LC, Wagner TM, Devilee P, Vijg J, Szabo C, Tavtigian SV, Nathanson KL, Ostrander E, Frank TS (2001). Interpreting epidemiological research: blinded comparison of methods used to estimate the prevalence of inherited mutations in BRCA1. J Med Genet 38: 824–833.
11. Underhill PA, Jin L, Lin AA, Mehdi SQ, Jenkins T, Vollrath D, Davis RW, Cavalli-Sforza LL, Oefner PJ (1997) Detection of numerous Y chromosome biallelic polymorphisms by denaturing high-performance liquid chromatography. Genome Res 7: 996–1005.
12. Xiao W, Oefner PJ (2001) Denaturing high-performance liquid chromatography: A review. Hum Mutat 17: 439–474.
13. Breast Cancer Information Core, BIC website: http://www.nhgri.nih.gov/Intramural_research/Lab_transfer/BIC
14. Meindl A on the behalf of the German Consortium for Hereditary Breast and Ovarian Cancer (2002) Comprehensive analysis of 989 patients with breast or ovarian cancer provides BRCA1 and BRCA2 mutation profiles and frequencies for the German population. Int J Cancer 97: 472–480.

9

MELTMADGE: ECONOMICAL VERY HIGH-THROUGHPUT MUTATION SCANNING USING THERMAL RAMP ELECTROPHORESIS IN CONJUNCTION WITH MADGE (MICROPLATE ARRAY) GELS

Ian N. M. Day, Mohammed A. Aldahmesh, Lema Haddad, Khalid K. Alharbi, Xiao-he Chen, Hamid Rassoulian, Steve E. Humphries, Emmanuel Spanakis, and Diana Eccles
Division of Human Genetics, Southampton University Hospital, Southampton, UK

■ MICROPLATE ARRAY DIAGONAL GEL ELECTROPHORESIS (MADGE), 96-WELL AND HIGHER DENSITY GEL FORMATS FOR SHORT-TRACK LENGTH, HIGH-THROUGHPUT ELECTROPHORESIS

The well-known industry-standard 96-well microplate has evolved to form the basis of many high-throughput approaches in the biosciences over the past 30 years. However, microplates are only permissive of liquid-phase reactions or solid-phase (binding) separations, whereas electrophoresis can derive information about parameters such as size, shape, and charge of molecular moieties as well as acting as a

Guide to Mutation Detection, Edited by Graham R. Taylor and Ian N. M. Day.
ISBN 0-471-23444-3

highly resolving separation approach for complex mixtures. We described a generalized system, "MADGE" (microplate array diagonal gel electrophoresis) in 1994 that has a number of advantages for laboratory molecular genetic analysis of population samples (1, 2). This system challenges the traditional wisdom that electrophoresis should be avoided, and liquid-phase analyses developed, where higher-throughput applications are contemplated. The system combines liquid-phase microplate compatibility (laboratory transfers and informatics), convenient setup and usage of polyacrylamide gel (PAGE; which has higher resolution than agarose post-PCR). Short track length/short run time, compactness, and scalability with hand-sized, robust slab gels accessible for direct human interaction, that is, mini not micro scale, are additional features. Start-up costs are minimal, so "third world" laboratories can also apply the system. Essentially, any analysis that can be reduced to short tracks (e.g., <3 cm) gains the throughput advantages. One technician can process 10–100 gels (1000–30,000 tracks) per day depending on the application and the MADGE implementation in use. MADGE is applicable to many categories of DNA analysis (originally PCR-RFLP and allele-specific PCR analyses, but more recently for minisatellite and microsatellite sizing and identification of unknown mutations) and potentially to many other categories of biomolecule (protein, etc.), and we can only speculate on future utilization. Complete genome mapping (simple but very high-throughput PCR checking gels from radiation hybrid cell lines) has also been implemented (3).

In 1807, Ferdinand Frederic Reuss observed under a microscope the migration of colloidal particles in an electric field, perhaps the first electrophoretic separation. During the early 1970s both agarose and polyacrylamide electrophoresis gel usage evolved to slab gel formats (4, 5) much as they are still found in the laboratory today. Throughput requirements were not an issue at that time. Indeed, slabs were an advance from tube gels. For both tubes and modern capillaries, each "lane" is separate from any other and cross-referencing requires reliance on either an external frame of reference (e.g., time of elution) or an internal frame (e.g., coelectrophoresed mobility standards). Slabs give convenient alignment of tracks. MADGE, in the process of achieving microplate compatibility, uses 96 (higher density in newer formats) track origins (wells) with 8 × 12 array locations at the 9-mm pitch, identical with microplates devised by Dynatech and now a highly established industry standard for liquid-phase operations. The long axis of the array is at an acute angle, for example, 18.4°, relative to the direction of electrophoresis (see, e.g., Fig. 9.1a), so that a track length of 25–30 mm is available. It should be noted that each track is specifically isolated (like a set of tube gels or capillaries) but the tracks are in a slab. The two-piece kit dimensional gel former and piece of backing glass or plastic (Fig. 9.1b) for the gel both excludes air and gives a perfectly flat open face to the resultant gel, simplifying the production of agarose, PAGE, and other gel matrix or composite gels. The combination of features in the MADGE gel design gains substantial advantages (as described above and below), but the features of this system can be used in isolation, for example, the simplified production of open-faced horizontal PAGE gels (1) but with standard row(s) of wells.

Electrophoresis of DNA has been performed traditionally either in agarose or polyacrylamide gel matrix. Much effort has been directed to improved-quality agaroses capable of high resolution, but for small fragments, such as those from PCR and post-PCR digests, polyacrylamide still offers the highest resolution. Although agarose gels can easily be prepared in an open-faced format to gain the

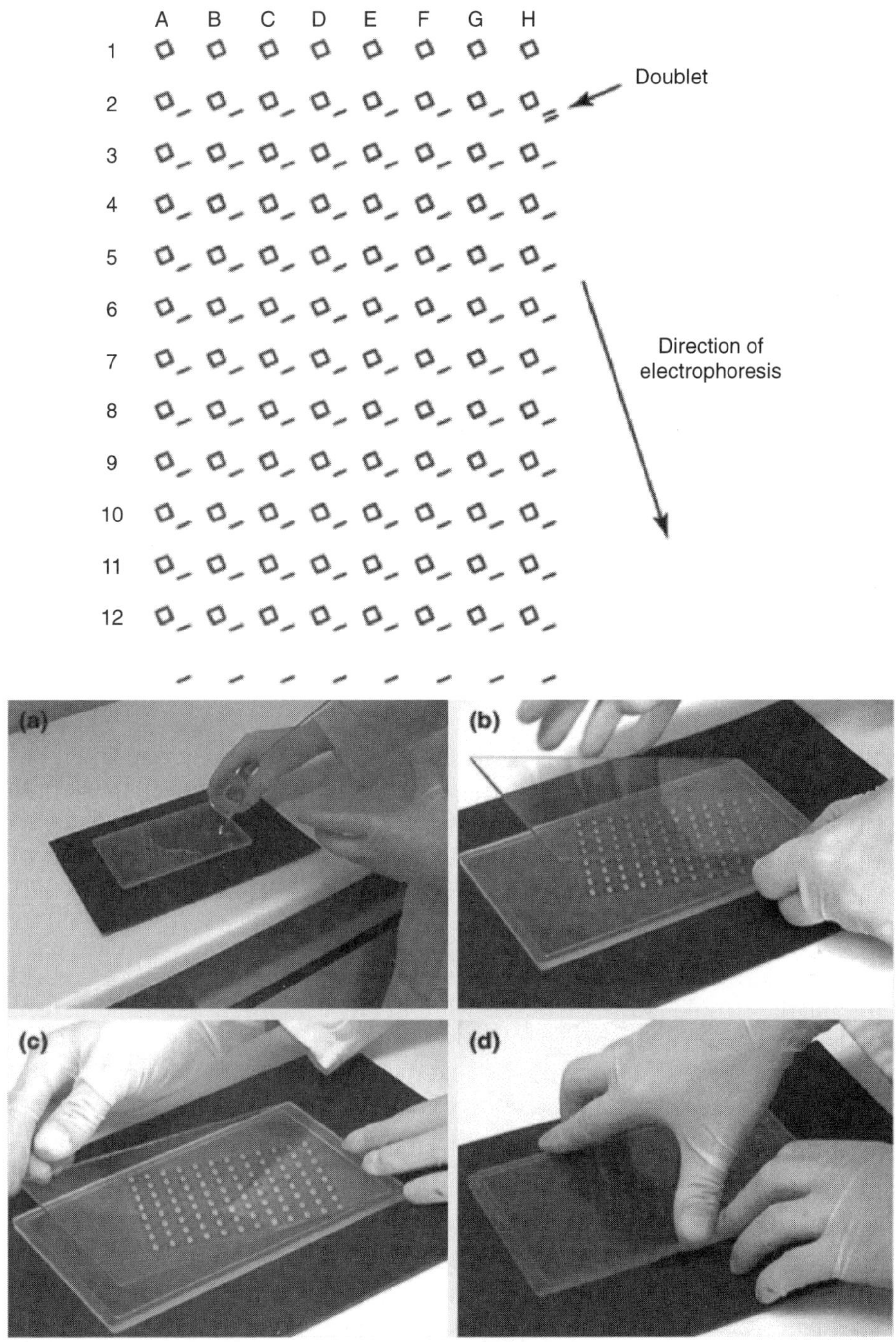

FIGURE 9.1. Microplate array diagonal gel electrophoresis (MADGE). A: Schematic diagram of a (MADGE image. The squares represent the wells, and the lines represent the bands in the lanes. With ethidium bromide staining, the wells are only visible as a dark image in a dark background, whereas the bands stand out brightly in a regular array. Although this is seemingly complex, the human eye and brain are efficient at pattern recognition. For example, here, only track H1 contains a doublet. The setup is such that well A1 is placed nearest the cathode. B: Preparation of a MADGE gel, original and simplest format. The plastic former is laid horizontal with teeth upward. Acrylamide gel mix is poured (a) into the "swimming pool," which contains the teeth. A sticky silane-coated glass plate is laid over (b–d). Once the gel has set (1–5 min if higher concentrations of polymerization agents are used), the glass plate is prised off, bearing an open-faced microplate-compatible 96-well PAGE gel suitable for submersible or semidry use.

conveniences of horizontal electrophoresis, acrylamide does not polymerize in the presence of air and the usual configurations for gel preparation lead to electrophoresis in the vertical dimension.

The original MADGE format (1) uses a two-dimensional plastic former in conjunction with one glass plate coated with γ-methacryloxypropyltrimethoxysilane (Sticky Silane). The plastic former contains a 2-mm deep, 100 × 150-mm rectangular "swimming pool." Within the pool, there are 96 2-mm cubic "teeth" (well formers) in an 8 × 12 array with 9-mm pitch directly compatible with 96-well plates. The array is set on a diagonal of 18.4° relative to the long side of the "pool," which is parallel to the eventual direction of electrophoresis, giving gel track lengths of 26.5 mm (Fig. 9.1a). The gel former is placed horizontally, acrylamide mix is poured into the pool, and the glass plate is overlaid (Fig. 9.1b). After the gel has set, the glass plate is prized off, bearing its open-faced 96-well gel. The original description of the apparatus and the gel setup is in Reference 1. It should be noted that, although the system may sound complex, many users have stated that this is the simplest gel system they have ever used. Additionally, the human eye/brain is extremely good at pattern recognition and the small-scale user, once accustomed, is not obliged to use image analysis software. A 26.5-mm track length (2 × 4 diagonal of array), such as in Figure 9.1a, or greater, such as the 2 × 6 diagonal in Figure 9.5 (see section on higher-resolution MADGE), using a polyacrylamide gel matrix, is efficient for many post-PCR analyses involving simple band pattern recognition. Because the simplest situation for post-PCR analysis is to load PCR product directly onto the gel, the question arises of the effects of carryover buffers and salts in the sample solutions on separation in MADGE. Our typical PCRs contain 50 mM KCl, and loading 5 ml leads to a reasonable resolution of patterns where relative mobility differences between bands are 5% or greater. Extremely high-salt digests or other less typical solutions may occasionally present a problem. Rather than resorting to laborious ethanol precipitations of many microplates of amplicons, or other, expensive purification procedures, we have usually been able to "dilute away" the solute problem by diluting the sample in the gel buffer without any compromise to band detection. If ethidium bromide turns out then to provide insufficient sensitivity, order of magnitude improvements in sensitivity are readily achievable with newer intercalating dyes such as Vistra Green. Indeed, poorer resolution of doublets with less than 5% mobility differences has often simply been the consequence of band overloading. The thicker (2 mm) gel used with transverse imaging would also offer lower resolution if bands were significantly tilted in orientation from glass toward gel surface. This is no greater a problem than for horizontal submerged agarose gels, and generally this has not been a problem at all in our laboratory. Attention to detail of buffers and ions in gel and sample (6) may improve results further, but for the typical applications described it must be stressed that direct loading of samples usually gives perfectly readable results. Gels are robust, can be reused, are directly stackable for storage or use, and are fully compatible with industry-standard PCR microplates thus minimizing procedural information transfers and enabling direct 96-channel pipetting. However, considerable advantage is gained even with 8- and 12-channel pipettes.

We have also described higher-density formats, including 192-well (7) and 384- and 768-well (8)gels mainly for single nucleotide polymorphism typing, but with the exception of 192-well gels, we doubt whether these other formats will prove suitable for combination with the temporal thermal ramp approach of melt-MADGE described below.

MELTMADGE: MUTATION SCANNING USING THERMAL RAMP ELECTROPHORESIS IN CONJUNCTION WITH MADGE GELS

De novo mutation detection is an important component of research in genetic disease. For mutations not known in advance, the choice at present is between direct sequencing, de novo scanning techniques to target sequencing more efficiently to appropriate regions, and "chip" technologies, equivalent with direct resequencing by multiple parallel oligonucleotide hybridization (9). For the identification of rare variants in high-throughput centers and for population studies, important criteria are the degree of parallelism of samples possible, the number of bases analysed per "run," and the total number of bases an individual or automated system can scan per day. For heterozygote sequencing in humans, the difficulties of accurate base calling (essential for every base to avoid excessive numbers of false positive calls) and high reagent and hardware costs significantly restrict sequencing as a first-line approach (10). Chip technologies (11), although promising, have a low capability for sample parallelism; a chip is expensive ($5000–10,000) for any selected sequence and also requires an optimization process; and the approach demands high capital expenditure on chip reading hardware.

Scanning technologies are much used, but the intense activity to develop improved approaches reflects the importance and shortfalls in this field. Such technologies generally rely on electrophoresis and give some information about the mutation from the resultant band pattern. SSCP has been favored by research laboratories; it is easy to implement, as evidenced by thousands of Medline references, although sensitivity to all base changes may range from only 30% up to 90% (12, 13). However, the average sensitivity of SSCP to base changes remains uncertain. DGGE has been favored by diagnostics laboratories: It is laborious to set up but generally more sensitive and robust (14). An objection to DGGE raised by some investigators is the additional cost of oligo clamp synthesis and the occasional adverse influence of the clamp on the PCR reaction itself. The additional cost element corresponds to one or two additional oligos, or to a fluorescence end label. It is small compared with the costs of sequencing many templates and seems a reasonable compromise to achieve a "PCR and run" technique avoiding other intermediate steps and costs. It has not been our experience that GC clamps compromise many PCRs. We have previously reported modifications of SSCP to achieve scanning rates of approximately 20,000 bases per day per person (15, 16). However, SSCP has limitations related to the unpredictable effect (if any) of a base change, with resulting inability to predict, control, and maximize band resolution, as well as the relative inconvenience of detection systems suitable for single-stranded DNA. The folding and tertiary structure and hence electrophoretic mobility of a single strand depends on chance internal base pairing and other influences of bases on overall tertiary structure. The effectiveness is therefore a matter of the chance with which a particular single strand sequence will interrogate all its own internal bases. By contrast, in DGGE the base sequence of one strand is being interrogated in a predictable fashion by the complementary strand.

The principle of denaturing gradient gel electrophoresis (DGGE) (17) is that homoduplexes differing by one base (usually one hydrogen bond) and heteroduplex molecules containing a base mismatch melt and alter their electrophoretic mobility under slightly different conditions of denaturation encountered at some point in the space of the gel track down which they are being electrophoresed. This is achieved by using a gel with either a chemical denaturing gradient (DGGE) or a thermal gradient (TGGE) from the row of wells of origin down the length of the tracks. However, the use of a spatial gradient has several disadvantages. First, preparing gradient gels is cumbersome and inflexible and sophisticated (nonlinear) gradients are practically impossible to prepare. Second, the spatial gradient introduces the constraint of a single, rather than multiple, rows or arrays of wells, which severely limits throughput. Finally, the setup does not readily allow the use of very short track lengths, which further reduces the possibility of utilizing high-density arrays of tracks.

The melt-MADGE system that we have developed uses the same principle as DGGE or TGGE but **exchanges the dimension of the gradient, so that it is a thermal *ramp* in time rather than a *gradient* in space.** Thus the PCR products for analysis are loaded on a homogeneous gel, and the temperature of the entire gel is raised during the course of the electrophoresis. DNA fragments that form a melted domain at a lower temperature (e.g., heteroduplexes) will display reduced mobility at an earlier stage of the run and hence will not migrate as far as their more thermostable cognate homoduplexes. A meltMADGE proof-of-principle experiment is shown in Figure 9.1. The reconfiguration of the denaturing gradient to have temperature as the dependent variable and time as the independent (controlled) variable opens a range of important advantages:

(i) It creates the option for high-density arraying of wells and tracks (e.g., MADGE arrays), hence enabling high throughput of analyses.

(ii) Programmability of the ramp, which can be absolutely arbitrary and uses temperature rather than chemicals, is achievable easily in time. Arbitrary gradients (rather than simple linear gradients) would be very difficult in space. Gradients as an integral feature of the gel demand special gels. Programmability gives flexibility and convenience.

(iii) Potential for arbitrary ramps enables very short track lengths, and short run time (e.g., 1 h) further enables high track density and hence throughput.

(iv) The use of ordered arrays (e.g., microplate format of MADGE) much simplifies data handling.

(v) Arbitrary programming enables arbitrary resolution, thus reducing the analysis to simple pattern recognition rather than detailed relative mobility measurements. Pattern recognition evades the need for closely juxtaposed tracks or for internal markers. By contrast, the resolution of SSCP is often small, cannot be programmed and hence cannot be arbitrarily great. Thus advantages (i)–(v) cannot be realised with SSCP.

(vi) For PCR product analysis for sequence variations (known or de novo), the system will usually enable the user to **"PCR and run."** This concept represents an operational standard that can only be surpassed by evading electrophoresis altogether, and then only if the liquid phase procedures are simpler, more economical, or more sensitive in some way.

meltMADGE has a resolution theoretically greater than DGGE, because ramps of any complexity can be planned, including, for example, steep initial ramps to resolve heteroduplexes, followed by shallow subsequent ramps to enhance resolution of homoduplexes. Discontinuous ramps to combine analysis of more than one PCR product per run, where their T_m are quite different, will also be feasible. Sequencing can be used as first-line analysis. However, parallel sequencing of thousands of samples simultaneously would still be a very substantial and expensive task, and to be successful, no base call in any sequencing reaction should leave any (false) possibility that there could be heterozygosity at that position. In practice, the latter is difficult to achieve (10). By contrast, using one strand to interrogate another in a melting assay is an accurate, robust, and simple test of bases for perfect match (18). The feasible track usage rate for the first meltMADGE prototype is 1000 per 2h, which would correspond with 20–80 manual or automated instrumentation sequencing gels depending on whether a one track/four label or a one track/one label sequencing method were employed. Sample parallelism is also 100- to 1,000-fold greater than with a chip, in which one PCR product can be analyzed every 15min. Additionally, chip techniques require post-PCR procedures such as in vitro transcription and fragmentation before chip hybridization (9), rather than being PCR-and-run. Sample parallelism is also problematic with capillary electrophoresis instrumentation, although possible at high equipment cost (19). The design of a "meltMADGE machine" also means that its final cost is likely to be much less than a chip and "chip reader": The former requires electrophoresis in equipment otherwise akin to a PCR thermal cycler; the latter requires photolithography, an oligonucleotide synthesizer, and a purpose-built confocal microscope.

These considerations suggested to us that for high-throughput laboratories, such as our own, engaged in population-based molecular genetic epidemiological research, or central reference facilities for diagnostics or research (on the genome of any species), the concept of "meltMADGE" might better meet the criteria of economy, sample parallelism, convenience of setup, and accessibility. However, like the original MADGE technology, the low hardware costs and running costs render the technology accessible to large and small laboratories alike.

■ APPARATUS

The prototype apparatus was purpose-built (20). In brief, a 10cm × 10cm × 15cm electrophoresis tank contains buffer that is continuously mixed to ensure spatial thermal homogeneity, and the temperature is varied in real time by programmable software controlling heating and cooling systems and receiving feedback from temperature sensors in the tank. Overall, this is little more complex than a PCR machine, although for PCR, low thermal mass for rapid temperature shifts is the objective, whereas for meltMADGE, the aim is high thermal mass for stability of small temperature shifts. The initial prototype accommodates 10–12 horizontal MADGE gels spaced in a carrier stack. Each 2-mm-thick gel is adherent to and supported by a 2-mm glass plate and is also covered by 2-mm glass, with spacing between each glass/gel/glass sandwich for buffer circulation. Because buffer, and hence thermal, circulation is vigorous, the glass is both a good thermal conductor and also quite thin, and the thermal ramps are relatively slow and shallow, reasonable spatial thermal homogeneity is achieved, at least to within 0.1°C as determined by a reference platinum resistivity thermometer. This basic system enables analysis of 1000 PCR products with 1- to 2-h run time in a 2-liter tank, using 10–12 palm-sized MADGE gels.

We have since simplified the construction of apparatus to make the system more accessible to any laboratory until an established commercial supply becomes available. The simplification combines any off-the-shelf recirculating water bath that is programmable to deliver a thermal ramp; an electrophoresis tank and gel rack as in the original system, which has electrodes at the ends of the tank and a motorized impellor to ensure complete thermal homogeneity of the buffer in which the gels are submerged; and a glass serpent as in the original design, which sits in the base of the tank and carries water circulation from the water bath. A larger (10–15 liter) tank system with a purpose-built meltMADGE cassette is now available from Ingeny, Holland, which has been used in our laboratory to date mainly for thermostatted MADGE applications such as improved resolution band sizing for microsatellite marker allele sizing (21).

GENERAL APPROACH TO MELTMADGE ASSAY SETUP

The most thorough approach to assay setup for a specific region embodies the following steps

(i) Assay and primer design; may also include design of an artificial mutant control
(ii) PCR optimization
(iii) Constant temperature checks of melting
(iv) Trial and validation of a suitable thermal ramp

Shortcuts (at risk of confusion) can include omission of artificial controls (where polymorphic alleles or known mutations are available as genomic DNA) for PCR amplification; direct use of a standard set of conditions for step (ii); and omission of step (iii).

Assay design employs the same principles and programs as DGGE (22). The region for interrogation is identified. Thermal equality of melting the whole domain is examined by a predictive computer program based on empirical and theoretical parameters of strand melting. A long GC clamp is introduced to one end of the planned amplicon, which improves the thermal equality of melting and prevents complete strand dissociation during electrophoresis. A short GC clamp introduced to the other end of the domain may further improve melting characteristics. Suitable PCR primers are planned and checked in one of many PCR primer planning programs. A suite of tools (Tixis, Greek for "melt") written by E. Spanakis to run as macros integral to Microsoft Office and built around Leonard Lerman's Melt programs, is in use in our laboratory. Making a longer (5–10 nt longer) third PCR primer with an introduced base mismatch relative to native template enables PCR creation of a mutant allele. This can be coannealed (e.g., heat to 95°C, cool quickly to 30°C) with "wild-type" amplicon to create an artificial heterozygote as a positive control. Ensuring random strand annealing and thus heteroduplex formation, rather than all strands potentially representing new homoduplex strand synthesis at the end of each PCR, is important for all templates.

Usually we optimize PCRs on a thermal gradient PCR block by using a pooled DNA template and a different magnesium chloride concentration (e.g., 1–4.5 mM) in each row of a 96-well microplate. Gel checks in a nondenaturing dry-gel MADGE (8), using a short (5–15 min) electrophoresis are used to determine the conditions under which a single band obtains. If this fails, primers are redesigned.

This same set of amplicons can then be checked in constant-temperature meltMADGE runs, each for the same amount of time, temperatures for example at 10,

5, and 0° below the predicted melting temperature of the native homoduplex, taking account of the effect of urea in the gel in lowering DNA melting temperature (2 degrees per molar unit). Alternately, individual templates may be used for step (iii), including any available known natural or artificial positive controls, or a sufficient number of samples to ensure that any known polymorphism should be represented in the heterozygous state. H-PAGE gels with one or more conventional rows of wells, rather than MADGE arrays, may be used at this stage. Occasional amplicon designs turn out to show unexpected artifactual bands during melt analyses, which were not apparent on simple checking gels: In these cases, a redesign may be necessary. The whole design and validation procedure can be completed in 1–2 days (for many amplicons in parallel, if necessary) on account of the ease of setup and short run times for each experiment.

A thermal ramp linear with time, starting 3.5° below the native duplex melting temperature and rising over 90 min to 0.5° about melting temperature, the latter temperature to take account of mutations that add a hydrogen bond (T to C and A to G transitions), usually proves satisfactory. However, there is considerable flexibility for this precise range, provided that the ramp will be resolving at least one of the heteroduplexes (or homoduplexes) for part of the time of the run.

EXAMPLES FROM DIAGNOSTICS AND RESEARCH

We show here some illustrations from both a research and a diagnostic context, for the following human genes: *LDLR* (Fig. 9.2); *APOE* (Fig. 9.3); *BRCA1* (Fig. 9.4); *MC4R* (Fig. 9.5). *LDLR* (early work of L. Haddad, recent work of X. Chen) exon 3 was used for initial proof-of-principle experiments. From first principles, sensitivity to base changes should be very comparable with DGGE, estimated to be 90–100% (14). Our present evidence is that most known heterozygotes for single base changes and small indels are identified, considerably more than were identified by SSCP for the *LDLR* study, and no known single nucleotide polymorphism yet having proven undetectable: Our almost complete blind analysis of *BRCA1* will be our largest comparative study.

For *APOE*, the exon 4 region was set up as a proof of principle that a pair of common polymorphisms in a very high-percentage GC-rich region (71% for the exon) could be "rediscovered" by meltMADGE and that the three alleles (E2,3,4) and six genotypes could be genotyped directly from meltMADGE (work of E. Spanakis, to be submitted). Model human *APOE* homozygotes and heterozygotes for the common alleles ε2, -3, and -4 (differentiated at two codon positions, C112R and C158R) were loaded alternately at various positions of a meltMADGE (real size and detail shown) to show that the gel offers the same resolution throughout its surface. The alleles are identified by distance run (if not by eye). This *APOE* model also demonstrates that a variation can be detected with the same efficiency independently of the position, along the sequence, where it occurs.

As an example of a gene where high-throughput mutation scanning would be useful in a diagnostic setting, 60 amplicons covering all exons of *BRCA1* have been optimized in our laboratory for *BRCA1*, for a large mutation scanning methods comparative study with SSCP, PTT, and dHPLC (paper to be submitted, work of M. A. Al-Dahmesh and D. Eccles). We have not yet broken the sample blinding codes for this study but note that numerous different mutations have been observed, all

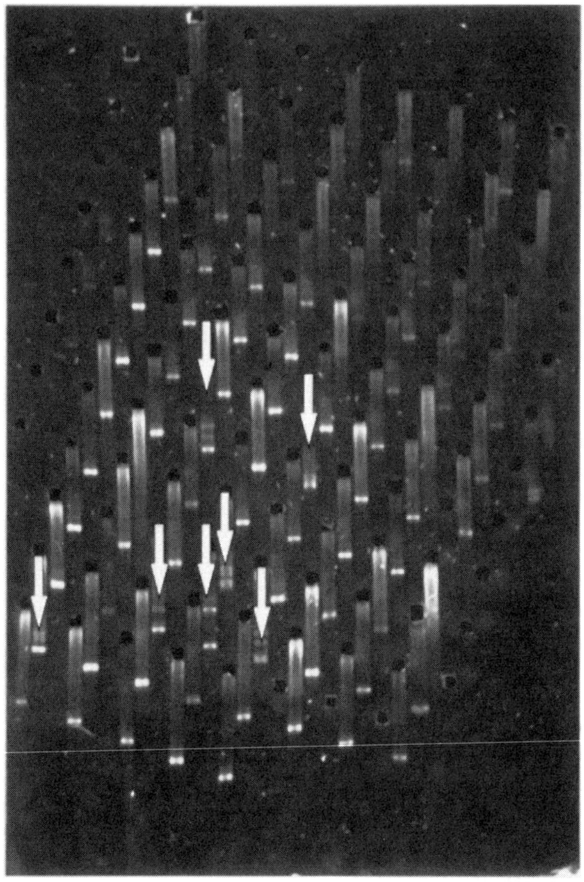

FIGURE 9.2. Image of a meltMADGE analysis. With a purpose-built temporal thermal ramp electrophoresis apparatus meltMADGE, aDGGE-like analysis of GC-clamped PCR products derived from familial hypercholesterolemic patients' LDLR gene exon 3 was undertaken in MADGE format. A single band indicates (as in DGGE) no base variation, whereas band splitting during electrophoresis (resolving up to four bands, two homoduplexes and two heteroduplexes) indicates heterozygosity for a mutation in this exon. Several different patterns representing different mutations are evident. Run time was 1 h, and 10 MADGE gels can be analyzed simultaneously in one 2-liter tank. In the absence of partial melting, each PCR product appears as the same-sized single band on MADGE.

expected common polymorphisms have appeared in the assays, and one new polymorphism has been observed.

Finally, as a research project concerned with implementing mutation scanning for very large-scale studies, we have optimized five meltMADGE amplicons covering all *MC4R* exons and examined 1100 unselected elderly subjects for unknown rare mutations (work of K. K. Alharbi, to be submitted). *MC4R* is the only obesity gene so far identified for which mutations act in the dominant rather than recessive mode and is therefore a good candidate to display rare mutations of milder effect manifesting as moderate rather than severe obesity in an unselected population. Two rare variants, one previously described and the other previously unknown, were observed. Additionally, all artificial mutations designed for assay validation have

A

B

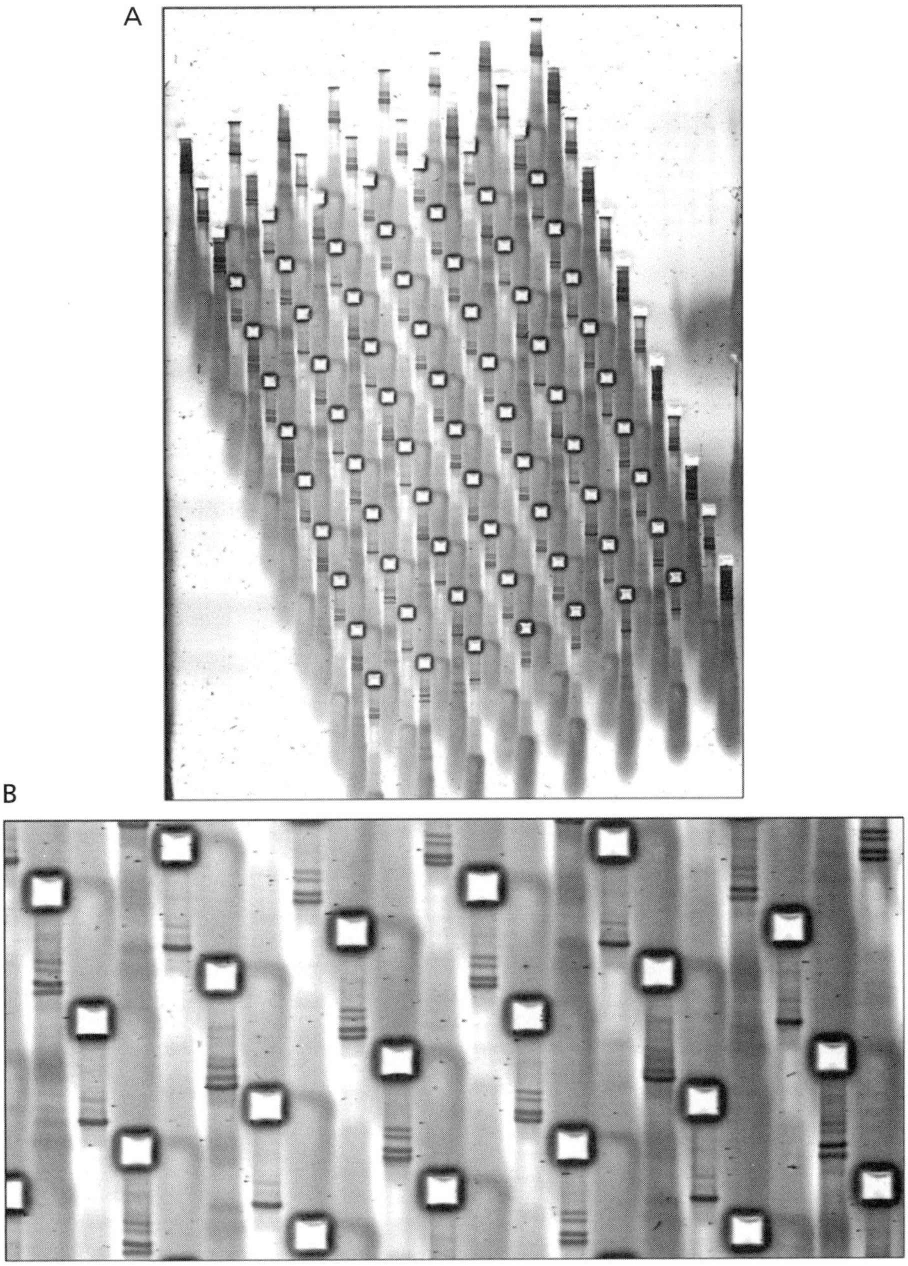

FIGURE 9.3. *Ab initio* recognition of common alleles 2, 3, and 4 in the very %GC-rich APOE gene, using meltMADGE. Model human *APOE* homozygotes and heterozygotes for the common alleles ε2, -3, and -4 (differentiated at two codon positions, C112R and C158R) were loaded alternately at various positions of a meltMADGE (real size and detail shown) to show that the gel offers the same resolution throughout its surface. The alleles are identified by distance run (if not by eye). Our *APOE* model also demonstrates that a variation can be detected with the same efficiency independently of the position, along the sequence, where it occurs. Loading: Wells are as in a microplate array: rows A, B, . . . , H are from right to left and columns 1, 2, . . . , 12 from top to bottom. Columns 1, 3, 5, 7, 9, and 11: A = ε3/ε4, B = ε2/ε4, C = ε3/ε3, D = ε2/ε3, E = ε3/ε4, F = ε4/ε4, G = ε3/ε3, H = ε2/ε3. Columns 2, 4, 6, 8, 10, and 12: A = ε2/ε3, B = ε3/ε3, C = ε4/ε4, D = ε3/ε4, E = ε2/ε3, F = ε3/ε3, G = ε2/ε4, H = ε3/ε4. A: complete meltMADGE gel. B: enlargement of part of the gel. Alignment of bands (e.g., the reader can place a ruler over the image) reveals three mobilities of homoduplex, of which only two homoduplexes will be present in any heterozygote. Mobilities are determined by total number of hydrogen bonds in the homoduplex: n(E4) = n(E3) + 1 = n(E2) + 2, representing, respectively, isoforms ArgArg, CysArg, and CysCys.

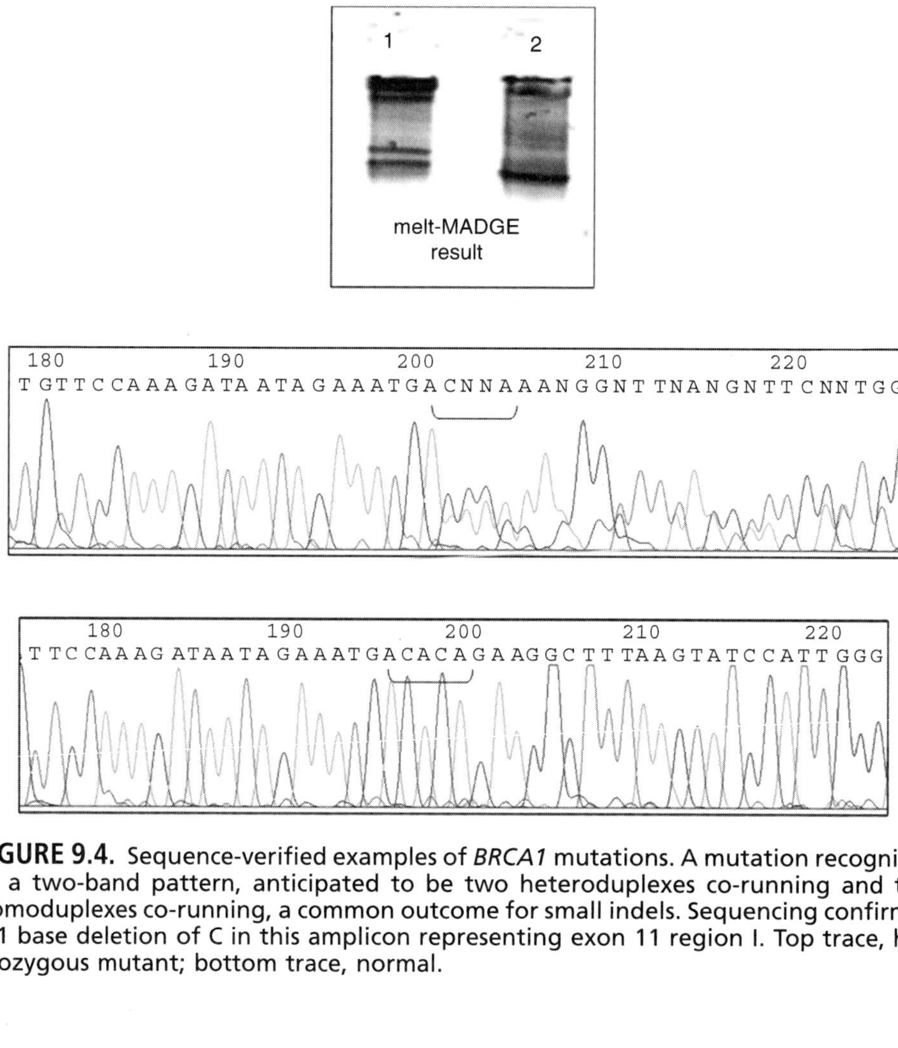

FIGURE 9.4. Sequence-verified examples of *BRCA1* mutations. A mutation recognized as a two-band pattern, anticipated to be two heteroduplexes co-running and two homoduplexes co-running, a common outcome for small indels. Sequencing confirmed a 1 base deletion of C in this amplicon representing exon 11 region I. Top trace, heterozygous mutant; bottom trace, normal.

shown appropriate band patterns when used to create artificial heterozygous amplicon.

■ THE FUTURE OF MELTMADGE

Most laboratories have not so far needed access to high-throughput mutation scanning for rare variants. Complex trait genetic research has so far focused mainly on common single nucleotide polymorphisms. However, there may be a convergence between the field of rare mutations of severe effect (i.e., clinical molecular genetics diagnostics and research) and the field of common variations of milder effects (i.e., complex trait genetics). It remains possible that many rare mutations of moderate effect, each individually rare but cumulatively quite common, impact on the population burden of disease. Diagnostics laboratories are faced with an ever growing range of disease genes for which to provide mutation scanning; and research has yet to explore systematically the hypothesis that there may be many rare mutations of

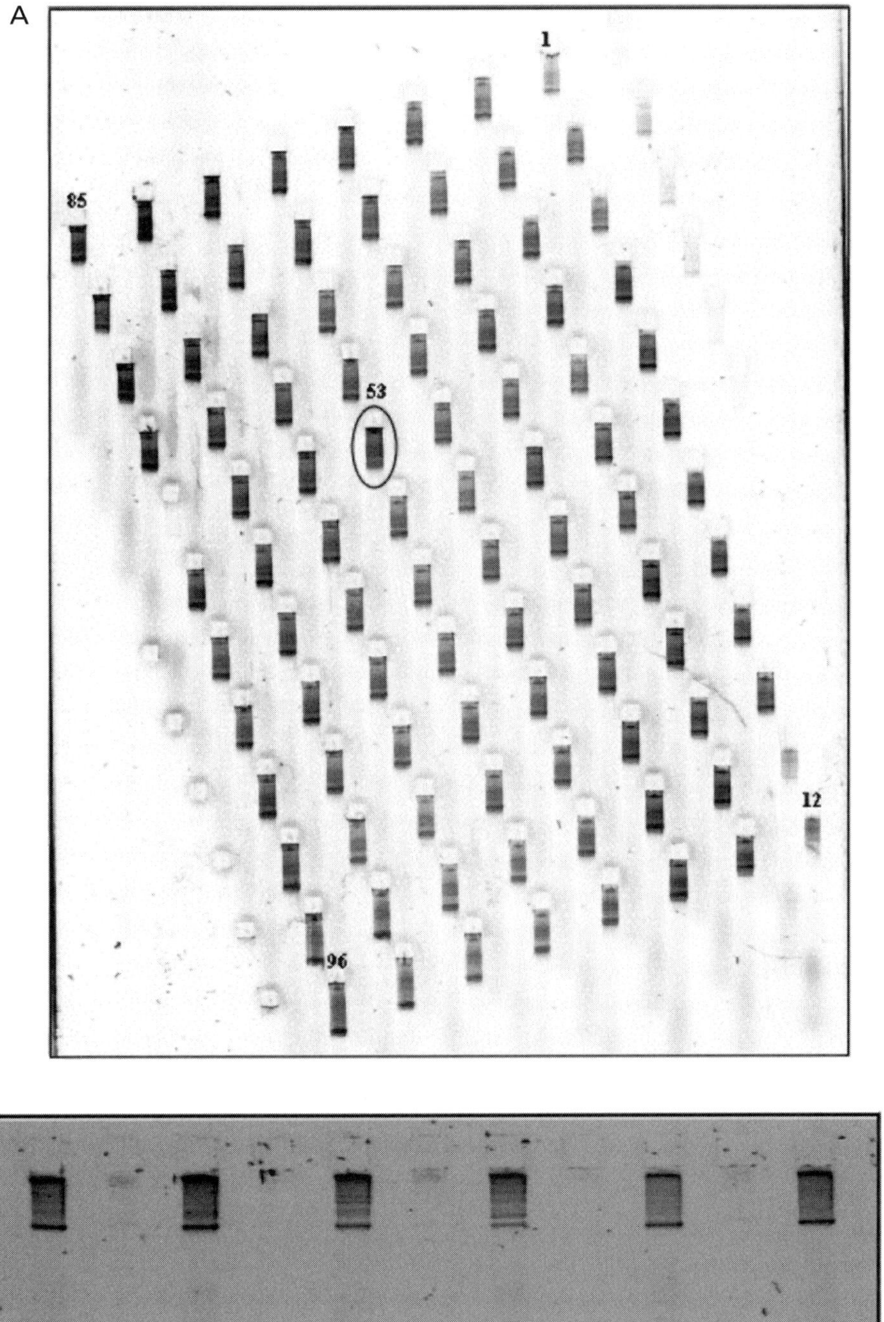

FIGURE 9.5. Scanning a population sample for possible unknown *MC4R* mutations. A: One of twelve simultaneously run meltMADGE arrays, each well of a 96-well plate representing a separate subject from a cohort study. The GC-clamped amplicon is one of five amplicons representing the long coding exon of *MC4R*. In GenBank sequence S77415, primers were: forward primer: 551–569 bp (+3 bp GC-rich short clamp); reverse primer: 771–791 bp (+40 bp GC-rich long clamp). Linear melt ramp at $T = 58$–62°C over 1.5 h in 6% denaturing polyacrylamide gel with 4 M urea. A single mutant is observed at array position 53, more readily visible on an enlarged H-PAGE melt run (panel B). B: V103I polymorphism with base change at nucleotide (A307G) based in MC4R amplicon "MR2." In order, the tracks show wild-type melting pattern, artificial (primer generated) mutant, coannealed mix of wild-type and artificial mutant amplicons, natural sequence variant (a rare polymorphism) V103I, and two further wild-type amplicons run under similar conditions as panel A.

moderate effect in the unselected population. It remains to be determined whether meltMADGE is less or more sensitive than, for example, dHPLC, but its features of very high throughput including fast turnaround of assay setup experiments, at low cost of laboratory setup and low reagent costs should at a minimum make it very suitable for the research laboratory. Not all the features of meltMADGE need be used, and in low-throughput mode it also represents a convenient form of DGGE, a feature that may be of much more interest to laboratories other than our own.

■ ACKNOWLEDGMENTS

I. N. M. Day was a Lister Institute Research Professor during most of these developments, initially in the laboratory of S. E. Humphries, who is supported by the British Heart Foundation (RG95007 and RG2000015). Proof-of-principle experiments were by L. Haddad, who was a British Heart Foundation PhD student. M. A. Aldahmesh holds a PhD studentship from the Ministry of Higher Education of Saudi Arabia, and additional support has been from a grant from the UK Breast Cancer Campaign to D. Eccles. K. K. Alharbi holds a Scholarship from the Department of Applied Medical Science, King Saud University, Saudi Arabia. The MRC is thanked for project grants G9605150MB (postdoctoral fellowship of E. Spanakis 1996–9) and a earlier ROPA award that led up to this development. E. Spanakis was a Wessex Medical Trust Senior Research Fellow, 1999–2001. Most recent work is funded by a UK Department of Health National Genetics Reference Laboratories Award (postdoctoral fellowship of X. Chen). MADGE is subject to patent.

■ REFERENCES

1. Day IN, Humphries SE (1994) Electrophoresis for genotyping: microtiter array diagonal gel electrophoresis on horizontal polyacrylamide gels, hydrolink, or agarose. Anal Biochem 222(2): 389–395.
2. Day IN, Humphries SE (1994) Electrophoresis for genotyping: Devices for high throughput using horizontal acrylamide gels (H-PAGE) and microtitre array diagonal gel electrophoresis (MADGE). Nature Product review[June 36-37].
3. Watanabe TK, Bihoreau MT, McCarthy LC, Kiguwa SL, Hishigaki H, Tsuji A et al. (1999) A radiation hybrid map of the rat genome containing 5,255 markers. Nat Genet 22(1): 27–36.
4. Studier FW (1973) Analysis of bacteriophage T7 early RNAs and proteins on slab gels. J Mol Biol 79(2): 237–248.
5. Sugden B, De Troy B, Roberts RJ, Sambrook J (1975) Agarose slab-gel electrophoresis equipment. Anal Biochem 68(1): 36–46.
6. Kozulic B (1994) Looking at bands from another side. Anal Biochem 216(2): 253–261.
7. O'Dell SD, Gaunt TR, Day IN (2000) SNP genotyping by combination of 192-well MADGE, ARMS and computerized gel image analysis. Biotechniques 29(3): 500–506.
8. Gaunt TR, Hinks LJ, Rassoulian H, Day IN (2003) Manual 768 or 384 well microplate gel "dry" electrophoresis for PCR checking and SNP genotyping. Nucleic Acids Res 31(9): e48.
9. Hacia JG, Brody LC, Chee MS, Fodor SP, Collins FS (1996) Detection of heterozygous mutations in BRCA1 using high density oligonucleotide arrays and two-colour fluorescence analysis. Nat Genet 14(4): 441–447.

10. Parker LT, Zakeri H, Deng Q, Spurgeon S, Kwok PY, Nickerson DA (1996) AmpliTaq DNA polymerase, FS dye-terminator sequencing: Analysis of peak height patterns. Biotechniques 21(4): 694–699.
11. Goffeau A (1997) DNA technology: Molecular fish on chips. Nature 385(6613): 202–203.
12. Sheffield VC, Beck JS, Kwitek AE, Sandstrom DW, Stone EM (1993) The sensitivity of single-strand conformation polymorphism analysis for the detection of single base substitutions. Genomics 16(2): 325–332.
13. Hayashi K, Yandell DW (1993) How sensitive is PCR-SSCP? Hum Mutat 2(5): 338–346.
14. Moyret C, Theillet C, Puig PL, Moles JP, Thomas G, Hamelin R (1994) Relative efficiency of denaturing gradient gel electrophoresis and single strand conformation polymorphism in the detection of mutations in exons 5 to 8 of the p53 gene. Oncogene 9(6): 1739–1743.
15. Whittall R, Gudnason V, Weavind GP, Day LB, Humphries SE, Day IN (1995) Utilities for high throughput use of the single strand conformational polymorphism method: Screening of 791 patients with familial hypercholesterolaemia for mutations in exon 3 of the low density lipoprotein receptor gene. J Med Genet 32(7): 509–515.
16. Day IN, Whittall RA, O'Dell SD, Haddad L, Bolla MK, Gudnason V et al. (1997) Spectrum of LDL receptor gene mutations in heterozygous familial hypercholesterolemia. Hum Mutat 10(2): 116–127.
17. Fischer SG, Lerman LS (1983) DNA fragments differing by single base-pair substitutions are separated in denaturing gradient gels: correspondence with melting theory. Proc Natl Acad Sci USA 80(6): 1579–1583.
18. Guldberg P, Guttler F (1993) A simple method for identification of point mutations using denaturing gradient gel electrophoresis. Nucleic Acids Res 21(9): 2261–2262.
19. Clark SM, Mathies RA (1993) High-speed parallel separation of DNA restriction fragments using capillary array electrophoresis. Anal Biochem 215(2): 163–170.
20. Day IN, O'Dell SD, Cash ID, Humphries SE, Weavind GP (1995) Electrophoresis for genotyping: Temporal thermal gradient gel electrophoresis for profiling of oligonucleotide dissociation. Nucleic Acids Res 23(13): 2404–2412.
21. Chen XH, O'Dell SD, Day IN (2002) Microplate array diagonal gel electrophoresis for cohort studies of microsatellite loci. Biotechniques 32(5): 1080–2, 1084, 1086.
22. Abrams ES, Murdaugh SE, Lerman LS (1990) Comprehensive detection of single base changes in human genomic DNA using denaturing gradient gel electrophoresis and a GC clamp. Genomics 7(4): 463–475.

10

HIGH-THROUGHPUT SNP GENOTYPING BY MINISEQUENCING PRIMER EXTENSION USING OLIGONUCLEOTIDE MICROARRAYS

Katarina Lindroos, Ulrika Liljedahl, Snaevar Sigurdsson, and Ann-Christine Syvänen
Department of Medical Sciences, Uppsala University, Uppsala, Sweden

A promising approach toward high-throughput genotyping of single nucleotide polymorphisms (SNPs) is to use arrays of immobilized oligonucleotides in miniaturized assays (1, 2). A significant advantage of performing the assays in microarray formats is that the costs of genotyping are reduced because many SNPs are analyzed simultaneously in each sample and because the reaction volumes employed in the microarrays are small. The three major reaction principles that are currently in use for genotyping SNPs, namely, hybridization with allele-specific oligonucleotide (ASO) probes (3), oligonucleotide ligation (4), and DNA polymerase-assisted primer extension (5) have all been utilized in microarray-based assay formats. Simultaneous genotyping of multiple SNPs by ASO hybridization is hampered by the poor specificity of ASO probes discriminating between SNP genotypes in large, diploid genomes (6, 7). Therefore, the enzyme-assisted methods are gaining acceptance as more specific alternatives than ASO hybridization for multiplex SNP typing (8, 9).

Guide to Mutation Detection, Edited by Graham R. Taylor and Ian N. M. Day.
ISBN 0-471-23444-3

We have developed a microarray system based on "minisequencing" single nucleotide primer extension that allows highly specific, multiplex genotyping of SNPs (5). In the minisequencing reaction, a DNA-polymerase is used to extend detection primers that anneal immediately adjacent to the sites of the SNPs with labeled nucleotide analogs that are complementary to the nucleotide(s) at the SNP sites (10) (Fig. 10.1A). In the microarray format of the method, detection primers specific for the SNPs to be analyzed are attached covalently to surface-activated glass microscope slides through an amino group in their 5′-end. The primers are applied tothe microscope slide with an array spotter in a configuration of 80 "sub-arrays" that may contain up to 240 primers each (11) (Fig. 10.2). This format allows the generation of several thousand genotypes per slide. The throughput of our genotyping method is limited mainly by the requirement of establishing and performing multiplex PCRs of the DNA regions spanning the SNPs before the genotyping reactions on the microarrays. The PCR products spanning the SNPs are allowed to anneal to the immobilized detection primers, followed by extension of the primers with fluorescent ddNTPs using a DNA polymerase (Fig. 10.1A). In this method ddNTPs labeled with a single fluorophore (TAMRA) are used to detect all four nucleotides in separate reactions for each sample (12). After the minisequencing reactions, the incorporated fluorescence is measured in a fluorescence scanner. Figure 10.2 shows an example of the result from genotyping one sample for 100 SNPs. We also use a modification of the method, in which cyclic minisequencing reactions with four fluorescently labeled ddNTPs are performed in solution using primers with 5′-"Tag" sequences, followed by capture of the reaction products on microarrays carrying oligonucleotides that are complementary to the "Tag"

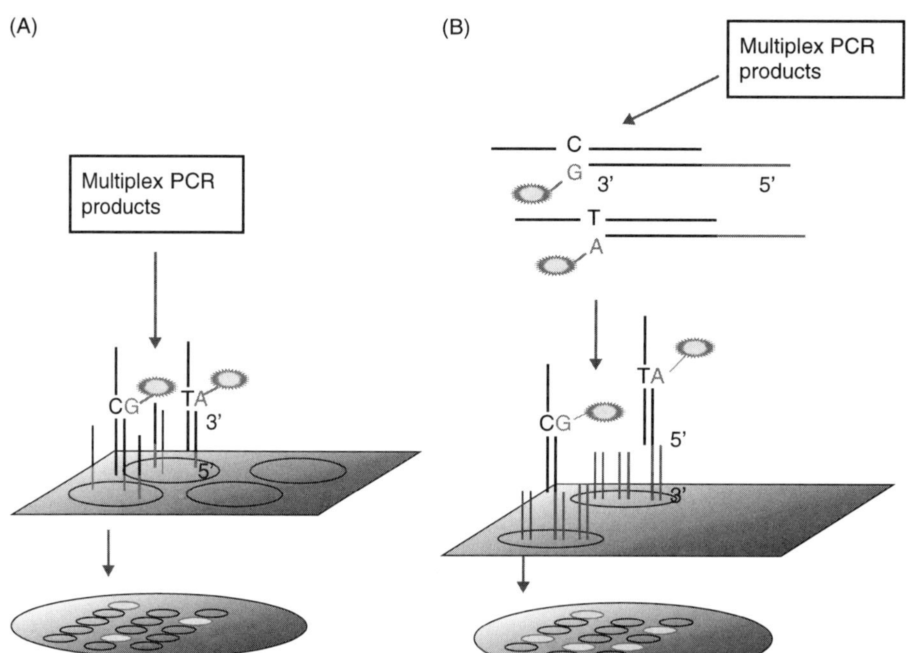

FIGURE 10.1. Principle of (A) minisequencing with immobilized primers, and (B) Tag-array-based minisequencing.

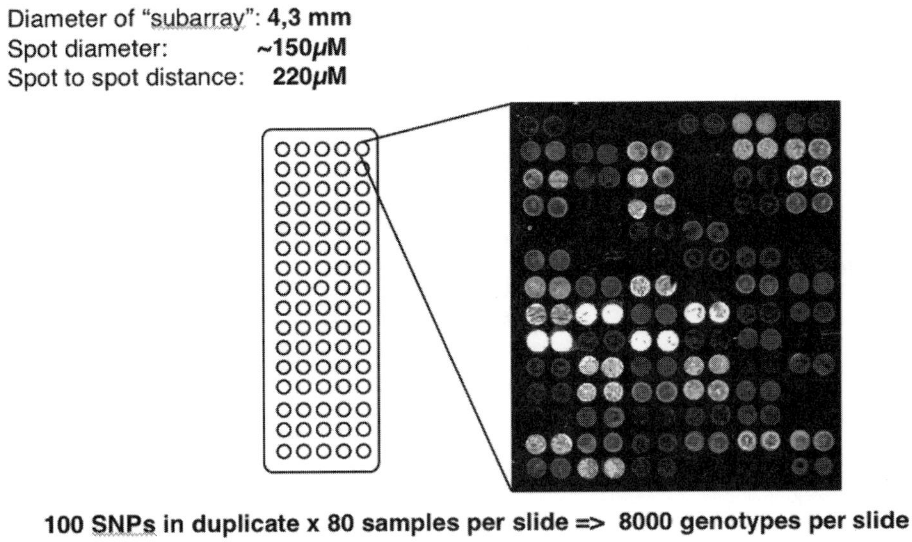

FIGURE 10.2. The "array-of-arrays" format of minisequencing.

sequences ("cTags") (Fig. 10.1B). The concept of using "tagged" PCR primers was first described for analyzing gene expression in yeast by PCR (13) and has since been applied to SNP genotyping by primer extension and capture on fluorescent microparticles (14), high-density oligonucleotide arrays (Affymetrix, GenFlex™ arrays, Ref. 15), and medium-density, custom-made oligonucleotide arrays (16). Our modification of the "Tag-array" method makes use of the format with 80 subarrays per microscope slide, to which the cTags are immobilized through a 3′- amino group as described above (21). Figure 10.3 compares the steps of the two procedures. All materials and equipment required for setting up and performing the assay are generally available from common suppliers. Thus the technology can be established in molecular biology laboratories wishing to increase their "in-house" throughput for genotyping SNPs.

Multiplex PCR amplification of more than 10 fragments reproducibly and successfully from multiple samples has proven to be difficult (3, 11). A "touchdown" PCR procedure (17) to circumvent differences in amplification efficiency due to differences in melting temperatures between the PCR primers in combination with universal 5′-sequences on the primers may be used (18) to unify the reaction kinetics of the primer annealing. To avoid complementary 3′ overlap of the primers, they can be designed with common terminal dinucleotides, for instance, AC, at their 3′ end (19).

The specific minisequencing primer should hybridize to the DNA-template with its 3′ end with the nucleotide adjacent to the variant nucleotide to be analyzed. The primers should be approximately 20 bases long, all with similar T_m values close to 57°C (excluding the 15 T residue). To avoid misincorporation of ddNTPs the primer should not form a 3′ hairpin-loop or other complementary structure. Such structures may be avoided by using a primer from the complementary DNA strand. The minisequencing primers to be immobilized contain 15T residues as spacers 5′ of the specific sequence and a 5′-amino group for attachment to the glass surface. Mini-

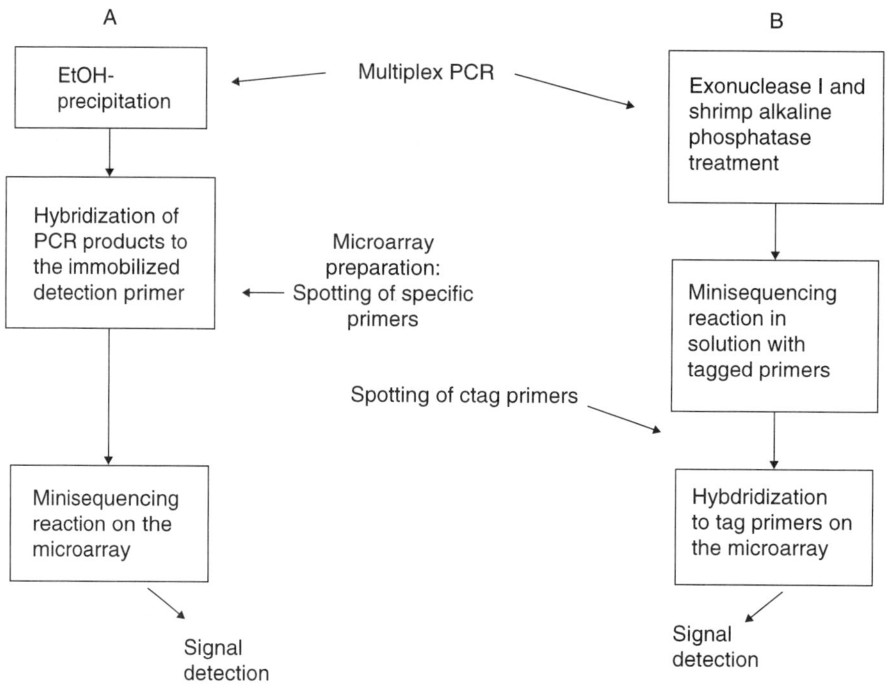

FIGURE 10.3. Steps of the procedure for (A) minisequencing with immobilized primers and (B) Tag-array-based minisequencing.

sequencing primers to be used in solution in the Tag-array assay are designed with a 5′-tag sequence. The tag-oligonucleotide sequences should be around 20–27 bases long, and their T_m should be approximately 55°C. Their self similarity and similarity to each other should be low (16). The complementary tag sequences to be immobilized contain 15T residues as spacers 3′ of the specific sequence and a 3′-amino group (see Fig. 10.1).

An advantage of the "Tag-array" system is that generic microarrays can be manufactured in advance and used for many different applications. A second advantage of the approach is the increased sensitivity obtained through the cyclic minisequencing reactions. A disadvantage is that two oligonucleotide primers (the tagged minisequencing primer and the immobilized "anti-tag capture" primer) are needed for each SNP compared to only one in the minisequencing reaction using specific immobilized primers. The latter approach may also be easier to automate and more advantageous for quantification of sequence variants.

REFERENCES

1. Hacia JG (1999) Resequencing and mutational analysis using oligonucleotide microarrays. Nat Genet 21: 42–47.
2. Southern EM (1996) DNA chips: analysing sequence by hybridization to oligonucleotides on a large scale. Trends Genet 12: 110–115.

3. Hacia JG, Sun B, Hunt N, Edgemon K, Mosbrook D, Robbins C, Fodor SP, Tagle DA, Collins FS (1998) Strategies for mutational analysis of the large multiexon ATM gene using high-density oligonucleotide arrays. Genome Res 8: 1245–1258.
4. Gunderson KL, Huang XC, Morris MS, Lipshutz RJ, Lockhart DJ, Chee MS (1998) Mutation detection by ligation to complete n-mer DNA arrays. Genome Res 8: 1142–1153.
5. Pastinen T, Kurg A, Metspalu A, Peltonen L, Syvanen AC (1997) Minisequencing: a specific tool for DNA analysis and diagnostics on oligonucleotide arrays. Genome Res 7: 606–614.
6. Wang DG, Fan JB, Siao CJ, Berno A, Young P, Sapolsky R, Ghandour G, Perkins N, Winchester E, Spencer J, Kruglyak L, Stein L, Hsie L, Topaloglou T, Hubbell E, Robinson E, Mittmann M, Morris MS, Shen N, Kilburn D, Rioux J, Nusbaum C, Rozen S, Hudson TJ, Lander ES et al. (1998) Large-scale identification, mapping, and genotyping of single-nucleotide polymorphisms in the human genome. Science 280: 1077–1082.
7. Cho RJ, Mindrinos M, Richards DR, Sapolsky RJ, Anderson M, Drenkard E, Dewdney J, Reuber TL, Stammers M, Federspiel N, Theologis A, Yang WH, Hubbell E, Au M, Chung EY, Lashkari D, Lemieux B, Dean C, Lipshutz RJ, Ausubel FM, Davis RW, Oefner PJ (1999) Genome-wide mapping with biallelic markers in *Arabidopsis thaliana*. Nat Genet 23: 203–207.
8. Fortina P, Delgrosso K, Sakazume T, Santacroce R, Moutereau S, Su HJ, Graves D, McKenzie S, Surrey S (2000) Simple two-color array-based approach for mutation detection. Eur J Hum Genet 8: 884–894.
9. Southern E, Mir K, Shchepinov M (1999) Molecular interactions on microarrays. Nat Genet 21: 5–9.
10. Syvänen A-C, Aalto-Setälä K, Harju L, Kontula K, Söderlund H (1990) A primer-guided nucleotide incorporation assay in the genotyping of apolipoprotein E. Genomics 8: 684–692.
11. Pastinen T, Raitio M, Lindroos K, Tainola P, Peltonen L, Syvanen AC (2000) A system for specific, high-throughput genotyping by allele-specific primer extension on microarrays. Genome Res 10: 1031–1042.
12. Raitio M, Lindroos K, Laukkanen M, Pastinen T, Sistonen P, Sajantila A, Syvanen AC (2001) Y-chromosomal snps in finno-ugric-speaking populations analyzed by minisequencing on microarrays. Genome Res 11: 471–482.
13. Shoemaker DD, Lashkari DA, Morris D, Mittmann M, Davis RW (1996) Quantitative phenotypic analysis of yeast deletion mutants using a highly parallel molecular barcoding strategy. Nat Genet 14: 450–456.
14. Cai H, White PS, Torney D, Deshpande A, Wang Z, Marrone B, Nolan JP (2000) Flow cytometry-based minisequencing: a new platform for high-throughput single-nucleotide polymorphism scoring. Genomics 66: 135–143.
15. Fan JB, Chen X, Halushka MK, Berno A, Huang X, Ryder T, Lipshutz RJ, Lockhart DJ, Chakravarti A (2000) Parallel genotyping of human SNPs using generic high-density oligonucleotide tag arrays. Genome Res 10: 853–860.
16. Hirschhorn JN, Sklar P, Lindblad-Toh K, Lim YM, Ruiz-Gutierrez M, Bolk S, Langhorst B, Schaffner S, Winchester E, Lander ES (2000) SBE-TAGS: An array-based method for efficient single-nucleotide polymorphism genotyping [In Process Citation]. Proc Natl Acad Sci USA 97: 12164–12169.
17. Don RH, Cox PT, Wainwright BJ, Baker K, Mattick JS (1991) "Touchdown" PCR to circumvent spurious priming during gene amplification. Nucleic Acids Res 19: 4008.
18. Shuber AP, Grondin VJ, Klinger KW (1995) A simplified procedure for developing multiplex PCRs. Genome Res 5: 488–493.
19. Zangenberg G, Saiki RK, Reynolds R (1999) Multiplex PCR: Optimization guidelines. In *PCR Applications*, Innis MA, Gelfand DH, Sninsky JJ (eds.), Academic Press, London, UK. pp. 73–94.

20. Guo Z, Guilfoyle RA, Thiel AJ, Wang R, Smith LM (1994) Direct fluorescence analysis of genetic polymorphisms by hybridization with oligonucleotide arrays on glass supports. Nucleic Acids Res 22: 5456–5465.
21. Lindroos K, Sigurdsson S, Johansson K, Rönnblom L, Syvänen A-C (2002) Multiplex genotyping in pooled DNA samples by a four colour microarray system. Nucleic Acids Res 30: e70.

11

MUTATION DATABASES

Rania Horaitis
Human Genome Variation Society, Genomic Disorders Research Centre, Carlton South, Australia

Richard G. H. Cotton
Genomic Disorders Research Centre, St. Vincent's Hospital, The University of Melbourne, Victoria, Australia

OVERVIEW

Importance of Documenting Mutations

Mutation databases of human genes are assuming an increasing importance in all areas of health care. Increasingly experts in the mutations and diseases of particular genes are curating published and unpublished mutations in locus-specific databases (LSDB). This information is especially needed by clinics and laboratories dealing with inherited disease and cancer, by biochemists interested in protein function, by technologists and evolution and migration researchers, and for use in forensic science. New methods for the detection of mutations and the completion of the human genome sequencing project have seen an exponential rise in variation information. To ensure future availability to health care professionals and others, this information must be collected, quality controlled, documented, and stored safely. Figure 11.1 shows the dramatic rise in number of mutations as documented in OMIM from its inception until December 1998. On 5 June 2002, OMIM contained 13,682 mutations, the rise still exponential as expected.

Brief History of Mutation Databases

The first mutation that caused human disease was described in 1949 in the hemoglobin gene. Since then, there have been almost 1300 genes in which at least one

Guide to Mutation Detection, Edited by Graham R. Taylor and Ian N. M. Day.
ISBN 0-471-23444-3

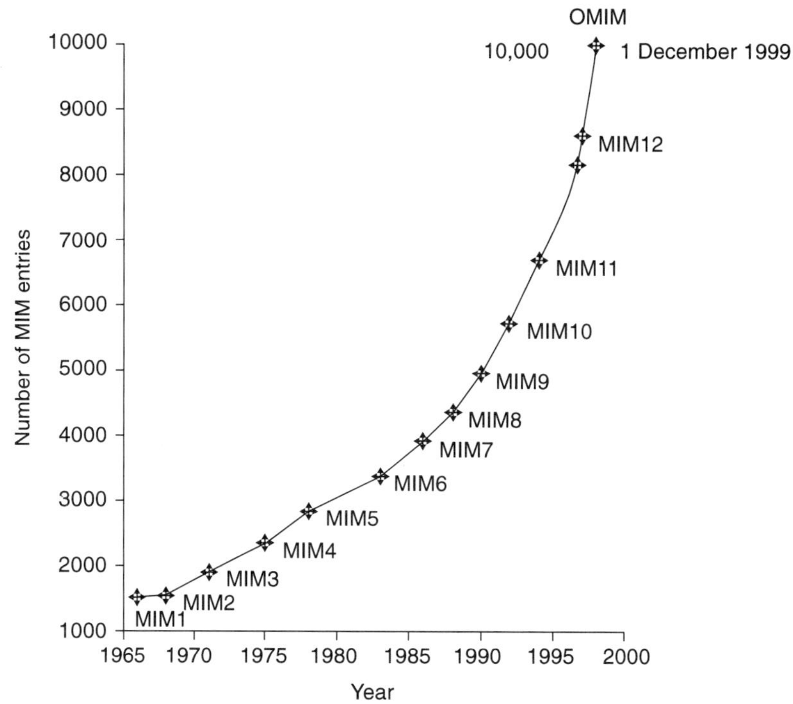

FIGURE 11.1. Number of entries in *Mendelian Inheritance in Man* (graph courtesy of OMIM).

known mutation is described. The first database of these mutations was begun in the 1970s by Prof. Victor McKusick and is known as MIM or "Mendelian Inheritance in Man." This database was in printed format, but now it is a useful online resource (OMIM). The first locus-specific database, that is, a database specific to a particular gene locus, was begun in 1976 when lists of hemoglobin mutations (200 variants) were published yearly in *A Syllabus of Human Hemoglobin Variants* by the Sickle Cell Anemia Foundation in the United States. Now this too is online, with around 700 variants.

Types of Mutation Databases

Currently two types of mutation databases are available: general mutation databases and locus-specific mutation databases.

General Databases

General mutation databases contain pooled information on variation across the whole genome. Some collect only mutations, others only single nucleotide polymorphisms (SNPs). These databases have developed tools for analyzing existing data collections while providing consistent user interfaces to all genes. The general databases are not sufficient in themselves for the needs in medical genetics because they are not run by experts in every gene but rather by bioinformatics experts who can present the data collected in a multitude of ways. Because of the effort involved in acquiring variation data on every gene described, only general information about published mutations can be included in these databases. Data that are unpublished, however, remain unreported in general databases.

The most commonly used freely available central mutation databases are:

(i) Online Mendelian Inheritance in Man; OMIM
http://www.ncbi.nlm.nih.gov/Omim/searchomim.html
OMIM documents inherited disorders and includes all known genes; however, only *selected* published mutations are included. These include the first mutation described in a locus, the most common mutations (or the first dozen or so), mutations of historical interest, unusual mechanisms of mutagenesis or unusual pathogenesis, or any other mutation of interest.

(ii) Human Gene Mutation Database; HGMD
http://archive.uwcm.ac.uk/uwcm/mg/hgmd0.html
HGMD became publicly available online in 1996. It is a collection of *all the published* germline mutations in nuclear genes along with their references and summaries of mutations in each gene. Maps and reference sequences are also included. There is, however, a lag from publication time to appearance on HGMD because of commercial contracts of approximately nine to thirteen months.

(iii) Human Genome Variation Database; HGVbase
http://hgvbase.cgb.ki.se/
HGVbase attempts to summarize all known sequence variations (i.e., mutations *and* SNPs, indels, simple tandem repeats, and other sequence alternatives) in the human genome and to facilitate research into how genotypes affect common diseases, drug responses, and other complex phenotypes. All sequence variations are presented with details of how they are physically and functionally related to the closest neighboring gene. At present, the majority of data is of the SNP type. HGVbase plans to start acting as a central depository for mutation collection soon for both published *and* unpublished variations in a new effort by the Human Genome Variation Society (see later).

(iv) dbSNP
http://www.ncbi.nlm.nih.gov/SNP/
The most common variations are single nucleotide polymorphisms (SNPs), and these occur approximately once every 100 to 300 bases. The term "SNP" is loosely used by this database and infers "minor genetic variation," thus dbSNP includes disease-causing clinical mutations as well as neutral polymorphisms, although it generally contains polymorphisms. There has recently been great interest in SNP discovery and detection as SNPs are expected to facilitate large-scale association genetics studies. Established in September 1998 by NCBI to serve as a central repository for SNPs, microsatellite repeats, and small insertion/deletion polymorphisms, this database now contains over 4 million human SNPs so far. Although not part of GenBank, the sequences of dbSNP records are expected to be contained within the sequences of one or more GenBank records. Submissions are invited from all laboratories; the major contributors are laboratories associated with the National Human Genome Research Institute (NHGRI) (USA) grants program.

Locus-Specific Mutation Databases (LSDBs)

LSDBs concentrate on variation within a single gene and are usually run by a consortium of collaborating researchers with scientific expertise in a particular gene or phenotype. Some curators are responsible for a number of LSDBs at a single website that may be specific for a particular disease or group of related genes. Because of the expertise of the curator, LSDBs often contain far more information on each mutation than general mutation databases, and some contain such extensive information that they have become known as knowledgebases. LSDBs also contain a greater number of mutations in each gene than general databases, because unpublished mutations are included as they are curated by the expert(s) in the gene presenting the database. As unpublished mutations can often equal published mutations, this is a huge amount of information to be obtained. The phenylalanine hydroxylase knowledgebase, for example, shows this:

	No. of Mutations	
Database	May 2002	September 2004
OMIM	67	67
HGMD	338	447
PAH knowledgebase	447	485

Over 270 LSDBs are now available on the WWW, and there are likely to be many more. Because of the different curators' expertise in informatics or ability/inability to spend time or money, LSDBs vary enormously in content and presentation (1). Some LSDBs are nothing more than a collection of mutations and their references, whereas others are fully interactive databases with a wealth of knowledge.

A list of LSDBs and links is available at:
http://www.genomic.unimelb.edu.au/mdi/dblist/dblist.html

The Problem with the Current System

Many mutations are discovered in the laboratory but remain unpublished or not submitted to any database; hence this information that could be vital is unavailable. Community health relies on the best access to current knowledge both to improve research questions and to provide maximal health care; therefore, a maximum number of mutations should be available. General databases such as OMIM and HGMD are unable to search and accumulate *all* mutations, and it is no longer possible to publish single mutations in journals (because of the high volume of new mutations being discovered). At present there is no complete systematic listing of *all* mutations (see Fig. 11.2).

A Consortium to Solve the Problem is Formed

In 1994 Richard Cotton organized a small meeting in Montreal to discuss this problem with Victor McKusick, Charles Scriver, Haig Kazazian, Jr., Douglas Wallace, Lap-Chee Tsui, and Aravinda Chakravarti. It was decided that to capture all mutations to work toward a system of up-to-date databases available via the Internet, a consortium of database curators would be needed because these are the experts in individual genes and they can curate the data most properly. This small group has now grown to around 600 members in 34 countries and until October

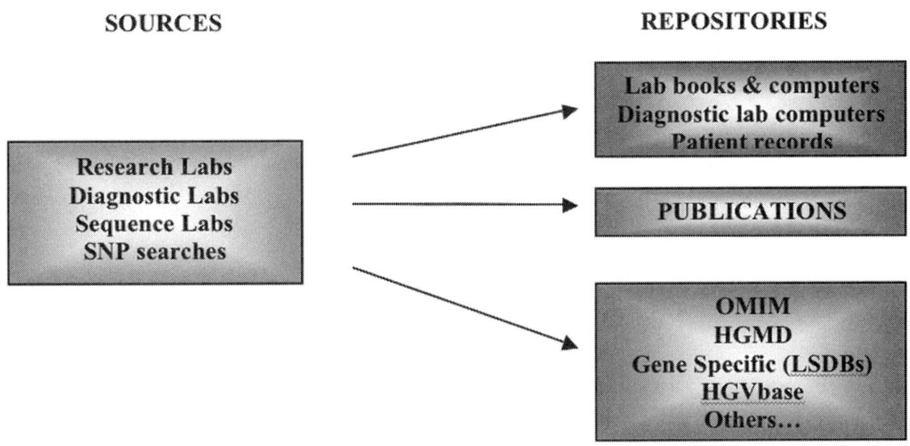

FIGURE 11.2. Origin of mutations and FLOWI.

2001 was known as the HUGO Mutation Database Initiative (MDI). In October 2001 it was decided that the group should form an official "Society" and it is now known as the Human Genome Variation Society (HGVS; www.hgvs.org); there are 61 members to date. The MDI members are still kept in the loop with newsletters and any important information.

A Proposed Solution to the Problem

A group of MDI/HGVS members have been working on a solution to the problem of capturing all mutations. As experts in particular genes are more in touch with new mutations, they are likely to be able to collect and curate all the mutations in any given gene. A new central database system is being proposed, one that utilizes the strengths of the LSDB curators and the strengths of the general databases in bioinformatics expertise and display of data: "HGVSYS." In summary, mutations will be submitted by filling in a quality control form and sent to a central depository (WayStation) via the WWW. The WayStation will receive, briefly check, and then send the submission for review to an expert in the gene, either an LSDB curator or other expert who has agreed to act as a "Gene Editor." If the expert agrees to accept this mutation provided all the necessary criteria are fulfilled, the WayStation is notified and assigns a *Human Mutation* reference, hence making this an online publication of the journal. If a novel mutation, NCBI will assign it a PubMed ID. Once accepted, the submission will be sent to HGVbase for storage and display as well as to any LSDB that may exist for the gene in question if the submitters desire it. We hope this system gets under way by the end of 2004, pending a grant application (Fig. 11.3).

Using Mutation Databases

I Have Found a Mutation(s). How Do I Submit It to a Database?

(1) Most general mutation databases search the literature on a weekly or monthly basis by a computerized search procedure to acquire new data. If you manage to publish your mutation, it will make its way to a general mutation database.

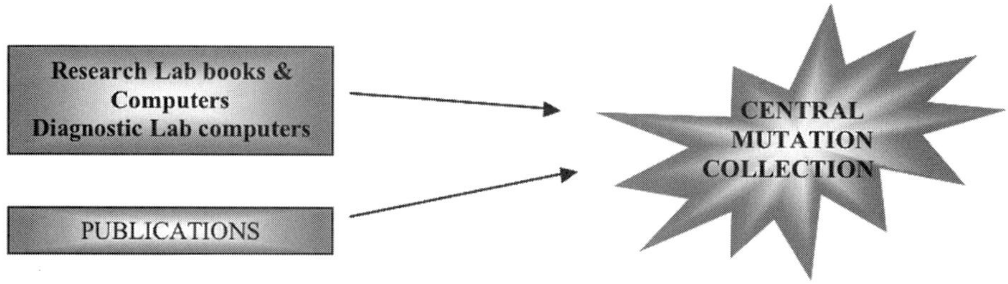

FIGURE 11.3. Simplified plan to collect mutations from *all* sources.

(2) If you do not publish your mutation, you will have to submit the data yourself to a LSDB if there is one for your gene. This is generally done either by directly contacting the curators themselves or by filling in an online form that is available on the website.

(3) We hope that by the end of 2002, you will able to submit your data to one place only, the HGVSYS WayStation briefly described above, where it will then be reviewed and, if accepted, sent to HGVbase for storage and display and any other LSDB that wants it. For the time being, you will have to follow step (2) above.

Are There Any Criteria for Data Submission I Need to Know?

Before material can be entered in a database, it must be ensured that the data is of the highest quality. Nomenclature must conform to standards, and quality control checks must be done.

(1) Nomenclature The HUGO MDI Nomenclature Working Group has proposed the following recommendations for naming of mutations. These have now become widely accepted.

(i) Antonarakis SE and the Nomenclature Working Group (1998) Recommendations for a Nomenclature System for Human Gene Mutations. Hum Mutat 11: 1–3.

(ii) den Dunnen JT, Antonarakis SE (2000) Mutation nomenclature extensions and suggestions to describe complex mutations: A discussion. Hum Mutat 15: 7–12.

These have already been discussed above.

(2) Quality Control A few points should be checked before submitting a mutation to ensure the highest quality of data in the database. When a mutation has been published it is likely that any problems were picked up by the reviewers, and so the quality should be high, although we have shown this is not necessarily true (2). The most important features to look for are:

PCR errors	Has the change been found in a 2nd PCR sample amplified from the patient's DNA?

Incomplete scanning	Even if 1 mutation is found, a 2nd mutation that may be disease causing may be there. If using a method that only detects 80% or less of the mutations, this may happen.
Phenotype	Is the change always present when the phenotype is present?
Present in 100 normals?	Ideally, a disease-causing mutation should be shown not to be present in 100 normal chromosomes.
Expression analysis	Ideally, this should be done to prove that the substitution affects activity.
Conserved residue	Is the base change in a conserved residue? If so, it will more likely be disease causing.
Polymorphism	Is the frequency of the base change reported in >1% of the normal population?
Reporting errors	Typographical errors Misreading a gel Incorrect assignment of base position Incorrect deduction of amino acid change

How Do I Search for a Mutation in a Database? The differences between general mutation databases and LSDBs have already been discussed. Searching for a mutation in either of these is a relatively simple and straightforward process that is obvious when logging on the database. It is not necessary to describe this here. You should familiarize yourself with OMIM, HGMD, and the LSDB of your gene of interest and see the differences in the information contained therein. OMIM, for example, contains information, especially clinical as text along with reference information, links to MEDLINE articles, and sequence information. HGMD, on the other hand, tabulates the information and categorizes it by mutation type (e.g., small deletions, small insertions, etc.) or phenotype with links to OMIM matching these phenotypes. Mutation maps and graphical displays of mutations in the coding regions, frequency of different types of mutations, and links to any LSDB and MEDLINE are also provided. A LSDB may have an enormous amount of information or very little depending on the gene. Examples of other information to be found/searched for are ethnic or geographic origin of patient, detection method, restriction enzyme sites, and clinical information.

Creating a Locus Specific Database

I Have Collected a Large Number of Mutations, But There is no Database for My Gene. What Do I Do With Them? Can I Create My Own Locus-Specific Database?

(1) Contact Rania Horaitis to determine whether someone else is already working on a new database for your gene.

(2) Form a consortium of interested researchers and clinicians to spread the load and ensure longevity of the database by having a community to find a new curator, if necessary; you may not be involved in this gene in a few years.

(3) Collect all mutations of your own and any others including polymorphisms and name them according to the recommendations (above). You may wish to publish this data as a review for a specialist journal such as *Human Muta-*

tion or another journal. This is a good incentive to collect the relevant information because a useful publication will be produced.

(4) Include "Core" information in your database to ensure usefulness.

What is the "Core" Information for Mutation Databases?

Recommendations have been made by the content working group of the MDI (3, 4) as to the bare minimum of data for cohesiveness. To make your database more meaningful and useful you should include all these elements. Of course, other data particular to your disease or gene may be included, and you are encouraged to include this.

Core Information (Essential Content)

(1) A unique identifier must be issued by the database for each allele.
(2) The source/report of the data
(3) The context of the allele; Ref. Seq. Acc. No.; DDBJ/EMBL/GenBank
(4) The name of the allele using the guidelines above, e.g.. 1425G → A
(5) Description of the change as fully as possible
(6) A document showing the "biography" of the database and a map of its details
(7) A minimum set of links to other databases should be included for mutual compatibility, e.g., OMIM
(8) The date when the database was last updated

Other Information to Include The list can vary depending on the gene/s and your interests; however; here are a few items to include that would make a variation database of more interest.

HUGO-approved gene symbol
Chromosome region
Gene information
Location exon/intron
Phenotype (if any)
RNA effect
Protein effect
Detection method
Clinical information about disease for clinicians and patients
Geographic origin
Ethnic origin
Expression studies
List of contributors to db
List of associations and organizations

What Software Can I Use?

You may choose to:

(1) Use commercially available packages: Access, Excel, FoxPro, FileMaker Pro, etc. Not recommended.

(2) Create your own: You may be good at informatics or have access to informatics expertise at your institute.

(3) Use software provided by HGVS Members:

Mutation database-specific software currently made available by HGVS members is listed below. You must contact the curators/creators yourself if you want to use it. Look at the databases running on this software to see if it is appropriate for you. Contact information is on the website.

Software Type	URL
MuStar®	www.hgu.mrc.ac.uk/Softdata/Mustar/
MUTbase	www.uta.fi/imt/bioinfo/MUTbase/
Universal Mutation DB	www.umd.necker.fr/

REFERENCES

1. Claustres M, Horaitis O, Vanevski M, Cotton RGH (2002) Time for a unified system of mutation description and reporting: A review of locus specific mutation databases. Genome Res 12: 680–688.
2. Cotton RGH, Horaitis O (2000) Quality control in the discovery, reporting, and recording of genomic variation. Hum Mutat 15: 16–21.
3. Scriver CR, Nowacki PM, Lehvaslaiho H (1999) Guidelines and recommendations for content, structure and deployment of mutation databases. Hum Mutat 13: 344–350.
4. Scriver CR, Nowacki PM, Lehväslaiho H (2000) Guidelines and recommendations for content, structure, and deployment of mutation databases: II. Journey in progress. Hum Mutat 15: 13–15.

12

MASS SPECTROMETRY FOR HIGH-THROUGHPUT GENOTYPING

Ivo Glynne Gut
Centre National de Génotypage, Evry, France

BACKGROUND

In the past few years several methods for single nucleotide polymorphism (SNP) genotyping have been presented. However, only very few of these methods have been integrated with automation and informatics to perform at high throughput. Key to the operation of a high-throughput genotyping system is the organization of the process. At CNG we have established a high-throughput SNP genotyping platform based on matrix-assisted laser desorption/ionization time-of-flight mass spectrometry (MALDI) detection. Accuracy, speed of data accumulation, multiplexability, and data structure are the major features of SNP analysis by MALDI. The method for allele distinction was developed in house and is called the GOOD assay. The major methods for SNP genotyping with primer extension and MALDI mass spectrometric detection and their application for high throughput are reviewed here.

INTRODUCTION TO MALDI

Mass spectrometry provides an attractive solution for SNP genotyping, mainly because it enables direct and rapid measurement of DNA products rather than reading a tag (fluorescent or radioactive) (1). Results can easily be scored by automated software.

In particular, MALDI has revolutionized the analysis of biomolecules. A matrix, usually a low-molecular-weight organic acid with a strong absorption at the laser

Guide to Mutation Detection, Edited by Graham R. Taylor and Ian N. M. Day.
ISBN 0-471-23444-3

wavelength, containing the analyte molecules is desorbed with a short laser pulse (Fig. 12.1). The ensuing physical processes result in the predominant formation of either singly positively or singly negatively charged ions. Ions are extracted with an electric potential and separated as a function of their masses and charge by the time of flight to a detector. The resolution of the current generation of MALDI mass spectrometers allows easy distinction of nucleobase substitutions in the mass range of 1–7 kDa, which corresponds to DNA sizes of 3–25 nucleobases.

Initially, MALDI was mainly applied for the analysis of proteins and peptides, and only more recently for nucleic acids. Its main advantage over conventional DNA analysis methods is its speed of signal acquisition (about 100 μs for one complete trace) and the result of the experiment being the molecular weight, an intrinsic physical property of each molecule. This precludes the use of size standards. Conventional electrophoretic methods for separating and detecting DNA are significantly slower and give less resolution. A serious limitation of MALDI is the size of the DNA products that can be analyzed. There is a dramatic drop-off of signal intensity with increasing mass and a problem of integrity of large DNA molecules. One contributing factor to this is the size-dependent tendency of the phosphodiester backbone of the DNA to fragment during the desorption process. Currently a practical limit is at DNA products of about 100 bases.

As complete automation from sample preparation to the acquisition and processing of data is possible, MALDI is generally considered an ideal analysis method for high-throughput applications like SNP genotyping, where product sizes are in the 3–25 base range.

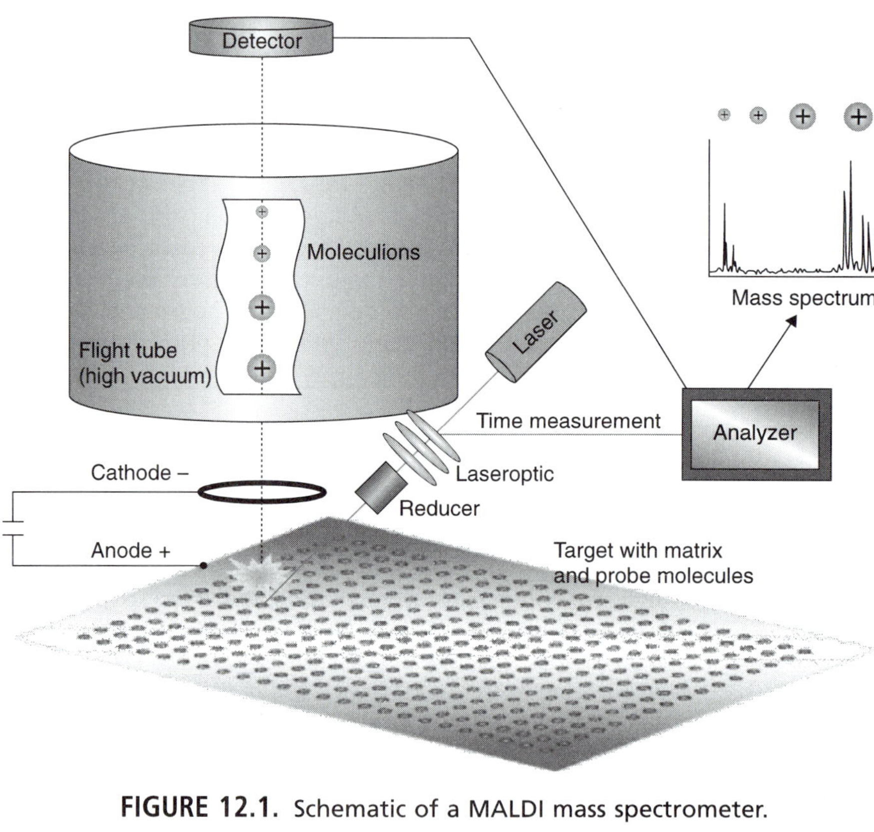

FIGURE 12.1. Schematic of a MALDI mass spectrometer.

PROBLEMS OF MALDI ANALYSIS OF DNA

It was observed that DNA analysis by MALDI was very inefficient, for example, 100 times more DNA has to be used in a preparation to achieve similar signal intensities compared to peptides of the same mass. The main problem in analyzing native DNA by MALDI stems from its negatively charged sugar-phosphate backbone. With native DNA, the phosphate residue provides a site of negative charge in solution and each DNA molecule carries as many negative charges as phosphate residues. The affinity of the phosphate residues for alkali counterions, such as sodium and potassium, is high but not high enough to result in complete saturation. These ions interfere in the ionization process by inducing adducts and thereby significantly reducing the signal intensity. There are marked drawbacks in terms of perturbation of data accumulation by impurities, calling for very stringent purification procedures. Several purification strategies have been implemented and integrated into a number of different protocols for allele identification. The most popular procedures for stringent purification include magnetic bead separation, reversed-phase column purification (ZipTips, Poros), gel filtration (G50), and ethanol precipitation. In general, these techniques tend to be cumbersome for automation and/or are quite expensive. This is the main disadvantage of most SNP genotyping methods using MALDI.

SNP GENOTYPING METHODS USING PRIMER EXTENSION AND MALDI DETECTION

Primer extension is a powerful and reliable way to generate allele-specific products of SNPs. Two prominent assays using primer extension with MALDI detection are the PROBE assay and the PinPoint assay (2, 3). These strategies use similar molecular biological procedures. A DNA polymerase extends a primer upstream of the SNP with a set of dNTPs and/or ddNTPs on a PCR amplicon, resulting in allele-specific products for MALDI detection (Fig. 12.2).

The DNA polymerase extends the 3′-end of the primer by specifically incorporating nucleotides that are complementary to the DNA template. The extension reaction terminates at the first nucleobase in the template where a nucleotide occurs that is complementary to one of the ddNTPs in the reaction mix. Generally, a thermostable DNA polymerase in a temperature-cycled reaction is used, leading to a linear amplification. An important point to note is that residual dNTPs and primers have to be removed before the primer extension reaction. This is achieved either by purification, for example, with magnetic beads, or by treatment with shrimp alkaline phosphatase and exonuclease I. Some protocols use primers

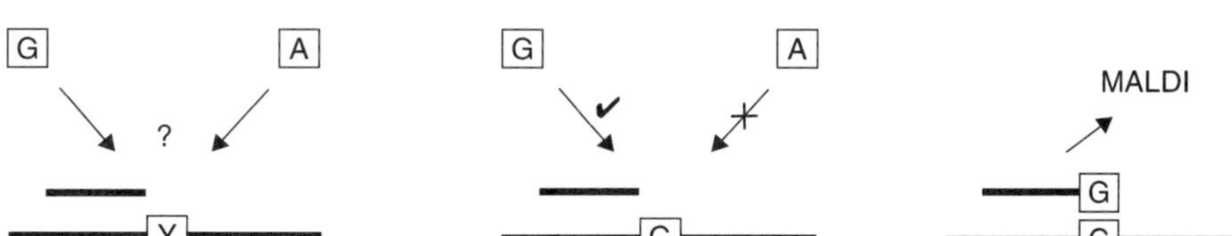

FIGURE 12.2. Schematic of primer extension.

or ddNTPs containing mass-tags that increase the mass differences between the allele products.

In a variation of the PROBE assay, the primer extension is implemented into an automated system called "MALDI on a chip technology" or "MassArray." A few nanoliters of purified and concentrated sample from the molecular biological reaction are piezo-pipetted onto a silicon chip, which is then inserted into the MALDI mass spectrometer, where each sample spot is measured automatically (www.sequenom.com).

A variation of the PROBE assay was recently described. To shift the masses of DNA products of the primer extension reaction into a more convenient mass range than 5000 Da, a chemically cleavable base was introduced close to the 3′-end of the primer. After primer extension allele-specific products are cleaved and analyzed in the 1000- to 2000-Da mass range.

Whereas for this assay, as for the PROBE assay, magnetic bead separation of products is applied before MALDI analysis, the PinPoint assay uses reversed-phase material or gel filtration. In a recent development of the MassArray system an ion-exchange material is used to de-salt reactions.

At thc CNG we developed a strategy that works around the problem of purification and have developed a purification-free SNP genotyping assay with MALDI detection—the GOOD assay (4) (Fig. 12.3). Currently a number of variants of this assay are used in high-throughput SNP genotyping production.

The regular GOOD assay consists of PCR, shrimp alkaline phosphatase digestion, primer extension, phosphodiesterase digestion, and alkylation. Simple dilution instead of purification of the reaction products before MALDI analysis suffices, allowing streamlined single tube preparation. A drawback is the potentially toxic reagent that is applied for alkylation. A simplified version of the GOOD assay using methylphosphonate-containing primers and a DNA polymerase that preferentially incorporates ddNTPs over dNTPs has been developed (5). For the simplified GOOD assay, because of a particularly selective DNA polymerase, Tma31FS, the

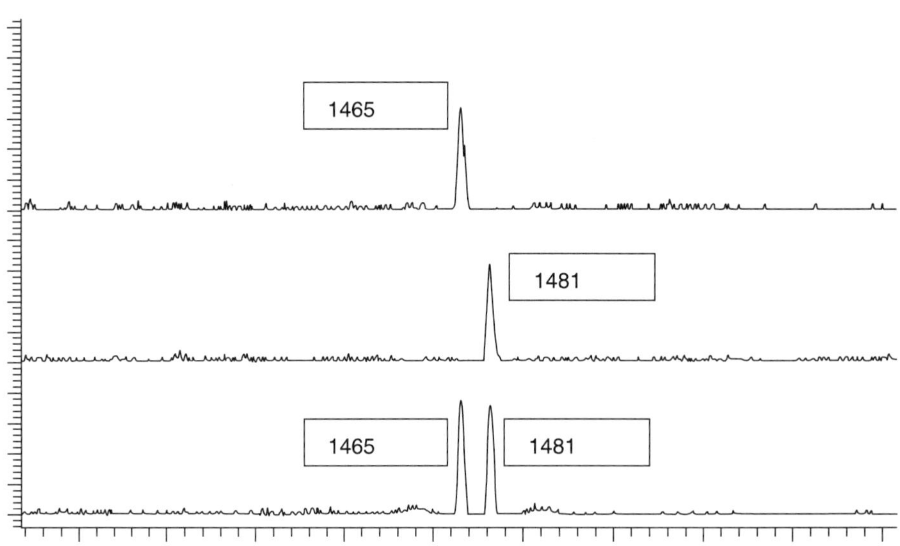

FIGURE 12.3. Result of a simple SNP genotyping result by the GOOD assay.

primer extension mix can be added immediately after PCR without the requirement of removing residual dNTPs or primers from the PCR. Also, the potentially toxic reagent of the alkylation is avoided. A fluid reaction sequence with three stages leads to allele-specific products, and again no purification is required before MALDI analysis.

■ HIGH-THROUGHPUT SNP GENOTYPING BY THE GOOD ASSAY

Because of the homogenous format of handling, the GOOD assay was straightforward to implement as an automated system. DNA samples are dispensed into 384-well microtiter plates, and all subsequent reagents are added into this plate. The PCR mix and primer extension mix additions are followed by a thermocycling step, whereas all other steps are followed by simple incubations. As samples are never transferred to a new vessel, there is no risk of sample mix-up. After the reaction sequence, the samples are transferred in format onto a Bruker MALDI target plate with the same liquid handling robot (BasePlate from the Automation Partnership) that is used for all other liquid additions. Targets are introduced into the MALDI mass spectrometers with a robot. Data are accumulated and interpreted automatically. Several quality control mechanisms have been integrated (6). Studies that are currently in progress include a cardiovascular target gene project, a DNA repair target gene study in a cancer DNA collection, several animal genotyping studies for traceability (cattle), and a plant genotyping study (corn, wheat, arabidopsis, and rapeseed).

■ VARIATIONS OF THE GOOD ASSAY

The GOOD assay has been adapted for a few very specific applications: haplotyping and epigenotyping.

Haplotyping: the information content of a single SNP is limited because of the biallelic nature of this marker. We have extended the GOOD assay to measure the phase of a number of SNPs in proximity (7). In the absence of family information the physical haplotypes can be measured in this way. A highly allele-specific PCR is integrated that enables the separation of segments of the maternal and paternal chromosomes as a function of heterozygous SNPs.

Epigenotyping: DNA methylation regulates gene expression in a dynamic fashion. It can be indicative of early events in tumorigenesis. We have adapted the GOOD assay for the analysis of DNA methylation (8). Genomic DNA is converted by treatment with bisulfite. Methylated cytosines are not affected by this procedure, whereas nonmethylated cytosines are converted to uracil. The region of interest containing the potentially methylated cytosines is amplified with primers specific for the DNA sequence after bisulfite conversion. Specific cytosine methylation is now analyzed by genotyping C/T, where C is representative of methylation on that particular position in the genomic DNA and T represents nonmethylation at that position. As methylation is a dynamic process, it can be in the range of 0–100%. The GOOD assay had to be established for quantitative treatment of results. It was determined that the degree of methylation can be determined within 5% accuracy.

■ REFERENCES

1. Tost J, Gut IG (2002) Genotyping single nucleotide polymorphisins by mass spectrometry. Mass Spectrom Rev 21: 388–418.
2. Little DP, Braun A, Darnhofer-Demar B, Köster H (1997) Identification of apolipoprotein E polymorphisms using temperature cycled primer oligo base extension and mass spectrometry. Eur J Clin Chem Clin Biochem 35: 545–548.
3. Little DP, Braun A, O'Donnell MJ, Köster H (1997) Mass spectrometry from miniaturized arrays for full comparative DNA analysis. Nat Med 3: 1413–1416.
4. Tost J, Kucharzak R, Lechner D, Gut IG (2004) The GOOD assay: A purification-free assay for genotyping by MALDI mass spectrometry. In *PCR Technology—Current Innovations* Second edition, Weissensteiner T, Griffin HG, Griffin A (eds), CRC Press, London, pp. 121–130.
5. Sauer S, Gelfand DH, Boussicault F, Bauer K, Reichert F, Gut IG (2002) Facile method for automated genotyping of single nucleotide polymorphisms by mass spectrometry. Nucleic Acids Res 30: E22.
6. Lechner D, Lathrop M, Gut IG (2002) Large-scale genotyping by mass spectrometry: Experience, advances and obstacles. Curr Opin Chem Biol 6: 31–38.
7. Tost J, Brandt O, Boussicault F, Derbala D, Caloustian C, Lechner D, Gut IG (2002) Molecular haplotyping at high throughput. Nucleic Acids Res 30: E96.
8. Tost J, Schatz P, Schuster M, Berlin K, Gut IG (2003) Analysis and accurate quantification of CpG methylation by MALDI mass spectrometry. Nucleic Acids Res 31: E50.

13

LOCUS-SPECIFIC MUTATION DATABASES AND NOMENCLATURE ANNEX

Christophe Beroud
Laboratoire de Génétique Moléculaire, Centre Hospitalier Universitaire (CHU) Montpellier, Institut Universitaire de Recherche Clinique (IURC), Montpellier, France

■ INTRODUCTION

Historically, collection of mutations and variations in human genes has taken place in the published literature. In the mid-1980s a few of these variations were available in various databases such as the Genome DataBase (GDB), GENBANK, EMBL, and SWISSPROT. Nevertheless, because of the structure of these databases, the extraction of relevant information about mutations was almost impossible. In this context, it became clear that specific software must be developed. A few laboratories therefore then started to build local databases to collect and document mutations in human genes. These creations remained confidential until the development of a new field: molecular epidemiology. In fact, the collection of p53 mutations involved in about 50% of human cancers allowed proof not only of the direct involvement of UV in the etiology of skin cancers but also the link between aflatoxin B1 exposure and increased frequency of hepatocarcinoma.

The field of genetic mutations involving a tumor suppressor gene was the best model for these emerging locus-specific databases (LSDBs). Today a LSDB is described as a collection of mutations from a single gene. LSDBs can have various structures and collect a wide range of data that will be detailed below.

Guide to Mutation Detection, Edited by Graham R. Taylor and Ian N. M. Day.
ISBN 0-471-23444-3

Because the number of somatic mutations is very important compared to germline mutations and because most of these are private, the largest number of available mutations from a single gene is today close to 13,000 and involve the p53 gene.

A few years later, LSDBs have been created to collect and organize data from genes involved in human genetic diseases. Since the beginning, new questions have emerged such as: "Is there a specific pattern of mutations associated with a phenotype?" and "Can we use these molecular data for predictive medicine including the response to a drug?" This needed a modification of the structure of the databases to move from a mutational event-based structure to a patient-based structure. The best, as always, lies in the happy medium to keep all tools developed for molecular epidemiology studies and add new ones for genotype-phenotype correlations and vice versa.

■ WHAT WILL BE THE AREA FOR LSDB DEVELOPMENT?

The human genome is thought to contain about 30,000–40,000 genes, and presently only 1/30th of these are known to be implicated in genetic diseases. Now that the entire sequence of the human genome is available, the development of new methods for mutation detection described in the other chapters of this book will lead to a huge increase in the identification of genes and their mutations associated with genetic diseases as well as cancers.

The collection of these mutations will be critical for researchers and clinicians to establish genotype-phenotype correlations. Other fields such as molecular epidemiology will also be developed by using these new data. Consequently, the future lies not in simple repositories of locus-specific mutations but in dynamic databases linked to various computerized tools for their analysis.

Because the field of LSDBs was growing so rapidly, the need to organize it and define recommendations and guidelines led to the development of the HUGO-Mutation Database Initiative (HUGO-MDI). For many years, this group of experts worked on various topics such as mutation nomenclature, ethics, and database and software structure.

The Human Genome Variation Society (HGVS) was inaugurated on 12 October 2001. This new society stemmed from the HUGO-MDI. The Society aims to foster discovery and characterization of genomic variations, including population distribution and phenotypic associations. We will promote collection, documentation, and free distribution of genomic variation information and associated clinical variations and endeavor to foster the development of necessary methodology and informatics.

■ LOCUS-SPECIFIC DATABASE VERSUS CORE DATABASE

Two main approaches were developed: one gene-one database (LSDB) and one database-all genes (core database).

Core Database

This approach is modeled by the Human Gene Mutation Database (HGMD), which comprises various types of mutation within the coding regions of human nuclear genes causing inherited disease. Somatic mutations and mutations in the mitochondrial genome are thus not included. Each mutation is entered only once, to avoid confusion between recurrent and identical-by-descent lesions. Mutations inferred from amino acid sequencing have been excluded because, in the absence of direct DNA analysis, some ambiguity may exist as to the DNA sequence changes involved. Silent mutations within the coding region that do not alter the encoded amino acid are also not recorded. If such mutations are known to adversely affect mRNA splicing or gene expression, or have been reported in significant association with disease, they may be included. HGMD does not usually include mutations lacking obvious phenotypic consequences, although a few such variants have been included where they could conceivably have some clinical effect (e.g., albumins, butyrylcholinesterases). Many published mutation searches identify more than one genetic change in a single patient. In such cases, the relationship between a given lesion and the clinical phenotype has not always been immediately clear, and the curators of HGMD have had to rely exclusively on the judgments of authors, peer reviewers, and journal editors. The possibility of unintentional inclusion of some lesions with little or no pathological significance can therefore not be ruled out.

Data are collected weekly by a combination of manual and computerized search procedures. In excess of 250 journals are scanned for articles describing germline mutations causing human genetic disease. The required data are extracted from the original articles and augmented with the necessary supporting data. Data included are mainly from the original published reports, although some data have been taken from "Mutation Updates" and review articles. Unpublished mutations and mutations reported only in abstract form are not included. Reports of such lesions can, however, be accessed for some genes via the locus-specific mutation databases.

Locus-Specific Database

In contrast to core databases, where the curators are not experts in the field of mutations from a specific gene, curators of a LSDB are always experts on detecting, analyzing, and interpreting the effects of mutations on gene function, and therefore data from a LSDB are usually of higher quality. Furthermore, the data collection covers a wider spectrum such as detailed clinical data for each patient and is more adapted to data mining such as genotype-phenotype correlation studies. In addition, a LSDB always collects all described mutations and not only the first report of a specific mutation. Because the curators are experts in the field of the LSDB they usually have numerous collaborations with research and diagnostic laboratories that allow the collection of unpublished data. In fact, it has been clear for some time that the collection of all mutations causing single-gene disorders into the literature and/or databases is difficult and thus usually incomplete. This point is emphasized by recent data showing that out of all the mutations in locus-specific databases curated by experts, 50% are unpublished. Considering that most diagnostic laboratories would not routinely publish mutations identified in the course of genetic diagnosis, this figure is likely to be an underestimate.

It is also apparent that print journals have in recent years become increasingly reluctant to publish mutations beyond those included in the first or second report that variations in a given gene are causative for a particular disease. Furthermore, it is highly unlikely that reports of such single mutations as the 520th mutation causing a particular disease have ever been accepted for publication.

The decrease in cost of DNA sequencing has led to its widespread availability and increasing use in clinical as well as research settings. There has thus been a concomitant increase in the detection of gene-related variations. The flow of these new data should be made available to the wider clinical and research communities, representing a significant and valuable opportunity. The availability of a core database will give a rapid answer to this problem with an immediate storing capacity for this new flow of information. Thu, a core database is particularly useful for genes with only few mutations reported.

When a network of experts is organized for a specific gene, a LSDB should be created. To organize published mutations, to acquire and document unpublished mutations, and to provide free public access to all, a number of LSDBs have already been created. Their number is expected to increase rapidly in the near future. From over a hundred LSDBs existing globally today (listing ~300 genes), the number is expected to double in a few years' time.

Two main restrictions in the development of LSDB must be considered: First, they have to be designed by computer scientists and be maintained for a long period of time; second, they need to be funded.

STRUCTURE OF A LOCUS-SPECIFIC DATABASE

LSDBs often originate as loosely organized compilations of data. Curators choose the implementation from available database management systems or create their own, depending on their needs and abilities. A common scenario is the locus-specific database, which has progressed from the initial compilation as a spreadsheet, then to several spreadsheets, and finally to a complex relational database with many tables.

Curators who do not wish to create a unique database design have the option of acquiring one of the generic softwares:

Software Name	Author	URL
MuStar®	Brown	www.hgu.mrc.ac.uk/Softdata/Mustar/
Mutation Checker	Lehvaslaiho	www.ebi.ac.uk/cgi-bin/mutations/check.cgi
MUTbase	Vihinen	www.uta.fi/imt/bioinfo/MUTbase/
Universal Mutation Database®	Béroud	www.umd.necker.fr/

Today no consensual structure has been validated for a LSDB. This is mainly due to the high degree of variation in the software development and the questions that are addressed for a specific gene. If no guideline is available, we can nevertheless give global recommendations.

A relational database should include the reference sequence of the gene and the human coding code. This will allow the development of an input tool that will help the curator to validate the data (the mutation name will be calculated according to the international nomenclature) and check the published information (10% of published mutations are wrong!). Various tables should also be created: one table containing the description of the mutation (many fields can be created according to the gene), one table for the sample (it is important to design an unique international identifier for the sample), and one table for the patient with clinical data (it is important to consider anonymization if you want to put these data on the web).

ADDITION OF INTEGRATED ANALYSIS TOOLS

The main difference between simple spreadsheets and relational databases is the availability of a programming language that allows the computer scientist to design advanced analysis tools. Ultimately, these tools should be accessible both at the local level and via the Internet. Today for most LSDBs only lists of mutations are available via the Internet, and one must get the data as a flat text file and create its own local data mining tools. Nevertheless, the best option is the availability of online advanced search and analysis routines. Below are listed few examples of these routines (extracted from the UMD software):

"Insertions and deletions analysis" searches whether flanking repeated sequences could be involved in the mutational process as described in Béroud et al. (1).

"Position" studies the distribution of mutations at the nucleotide level to identify preferential mutation sites.

"Potential stop codons" displays all codons from a specific exon that can be mutated in a stop codon by a single substitution.

"Mutational events" is comparable to "Position" but also indicates the type of mutational event. The result can be displayed either as a table or in a graphic representation.

"Frequency of mutations" allows one to study the relative distribution of mutations at all sites and to sort them according to their frequency. A graphic representation is also available and displays a cumulative chart of mutation distributions.

"Frequency of events" is similar to "Mutational events" but also indicates whether mutations are localized in a CpG dinucleotide.

"Distribution of mutations" and **"Binary comparison"** are two graphical routines that allow one to compare the distribution of mutations in up to eight groups of records and display the result either in one or two graphics.

"Stat exons" studies the distribution of mutations in the different exons. It enables detection of a statistically significant difference between observed and expected mutations.

"Distribution by exons" displays the partition of each type of mutation in each exon.

"Structure" studies the distribution of mutations both in the various structural domains of the protein and in the highly conserved domains.

"Phenotype-genotype analysis" gives an easy access to phenotype-genotype correlations. In a first step, the numbers of records with a specific symptom are sorted according to the description of this symptom. In a second step, the user can choose some of these symptoms to visualize the distribution of genotypes associated with each phenotype.

"Genotype-phenotype analysis" is similar to the previous function. The user can choose a particular genotype from a list and display the associated phenotypes.

All these routines can be applied to the overall database or to a subset of each database.

WHAT IS THE FUTURE OF LSDBS?

LSDBs are major databases because they are helpful for clinicians (phenotype-genotype correlations, etc.), geneticists (distribution and frequency of mutations, etc.), research biologists (structural domains, molecular epidemiology, genotype-phenotype correlations, etc.), and pharmacologists (genotype-drug response correlations, reconstructing pathophysiological circuit resulting in the identification of new targets for drug development, allele-specific gene-based drug therapy, etc.).

Their future will depend on a combination of many criteria: financial support, large networks of clinicians and biologists that will allow the collection of high-quality data, the availability of generic software, highly motivated curators, and recognition from the scientific community and publishers.

A few steps have already been made with the creation of the Human Genome Variation Society to federate the numerous existent mutation databases and the new possibility to getting a "PubMed ID" from online submission of new mutations. Many efforts will still have to be made to design the structure of the ultimate LSDB generic software.

ANNEX 1

Nomenclature for the Description of Sequence Variations (2)

Introduction

Recently, a nomenclature system has been suggested for the description of changes (mutations and polymorphisms) in DNA and protein sequences (3).

These nomenclature recommendations have now been largely accepted and have stimulated the uniform and unequivocal description of sequence changes. However, current rules do not yet cover all types of mutations, nor do they cover more complex mutations. This document lists the existing recommendations and summarizes suggestions for the description of additional, more complex changes (shown in italics) based on an article published in *Human Mutation* (4).

In this annex we have selected only a short description of the entire nomenclature system limited to the DNA mutation system.

General Recommendations

The term "***sequence variation***" is used to prevent confusion with the terms "***mutation***" and "***polymorphism***," mutation meaning "change" in some disciplines and "disease-causing change" in others and polymorphism meaning "non-disease-causing change" or "change found at a frequency of 1% or higher in the population."

The basic recommendation is to use ***systematic names*** to describe each sequence variation. For this, variations are described at the most basic level, that is, the DNA level, using either a genomic or a cDNA reference sequence. A genomic reference sequence is preferred because it overcomes difficult cases, including multiple transcription initiation sites (promoters), alternative splicing, the use of different poly-A addition signals, multiple translation initiation sites (ATG-codons), and the occurrence of length variations. When, as in most cases, the entire genomic sequence is not known, a cDNA reference sequence should be used instead.

- Sequence variations are described in relation to a reference sequence for which the accession number from a primary sequence database (Genbank, EMBL, DDJB, SWISS-PROT) should be mentioned in the publication/database submission (e.g., M18533),
- Tabular listings of the sequence variations described should contain columns for DNA, RNA, and protein and clearly indicate whether the changes were **experimentally determined** or only theoretically deduced.
- To avoid confusion in the description of a sequence change, precede the description with a letter indicating the type of reference sequence used:
- "g." for a ***genomic*** sequence (e.g., g.76A>T)
- "c." for a ***cDNA*** sequence (e.g., c.76A>T)
- *"m."* for a ***mitochondrial*** sequence (e.g., m.76A>T) (from David Fung, Camperdown, Australia)
- *"r."* for an ***RNA*** sequence (e.g., r.76a>u)
- *"p."* for a ***protein*** sequence (e.g., p.K76A)
- To discriminate between the different levels (DNA, RNA or protein), descriptions are unique:
- At DNA level, in capitals, starting with a number referring to the first nucleotide affected (e.g., c.76A>T)
- At RNA level, in lower case, starting with a number referring to the first nucleotide affected (e.g., r.76a>u)
- At protein level, in capitals, starting with a letter referring to the first amino acid (one-letter code) affected (e.g., p.T26P)
- A range of affected residues is indicated by a "_" character (underscore) separating the first and last residue affected (e.g., 76_78delACT)
- *NOTE:* current recommendations use the "-"-character (e.g., 76–78delACT)
- For deletions, duplications, or insertions in short tandem repeats, the most 3′ nucleotide is arbitrarily assigned as the nucleotide changed.
- Two sequence variations in one allele are listed between brackets, *separated by a " + " character* (e.g., [76A>C + 83G>C])
- *NOTE:* current recommendations use the ";" character as a separator (e.g., [76A>C; 83G>C])
- Sequence changes in different alleles (e.g., for recessive diseases) are listed between brackets, separated by a "+" character (e.g., [76A>C] + [87delG])
- *NOTE:* the current recommendation is [76A>C + 87delG]
- A unique identifier should be assigned to each mutation. The unique OMIM-identifier can be used; otherwise, database curators should assign unique identifiers

DNA Level

Nucleotides are designated by the bases (in upper case): A (adenine), C (cytosine), G (guanine), and T (thymidine)

Nucleotide Numbering:

- Nucleotide + 1 is the A of the ATG-translation initiation codon, the nucleotide 5′ to +1 is numbered –1; there is no base 0

- Non-coding regions:
  the nucleotide 5′ of the ATG-translation initiation codon is –1
  the nucleotide 3′ of the translation termination codon is *1
  intronic nucleotides
- **Beginning of the intron:** the number of the last nucleotide of the preceding exon, a plus sign and the position in the intron, e.g., 77 + 1G, 77 + 2T (when the exon number is known, the notation can also be described as IVS1 + 1G, IVS1 + 2T)
- **End of the intron:** the number of the first nucleotide of the following exon, a minus sign,, and the position upstream in the intron, e.g., 78-2A, 78-1G (when the exon number is known, the notation can also be described as IVS1-2A, IVS1-2G)
- For deletions, duplications, or insertions in single nucleotide (or amino acid) stretches or tandem repeats, the most 3′ copy is arbitrarily assigned to have been changed (e.g., ACTTTG**TG**CC to ACTTTGCC is described as 7_8delTG)

Description of Nucleotide Changes

Substitutions are designated by a ">" character

76A>C denotes that at nucleotide 76 a A is changed to a C

88 + 1G>T (alternatively IVS2 + 1G>T) denotes the G to T substitution at nucleotide + 1 of intron 2, relative to the cDNA positioned between nucleotides 88 and 89

89 – 2A>C (alternatively IVS2–2A>C) denotes the A to C substitution at nucleotide –2 of intron 2, relative to the cDNA positioned between nucleotides 88 and 89

NOTE: ***polymorphic variants*** are sometimes described as 76A/G, but this is not recommended!

Deletions are designated by "del" after the nucleotide(s) flanking the deletion site

76_78del (alternatively 76_78delACT) denotes a ACT deletion from nucleotides 76 to 78

82_83del (alternatively 82_83delTG) denotes a TG deletion in the sequence ACTTTG**TG**CC (A is nucleotide 76) to ACTTTGCC

IVS2_IVS5del (alternatives 88 + ?_923 + ? or EX3_5del) denotes an exonic deletion starting at an unknown position in intron 2 (after nucleotide 88) and ending at an unknown position in intron 5 (after nucleotide 923)

Insertions are designated by "ins" after the nucleotides flanking the insertion site, followed by the nucleotides inserted

NOTE: as a separator the "^" character is sometimes used but this is not recommended (e.g., 83^84insTG)

76_77insT denotes that a T was inserted between nucleotides 76 and 77

83_84insTG denotes a TG insertion in the TG-tandem repeat sequence of ACTTTGTGCC (A is nucleotide 76) to ACTTTGTG**TG**CC. Note that this sequence variation (a duplicating insertion) can also be described as a duplication, e.g. 82_83dupTG (*see duplications*)

Variability of short sequence repeats, e.g., in ACTGTGTGCC (A is nt 1991), are designated as 1993(TG)3-6 with nucleotide 1993 containing the first TG-dinucleotide which is found repeated 3 to 6 times in the population.

Insertion/deletions (indels) are described as a deletion followed by an insertion after the nucleotides affected

112_117delinsTG (alternatively 112_117delAGGTCAinsTG or 112_117>TG) denotes the replacement of nucleotides 112 to 117 (AGGTCA) by TG

Duplications are designated by "dup" after the nucleotides flanking the duplication site

77_79dupCTG denotes that nucleotides 77 to 79 were duplicated

Duplicating insertions in short tandem repeats (or single nucleotide stretches) can also be described as a duplication, e.g., a TG insertion in the TG-tandem repeat sequence of ACTTTGTGCC (A is nt 76) to ACTTTGTG**TG**CC can be described as 82_83dupTG (now 83_84insTG,)

Inversions are designated by "inv" after the nucleotides flanking the inversion site

203_506inv (or 203_506inv304) denotes that the 304 nucleotides from position 203 to 506 have been inverted

Translocations (no suggestions yet)

Changes in different alleles (e.g., in recessive diseases) are described as "[change allele 1] + [change allele 2]"

[76A>C] + [76A>C] denotes a homozygous A to C change at nucleotide 76

[76A>C] + [?] denotes a A to C change at nucleotide 76 in one allele and an unknown change in the other allele

Two variations in one allele are described as "[first change + second change]"

[76A>C + 83G>C] denotes an A to C change at nucleotide 76 and a G to C change at nucleotide 83 in the same allele

NOTE: current recommendations use the ";" character as a separator (e.g., [76A>C; 83G>C])

REFERENCES

1. Béroud et al. (2000) Hum Mutat 15: 86–94.
2. den Dunnen JT, Antonarakis SE (2001) Hum Genet 109(1): 121–124.
3. Antonarakis SE, the Nomenclature Working Group (1998) Recommendations for a nomenclature system for human gene mutations. Hum Mutat 11: 1–3.
4. den Dunnen JT, Antonarakis SE (2000) Mutation nomenclature extensions and suggestions to describe complex mutations: a discussion. Hum Mutat 15: 7–12.

Part II

METHODS

14

EXTRACTION OF NUCLEIC ACIDS

Graham R. Taylor
Regional DNA Laboratory, St. James's University Hospital, Leeds, UK

■ INTRODUCTION

DNA extraction can be applied to many different tissues, including whole blood, buccal cells, CVS samples, Guthrie blood spots, and paraffin-embedded fixed samples. There are commercial kits (e.g., Qiagen and Promega) for DNA extraction and automated DNA extraction systems (e.g., Whatman and Autogen). Sample tracking is a crucial part of DNA extraction, and it is important to take great care at any stage where samples are transferred to fresh tubes.

General Guidelines for all DNA Extractions

1. A sample extraction sheet is prepared.
2. Label all tubes to be used in the extraction procedure with the name of the patient and lab number.
3. Keep the tubes in the work racks in numerical order.
4. The names and lab numbers must be checked at all transfer stages throughout the extraction procedure to avoid sample mix-up.
5. Stagger the tubes in the rack to allow easy identification of sample transfers.

■ DNA EXTRACTION FROM WHOLE BLOOD

DNA Extraction by Salt Precipitation

Nucleated cells are separated from whole blood by lysing the red cells and then centrifuging to pellet the nuclei. After white cell lysis contaminating proteins are

Guide to Mutation Detection, Edited by Graham R. Taylor and Ian N. M. Day.
ISBN 0-471-23444-3

removed by precipitating with a high concentration of ammonium acetate. DNA is then precipitated with isopropanol followed by washing the DNA with ethanol. The DNA is then dissolved in Tris-EDTA and used for analysis.

Solutions

Red Blood Cell Lysis Solution (Autoclave)

155 mM ammonium chloride	8.3 g per liter
10 mM potassium bicarbonate	1 g per liter
1 mM EDTA	0.4 g per liter OR 2 ml 0.5 M EDTA per liter

White Cell Lysis Solution (Autoclave)

25 mM EDTA	9.3 g per liter OR 50 ml 0.5 M EDTA per liter
2% SDS	20 g per liter OR 200 ml 10% SDS per liter

Protein Precipitation Solution (use sterile water, do not autoclave)

10 M ammonium acetate	385.4 g per 500 ml

Ethanol

70% (v/v) ethanol, Analar	700 ml per liter

Isopropanol

Isopropanol, Analar

Glycogen

Glycogen	20 mg per ml

Store solutions at room temperature except glycogen, which is kept at –20°C.

Do not use new solutions for important samples until they have been proven reliable.

Red cell (RCL) buffer should be aliquoted into 500-ml bottles because a reduction in yields has been noticed when approaching the end of a single large bottle.

Equipment

1. Centrifuge with swing-out rotor (e.g., MSE).
2. 13-ml polypropylene centrifuge tubes (Sarstedt Ltd)
3. Plastic Pasteur dropping pipettes
4. Tube roller (Spiromix)
5. 37°C water bath
6. Fume cupboard
7. 2 ml microcentrifuge tubes (Sarstedt Ltd)
8. Microcentrifuge
9. 1.5-ml and 0.5-ml microcentrifuge tubes (Alpha Laboratories Ltd)
10. Pipette tips (Alpha Laboratories Ltd)
11. Rotary mixer (Gallenkemp)
12. Eye protection

Method

Whole Blood in EDTA or Lithium/Heparin

1. Check the order and alignment of the blood tubes and the tubes for transfer and initial the extraction sheet to indicate that this check has been made. Transfer 3 ml whole blood into a 15-ml polypropylene centrifuge tube.
2. Add 9 ml RCL solution (i.e., 3 volumes).
3. Spiromix for 10 min at room temperature.
4. Centrifuge at 3300 rpm (2000*g*) for 10 min in a swing-out rotor.
5. Pour off the supernatant—this will leave behind ~200 µl residual liquid.
6. Add 3 ml white cell lysis solution (i.e., 1 volume) to resuspended pellet—pipette up/down to lyse cells.
7. Invert several times—stable in this state at room temperature (RT) for 18 months.
8. Incubate for 30 min at 37°C.
9. Cool to below 21°C, 5 min on ice.
10. Add 1 ml protein precipitation solution.
11. Whirly mix for 20 s.
12. Centrifuge at 3300 rpm (2000*g*) for 10 min.
13. Pour supernatant into a clean 15-ml polypropylene centrifuge tube.
14. Add 3 ml isopropanol (propan-2-ol) and invert several times to precipitate DNA from the supernatant.
15. Centrifuge at 3300 rpm (2000*g*) for 10 min.
16. Gently pour off the supernatant, blot onto a paper towel.
17. Add 3 ml 70% ethanol to the DNA pellet, briefly mix.
18. Centrifuge at 3300 rpm (2000*g*) for 10 min.
19. Gently pour off the supernatant, blot onto a paper towel—ensure all liquid is removed.
20. Repeat 70% ethanol wash.
21. Check that DNA pellet is at the bottom of the tube.
22. Leave in fume cupboard inverted for approximately 30 min to dry.
23. Resuspend in 110 µl TE and leave at RT overnight.
24. Transfer to a DNA tube and measure the OD 260 nm.
25. Adjust the volume of TE to give a standard concentration of 500 µg/ml.
26. Leave on a tube rotator for a few hours or overnight to mix.

Phenol-Chloroform Extraction

Nucleated cells are separated from whole blood either by lysing the red cells (with either ammonium chloride solution or mild detergent) followed by centrifugation to pellet the nuclei, or by Ficoll gradient separation. These are then resuspended in buffered salt (STE) plus SDS and proteinase K to digest histones and other proteins. The DNA (and RNA) is then extracted in a two-phase system with phenol and chloroform extraction. At pH 7 and above, DNA selectively partitions into the aqueous phase (although some is lost in the phenol phase). DNA is recovered by adding 2 volumes of ethanol followed by spooling onto a Pasteur pipette. The DNA is then dissolved in Tris-EDTA and used for analysis. An alternative, more rapid method is to "salt out" (precipitate under conditions of high ionic strength) the proteins from the STE solution and spool out the DNA without phenol extraction.

Solutions

1 × STE plus 1% SDS (pH 8.0) (Autoclave)

100 mM NaCl	5.88 g per liter
10 mM Tris (pH 8.0)	1.21 g per liter
1 mM EDTA	0.37 g per liter OR 2 ml 0.5 M EDTA per liter
1% SDS	10 g per liter OR 100 ml 10% SDS per liter

Can be made by using 100 ml STE (10×) and 100 ml 10% SDS up to 1 liter with water.

TE Buffer pH 8.0 (Autoclave)

10 mM Tris (pH 8.0)	0.121 g per liter
1 mM EDTA	0.037 g per liter OR 0.22 ml 0.5 M EDTA per liter

Can be made by using 100 ml TE (10×) up to 1 liter with water.

Phenol Solution

Saturated solution in 0.1 M Tris from Northumbria Biologicals

Aliquot as 10 ml with a 0.1 M Tris overlay and store at –20°C.

Chloroform

Chloroform, Analar

Ethanol

Ethanol (100%), Analar

Proteinase K

20 mg/ml in water, store at –20°C

Equipment

See salt extraction method.

Method

Lysis of Fresh Blood If the blood is clotted, process only the unclotted part by this method. Centrifuge fresh blood at 3200 rpm for 10 min to separate the cells. Red cells will be at the bottom, white cells (buffy coat) will be a thin layer overlaying this, and plasma will be the top layer. Remove the upper layer to waste with plastic Pasteur. Remove the buffy coat, ensuring that you take a portion of the red cells to ensure all the buffy coat has been taken. Add 2.5 volumes of red cell lysis buffer to the buffy coat and spiromix for 10 min. The solution will go from a murky red to a clear deep red/maroon color. When blood clears, centrifuge at 3,200 rpm for 10 minutes. Tip off the buffer. A fawn-colored pellet of nuclei and white cells should be visible. It may be covered by some red cells, which can be gently washed off with red cell lysis buffer.

Lysis of Frozen Blood Lyse whole blood (or a stored buffy coat) with 2.5 volumes of red cell lysis buffer.

When blood clears, centrifuge at 3200 rpm for 10 min. Tip off the buffer. A fawn-colored pellet of nuclei and white cells should be visible. It may be covered by some red cells, which can be gently washed off with red cell lysis buffer.

The pellet from whole blood lysates is generally smaller and more contaminated with red cell debris. These pellets should be cleaned up by a second lysis step with ice-cold red cell lysis buffer B, followed by centrifugation at 3200 rpm once more. If the pellet is still discolored, do not use the salt extraction procedure; instead use phenol/chloroform.

Digestion of White Cells

1. Add 5 ml white cell lysis buffer (STE, SDS to 1%) and proteinase K to 50 μg/ml (50–100 μl of 20 mg/ml stock)
2. Mix gently. Solution will become very viscous.
3. Incubate at 37°C overnight or 55°C for 2 h.

Phenol/Chloroform Extraction

1. Add 5 ml phenol (use 10-ml tubes) and mix gently.
2. Centrifuge at 3200 rpm for 15 min.
3. Remove upper, aqueous phase with plastic Pasteur.
4. If necessary, cut the end of the Pasteur to widen the bore for easier transfer of viscous solution.
5. Remove aqueous phase and add 5 ml of chloroform, repeat centrifugation as above.
6. Remove aqueous phase and add 2–2.5 volumes of ethanol.
7. Mix gently. The DNA will appear as a white fibrous precipitate that can be recovered by winding around a bent glass Pasteur pipette.
8. Remove the Pasteur and snap off the end into a 2-ml Sarstedt tube, allow to dry in the fume cupboard.
9. Add 110 μl Tris-EDTA (TE) and dissolve by placing on a blood tube rotator for 1–3 days. If rapid analysis is needed, the DNA can be dissolved faster by heating to 55°C.

Alternative—Salt Extraction

1. This protocol can be used in place of the phenol/chloroform extraction method.
2. Add 0.5 volume of saturated NaCl or 7.5 M NH_4Ac to the STE and shake vigorously for 20 s.

3. Stand the tubes in ice/water for 15 min, then centrifuge at 4000 rpm for 20 min. A pellet of denatured protein should form at the bottom of the tube.
4. Gently remove the solution and transfer to a new 13-ml tube.
5. Add an equal volume of isopropanol (propan-2-ol) and spool out the precipitate.
6. Dry off the DNA and dissolve in 110 μl of TE buffer and dissolve by placing on a blood tube rotator for 1–3 days.

DNA EXTRACTION FROM PARAFFIN-EMBEDDED FIXED TISSUE BLOCKS

Wax block material is not the best material for DNA extraction for a number of reasons:

1. The samples are usually old, which increases the likelihood of DNA degradation.
2. The tissue samples have often been stored in formalin for extended periods before being embedded in wax, which increases the likelihood of DNA degradation. Because the samples are usually old it is difficult to find out exactly how the tissue was treated.
3. Degraded DNA is much more difficult to PCR and requires more amplification cycles.
4. DNA extracted from wax blocks can be contaminated with chemicals, for example, aldehydes or hemoglobin, which may inhibit PCR completely.
5. The amount of DNA extracted is usually low, and combined with degradation problems Southern blotting with wax block DNA is usually impossible.

Solutions

See Salt Extraction.

Equipment

See Salt Extraction.

Method

1. Use a scalpel blade to remove some of the tissue from the block, cut up in a clean weighing boat, and transfer into a 15-ml polypropylene tube. If shavings have been received, transfer them straight into the polypropylene tube.
2. Add 2 ml of white cell lysis solution and 20 μl of proteinase K. Vortex to mix.
3. Incubate overnight at 55°C.
4. Check the digestion in the morning. If the solution has gone cloudy and thick, proceed with the protocol. If not, add 20 μl more of proteinase K and leave the sample at 37°C until it has digested. If after 2 days the sample is still the same, proceed with the protocol.
5. Cool on ice for 5 min.

6. Add 700 μl of protein precipitation solution and vortex for 20 s.
7. Centrifuge at 3300 rpm for 10 min.
8. Transfer supernatant to a fresh tube. This is difficult to do, as the wax sometimes floats on the top.
9. Add 2 ml isopropanol and 10 μl glycogen (20 mg/ml). Mix.
10. Incubate at –20°C for 2 h.
11. Centrifuge at 3300 rpm for 10 min to pellet the DNA.
12. Carefully pour off supernatant and add 2 ml of 70% ethanol to wash the DNA.
13. Centrifuge at 3300 rpm for 10 min to pellet the DNA.
14. Pour off the supernatant and repeat the wash and centrifugation.
15. Pour off the supernatant and leave the DNA to dry in the fume cupboard.
16. Add 50–100 μl of TE and leave the DNA to resuspend in the usual way. The amount of TE added depends on the size of the pellet of DNA. The pellet will probably be contaminated with some wax, so only add 50 μl if uncertain. The sample can be diluted at a later stage if necessary.
17. Estimate the concentration of DNA with the spectrophotometer. Remember that there are probably contaminants in the sample, so the OD will be approximate.
18. Check the molecular weight of the DNA by electrophoresing 5 μl on an agarose gel. The DNA extracted is usually degraded, and it is unusual to see high-molecular-weight DNA.
19. The DNA is now ready to be used.

DNA CONCENTRATION ASSAYS

Some DNA tests are very sensitive to the initial DNA concentration. For this reason, DNA should be stored at a standard concentration (e.g., 500 μg/ml) where possible.

Solutions

DNA Standard

10 mg/ml DNA standard, salmon or herring sperm DNA

Equipment

UV/vis spectrophotometer and UV transparent cuvettes
Micropipette tips
Automatic pipettes

Method

Switch on the spectrophotometer and leave to warm up for 5 min.
Set the wavelength (λ) at 260 nm.
Done only once provided nothing in the spectrophotometer is changed.

Initial Blank and Calibration of the Spectrophotometer

Pipette 990 µl of water into cuvette and place in spectrophotometer with frosted side of glass facing toward you.

Always hold the cuvette on the frosted sides and ensure that the clear glass is clean.

Calibrate to zero absorbance.

The spectrophotometer is now ready to use to OD the DNA samples.

Calculation of DNA Concentrations if No Standard Curve is Available

1 OD = 50 µg/ml of double-stranded DNA.

The DNA sample is diluted by a factor of 100 in the cuvette. Therefore:

$$\text{DNA concentration } (\mu\text{g/ml}) = \text{OD reading} \times 5000$$

■ ASSESSING THE PURITY OF DNA

OD260/280 Ratios

Equipment

Spectrophotometer and cuvettes
Pipette tips
Calculator

Method

(a) Measure the OD of the sample, as described in Method in previous section, at a wavelength of 260 nm. Do not throw the sample away.

(b) Alter the wavelength to 280 nm.

(c) Zero the spectrophotometer at this new wavelength with a separate cuvette containing water (as described above).

(d) Measure the OD of the same sample, in your first cuvette, at this new wavelength.

To assess the purity of the DNA the ratio is of the two absorbancies can be calculated.

$$\text{OD at wavelength 260 nm/OD at wavelength 280 nm}$$

A range of 1.6–2.0 indicates sufficiently pure DNA for most applications.

Gel Analysis of DNA

Solutions

Agarose

0.8% Agarose	8 g per liter
10 × TAE (Tris acetate) buffer	100 ml per liter

Ethidium bromide solution (10 mg/ml)	100 µl per liter
1-kb ladder (Invitrogen)	

Loading Buffer

0.25% bromophenol blue	2.5 g per liter
0.25% xylene cyanol	2.5 g per liter
25% Ficoll 400	250 g per liter

in 10 × TAE (Tris acetate) buffer

Running Buffer (TAE Buffer)

0.4 M Tris	48.5 g per liter
0.19 M glacial acetic acid	11.4 ml per liter
10 mM EDTA	3.7 g per liter OR 20 ml 0.5 M EDTA per liter

Can be made by using 100 ml TAE (10×) made up to 1 liter with water

Equipment

1. Microwave oven
2. Horizontal agarose gel tank, comb, casting tray, and power pack
3. Pipette tips
4. Pipettes
5. 55°C water bath
6. Biohazard tape
7. Transilluminator and photographic equipment

Method

1. Mix the appropriate weight of agarose with TAE buffer.
2. Put in microwave at full power for 4 min 30 s; heat protecting gloves should be used when heating the agarose.
3. After this, microwave for 15- to 30-s intervals at full power until the agarose has dissolved, but be careful not to let it boil over.
4. Once fully melted leave in the microwave for 2 min to cool, as immediate removal can result in boiling over.
5. Place in 55°C water bath to allow it to cool without setting.
6. Seal the edges of a small gel casting tray with biohazard tape.
7. Add 100 µl (per liter) of ethidium bromide (caution, it is carcinogenic) to the cooled agarose solution and mix by gentle swirling.
8. Insert a fine comb into the casting tray.
9. Pour agarose to a depth of about 1 cm and allow to set.
10. Remove the biohazard tape.
11. Take 10 µl of a 500-µl DNA sample, add 2.5 µl loading buffer.

12. Use 1 µl of 1-kb ladder (1µg/µl, Invitrogen), 9 µl water, 2.5 µl loading buffer as a marker.
13. Fill tank to just above gel bed with TAE buffer (2 liters).
14. Place gel in tank, ensuring that the gel is just submerged.
15. Add the samples to the sample wells and run at 150 V for about 1 h, or until front dye is near the bottom of the gel.
16. Check that electrodes are bubbling so you know that the circuit is complete (or check that you have amps as well as volts by pressing the button on the power pack).
17. Photograph gel under UV (300 nm) transillumination.

15

POLYMERASE CHAIN REACTION AND ARMS

Mark D. Robinson and Graham R. Taylor
Regional DNA Laboratory, St. James's University Hospital, Leeds, UK

■ POLYMERASE CHAIN REACTION

The polymerase chain reaction (PCR) is an in vitro technique for the synthesis of specific sequences of DNA (including cDNA) with DNA polymerase. The specificity of the reaction is due to the requirement of all DNA polymerases for a double-stranded start site for de novo synthesis (1, 2). Two primers are used, which flank the region to be amplified. The primers are themselves incorporated into the reaction product. The reaction works by denaturing genomic or cDNA (or single-stranded DNA can be used) in the presence of excess (5–50 pmol/reaction) primers, dNTPs, buffer, and heat-stable *Taq* polymerase. The mixture is cooled to allow primer hybridization and incubated to allow polymerization of new strands. An almost exponential synthesis of DNA should result, producing nanogram to microgram amounts after 25–35 cycles. This can then be analyzed on gels directly after restriction digestion. It is standard practice in this laboratory to conduct hot-start PCR. This means that active reagents are only combined at high temperature and tends to increase the yield and specificity of the reaction. Hot start can be achieved either by physical separation of the reagents, for example, with a wax barrier or a two-stage setup, or by using enzyme formulations such as Clontech Advantage, Platinum Taq, or AmpliTaq Gold that are modified to require heat activation before use. In the case of AmpliTaq Gold, 10- to 15-min preheating at 95°C is required in the initial denaturation step.

PCR products should be analyzed by gel electrophoresis. Greater than 80% of the product should be as fragments of the expected size. The amplification reaction

Guide to Mutation Detection, Edited by Graham R. Taylor and Ian N. M. Day.
ISBN 0-471-23444-3

is conducted with an automated thermal cycler. Products are detected by staining using ethidium bromide or silver, by radioactive labeling, or by fluorescent labeling.

Reagents

Taq buffer is often supplied with the enzyme, but some *Taq* enzymes are not compatible with buffers from other manufacturers. This is probably because of the detergents used. AmpliTaq Gold will not be activated in high-ammonium salt buffers and should only be used in the buffer supplied. Because salt and magnesium ion concentrations vary in different buffers, annealing temperatures and PCR yields may need to be reoptimized if a buffer is changed.

Primers

Primers are routinely stored at –20°C in 50% glycerol-TE at 100 μM. These are used to make up working stocks.

PCR Buffer

Tris-Cl pH 8.4 10 mM
KCl 50 mM
$MgCl_2$ 1.5 mM
dNTPs 200 mM
Triton X-100 0.1%
BSA 0.01% (not essential)

dNTPs

These are obtained as ultrapure, neutralized solutions (100 mM) (from Amersham-Pharmacia, Promega, or Roche).

dNTP Solution 100 × (20 mM) Working Stock

50 μl of each dNTP 50 μl water

Store at –20°C.

For the incorporation of labeled dNTP, use stocks reduced for the hot dNTP.

Thermostable Polymerases

The polymerases used for the PCR fall broadly into two categories: those with proofreading (3′ → 5′ exonuclease) activity and those without. Most PCR amplifications use DNA polymerases that lack proofreading properties derived from *Thermus aquaticus* or other thermophilic eubacteria. A wide range of suppliers manufacture suitable enzymes, including Promega native *Taq* polymerase, AmpliTaq Gold (Applied Biosystems), and Clontech Advantage (Clontech). Unit activities are measured as Kunitz units, reflecting the amount of labeled precursor that is incorporated into acid-insoluble material, and enzyme is typically supplied at 5 units per microliter. Although this unit of measurement does not reflect the processivity of the enzyme (number of bases synthesized per enzyme binding event), final concentrations of between 1/50 and 1/200 dilutions usually work well. Different buffer formulations may produce annealing temperatures that vary by 1 or 2°C.

Equipment

PCR Machines (Thermal Cyclers)

There are several type of thermal cycler, and temperature ramp speed, accuracy, and uniformity may vary between machines. The HUGO courses have used MJ Research Tetrads, and other machines including the Applied Biosystems 9700 are equally effective.

Hot-start PCR Protocol with Heat-activated Polymerase

1. Add DNA (5–250 ng), primers (5–50 pmol), and enzyme/reagent mix to 10 µl. Always include a negative control.
2. Cover with oil if not using a heated lid.

TABLE 15.1. Stoichiometry of the PCR

Component of Reaction	Initial	After 10^6-Fold Amp.
Human DNA		
Weight	1 µg	1 µg
Moles	$5\,\mu\,10^{-7}$pmol	5×10^{-7}pmol
Concentration	5 fM	5 fM
Ratio to template	1:1	1:1
1-kb Target		
Weight	0.3 pg	0.3 µg
Moles	5×10^{-7}pmol	0.5 pmol
Concentration	5 fM	5 nM
Ratio to template	1:1	10^6:1
Primer		
Weight	65 ng	62 ng
Moles	10 pmol	9.5 pmol
Concentration	0.1 mM	0.095 mM
Ratio to template	2×10^7:1	2×10^7:1
Taq Polymerase		
Weight	0.1 µg	0.1 µg
Moles	0.1 pmol	0.1 pmol
Concentration	1 nM	1 nM
Ratio to template	2×10^5:1	2×10^5:1
dNTP (each)		
Weight	11.5 µg	11.4 µg
Moles	2×10^4 pmol	1.98×10^4 pmol
Concentration	200 µM	198 µM
Ratio to template	4×10^{10}:1	3.95×10^{10}:1

Source: After Ruano G, Brash DE, Kidd KK (1991) Amplifications, 7, p. 1–4.

This table is based on a 100-µl reaction assuming genome = 3.3×10^9 bp, 1 bp = 650 daltons, oligos as 20mers, 1 base = 325 daltons, no degradation due to thermal cycling, 1 U *Taq* = 50 fmol of enzyme, 1 dNTP = 575 daltons. For equal G,C,A,T template, $10^3\times2/4 = 500$ pmol each dNTP is consumed for each 1 kb product.

3. Place in thermal cycler, switch on, and start up for the initial 5-min denaturation step (15 min for AmpliTaq Gold).
4. Samples can be stored in the machine overnight.

■ ARMS

This is a modification of conventional PCR in which one of the primer pairs is designed to have the polymorphic base in the template at its 3′ position (3). *Taq* polymerase is unable to extend from a mismatched base, and so the generation of a PCR product only occurs if the 3′ base in the primer matches the template. The technique can be multiplexed to type 20–30 SNPs simultaneously.

Principle

The method employed by the CF(12)m PCR kit is based on the amplification refractory mutation system (ARMS). The principle is that an oligonucleotide with a mismatched 3′ residue will not function as a PCR primer. Primers are designed so that the 3′ base is at the site of a particular mutation, and amplification of this product only occurs in the presence (or absence) of this mutation.

The CF(12)m PCR comprises two ARMS multiplexes (Fig. 15.1):

Vial A detects the CF mutations 1717-1(G > A), G542X, W1282X, N1303K, ΔF508, and 3849 + 10Kb(C > T).

Vial B detects the mutations 621+1(G > T), R553X, G551D, R117H, R1162X, and R334W. In addition, this multiplex identifies genes in which the ΔF508 mutation is not present.

■TABLE 15.2. CF (12)m—Elucigene ARMS PCR Kit

Mutation	Incidence in a Northern English Population (% CF chromosomes)
ΔF508	79
621+1(G > T)	0.5
R117H	0.7
R334W	0.07
1717-1(G > A)	0.28
G542X	0.5
G551D	2.17
R553X	0.14
3849+10 kb (C > T)	0.07
R1162X	0.14
W1282X	0.07
N1303K	0.35
TOTAL	84

Astra-Zeneca used to supply this kit, which detects 12 of the more common CF mutations. An updated version of the kit is available from Orchid.

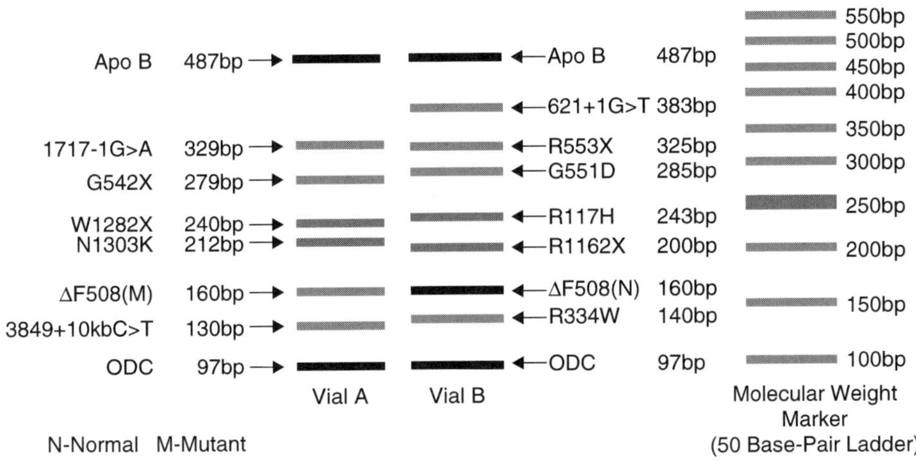

FIGURE 15.1. Scheme of mutations detected with a 2-tube, 12-mutation ARMS test.

As confirmation of amplification, both multiplexes contain control primers that amplify specific regions in the human apolipoprotein B (ApoB) and ornithine decarboxylase (ODC) genes.

Materials

Reagents Supplied

Vials A (pink) and B (clear) contain the appropriate primers and dNTPs for the reaction.

Each A/B tube contains 40 μl of reagent. To economize, we routinely split them into 5 tubes of 8 μl and set up a 10-μl PCR instead of the recommended 50 μl. Reaction buffer and *Taq* polymerase are supplied in separate tubes.

Genomic DNA

Controls

A normal DNA control is also supplied (at a concentration of 5 ng/μl).

We also have control DNA samples (representing each of the 12 mutations) diluted ready for use, which are stored with the ARMS kit and used in rotation.

Test DNA

DNA extracted by a routine method:

DNA is diluted to a concentration of 10 ng/μl for use with the ARMS kit.

For most DNA samples (which are resuspended at a concentration of 500 μg/ml) this is a 1/50 dilution.

Method

Maximum run size = 15 patients and 3 controls giving 18 tests in 36 tubes.

PCR Conditions

94°C, 20 min; (94°C 1 min; 58°C 2 min; 72°C 1 min) × 35
72°C 10 min
DO NOT SET TO COOL to below room temp.

1. Prepare diluted DNA for each patient sample to 10 ng/μL.
2. Label a pink tube A and a white tube B for each patient/control being tested.
3. Take the appropriate number of A and B vials from the kit (one tube of each for 5 patients) and microfuge the contents until the liquid is at the bottom.
4. Pipette 8 μl of primer mix A and B into your labeled pink and white tubes, respectively.
5. Pipette 1 μl of diluted DNA into tubes A and B for each patient/control.
6. For the negative control, add no DNA.
7. Prepare *Taq*/buffer solution. For 18 samples mix 39 μl sterile deionized water, 5 μl of ARMS buffer and 6 μl of *Taq* polymerase.
8. Add 1 μl of the *Taq*/buffer mix to each sample tube.
9. Add one drop of mineral oil to each tube.
10. Place on the PCR block and start the program running.

On completion, rapid cooling should be avoided, but the samples can be stored at room temperature overnight or in the refrigerator.

Agarose Gel Electrophoresis

1. 300 ml of 3% Nusieve 3:1 agarose (6.7 g Nusieve GTG low-melting-point agarose, 2.3 g normal agarose) is prepared in 1 × TBE. This is boiled and poured (with 25–30 μl of ethidium bromide) into a large gel tray with 2 combs, giving 40 wells.
2. Once set, cool the gel at 4°C for 15–30 min.
3. All 10 μl of each PCR product is loaded with 1.5 μl agarose gel loading buffer.
4. A 50-bp ladder (Pharmacia) is run in 4 lanes alongside the PCR products. 2 μl of ladder is used per lane.
5. The gel is run in 1 × TBE at 150 V for 3–4 h (the deep blue dye front (bromophenol blue) should be at the bottom of the gel).
6. The gels are visualized and photographed with the UV transilluminator.
7. The results can then be scored with a chart supplied in the instruction booklet.

Results

Negative control must show no amplification products.

The lower (ODC) and upper (ApoB) control bands must be visible in each sample lane for results to be accepted.

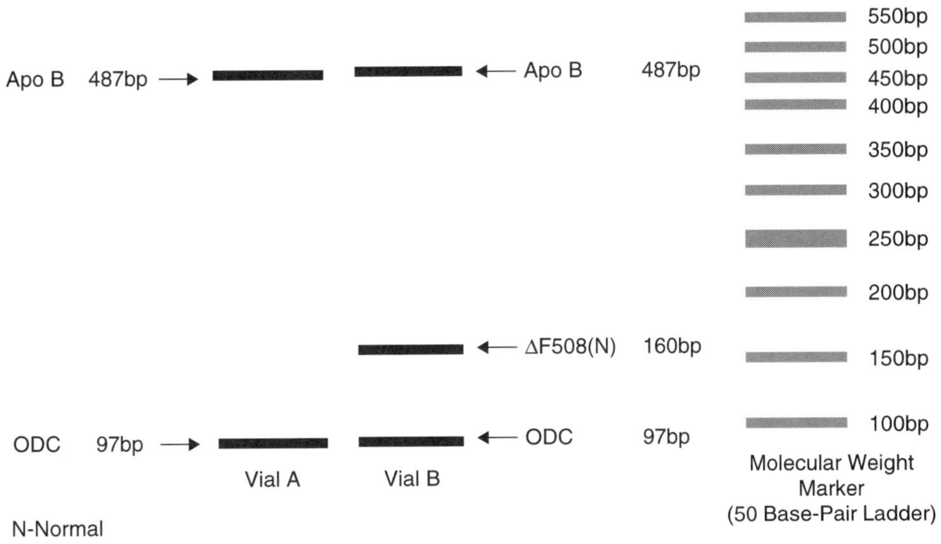

FIGURE 15.2. Expected band pattern of noncarrier.

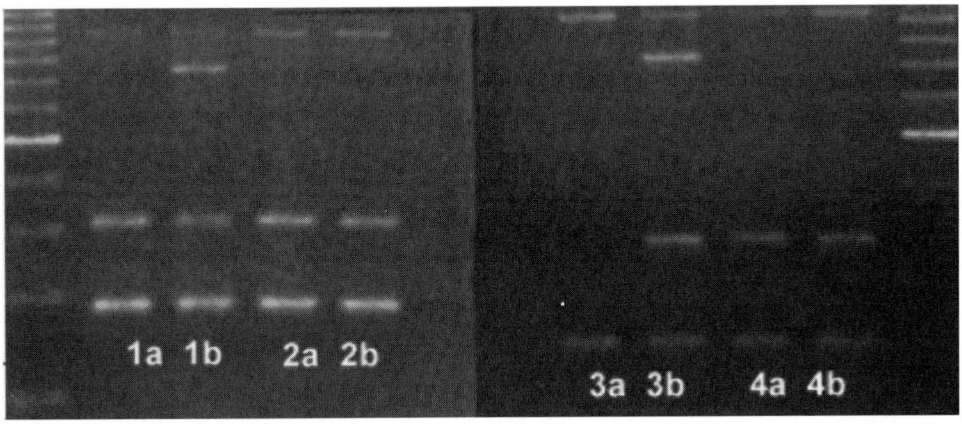

FIGURE 15.3. Typical genotype results on an agarose gel. First and last lanes are size standards; lanes a and b correspond to the a and b tubes; numbers refer to test cases.

For ΔF508 a band at 160 bp in track A represents a deleted chromosome. A band at 160 bp in track B represents a nondeleted chromosome. Heterozygotes and homozygotes can therefore be distinguished for this mutation.

PCR products for the other mutations will only be visualized when an individual has the mutation (Figs. 15.1, 15.2, and 15.3). The mutation can be identified depending on which track (A or B) it appears in and on its size relative to the 50-bp ladder. It is not possible to tell whether an individual is homozygous or heterozygous for these mutations.

REFERENCES

1. Kleppe R, Sekiya T, Loewen PC, Kleppe K, Agarwal KL, Buchi H, Besmer P, Caruthers MH, Cashion PJ, Fridkin M, Jay E, Kumar A, Miller RC, Minamoto K, Panet A, RajBhandary UL, Ramamoorthy B, Sidorova N, Takeya T, van de Sande JH, Khorana HG (1976) Total synthesis of the structural gene for the precursor of a tyrosine suppressor transfer RNA from *Escherichia coli*. 11. Enzymatic joining to form the total DNA duplex. J Biol Chem 251: 667–675.
2. Mullis K, Faloona F, Scharf S, Saiki R, Horn G, Erlich H (1986) Specific enzymatic amplification of DNA in vitro: The polymerase chain reaction. Cold Spring Harb Symp Quant Biol 51: 263–273.
3. Newton CR, Graham A, Heptinstall LE, Powell SJ, Summers C, Kalsheker N, Smith JC, Markham AF (1989) Analysis of any point mutation in DNA. The amplification refractory mutation system (ARMS). Nucleic Acids Res 17: 2503–2516.

16

MINISEQUENCING AND CAPILLARY ELECTROPHORESIS

Darren Cuthbert-Heavens
Mutation Detection Facility, Cancer Research UK, St. James's University Hospital, Leeds, UK

■ INTRODUCTION

In conventional sequencing deoxynucleotides are incorporated alongside terminating dideoxynucleotides to produce strings of fragments of varying length, enabling the complete sequence of a given fragment to be decoded. Minisequencing, however, is based on the dideoxy single nucleotide extension of an unlabeled oligonucleotide primer (Fig. 16.1). This makes the technique ideally suited for genotyping single nucleotide polymorphisms and is very useful where polymorphisms do not create or destroy restriction sites. By coupling this technique with fluorescent detection samples can be run in 25 minutes, allele calling can be automated, and with the possibility of multiplexing a number of polymorphisms can be investigated at once, resulting in significant increases in throughput (Fig. 16.2).

■ MINISEQUENCING PROTOCOL USING SNAPSHOT KIT

Solutions

SnaPshot Kit (Applied Biosystems)
POP4 polymer (Applied Biosystems)
Primers (e.g., Interactiva)
10× TBE (Invitrogen)

Guide to Mutation Detection, Edited by Graham R. Taylor and Ian N. M. Day.
ISBN 0-471-23444-3

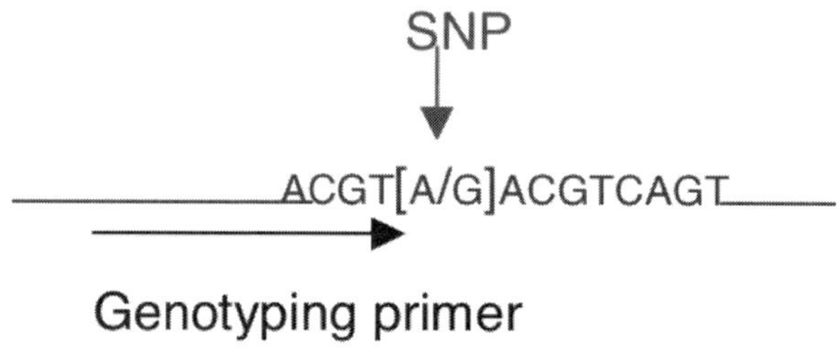

FIGURE 16.1. Principle of genotyping by single nucleotide primer extension.

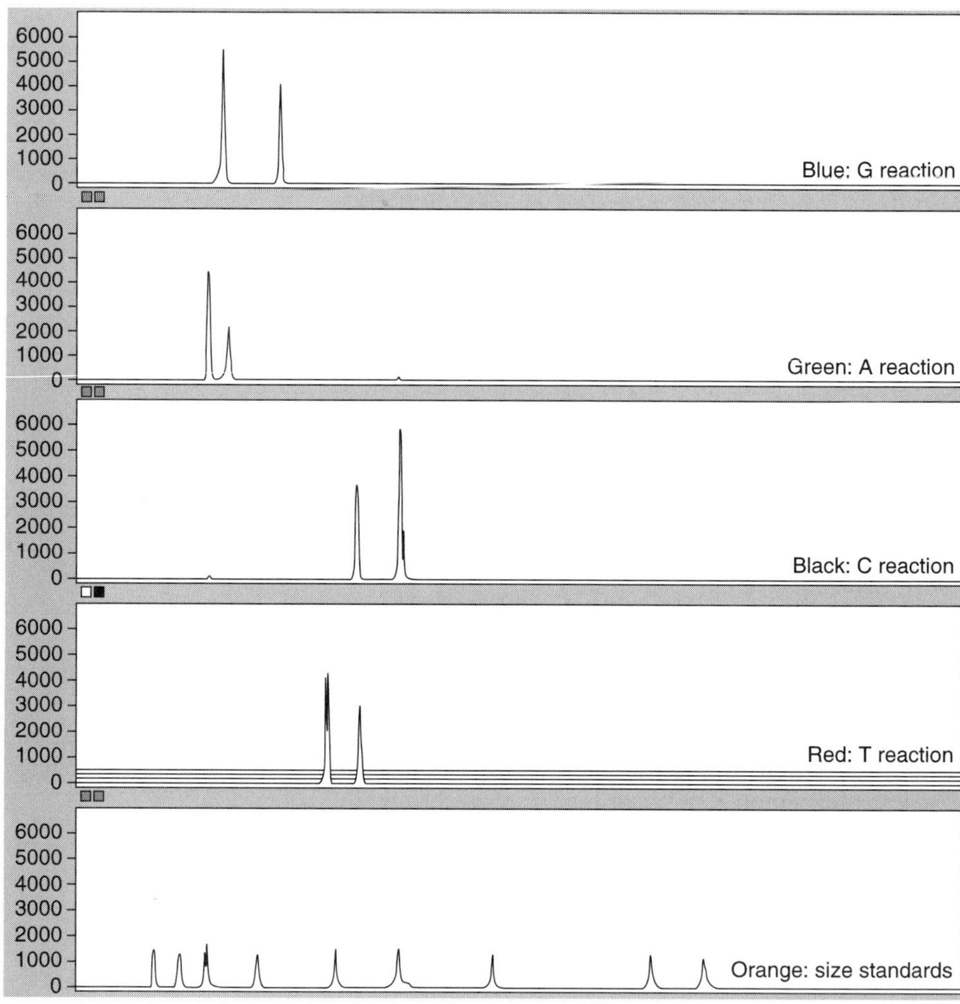

FIGURE 16.2. Example of a multiplex genotyping assay with the SnapShot system.

Method

Primer Design

Minisequencing primer design is limited because of the necessity to have the 3′ end of the primer adjacent to the polymorphic base. Primers can be designed against the sense or antisense strand to overcome potential problems caused by possible extendable hairpin structures and should be greater than 18 nucleotides in length.

Template Preparation

Genomic DNA is amplified via a conventional PCR in a 25-μl volume; primers should be designed to produce fragments around 150--200 base pairs in length with the polymorphism to be screened near the center of the fragment. After amplification, 5 μl of the reaction should be checked with an agarose gel.

Template Purification

PCR products need to be purified to remove excess primers and nucleotides using a digestion step involving shrimp alkaline phosphatase (SAP) and exonuclease I (Exo I).

SAP Exo I Method Combine the following in a 200-μl tube:

2 units SAP
2 units Exo 1
4 μl PCR product
Water to 10 μl

Incubate at 37°C for 1 h.
Incubate at 72°C for 15 min.

Minisequencing Reaction Combine the following in a 200-μl tube:

SNaPshot Ready Reaction Premix	5 μl
Minisequencing primer (0.15 pmol/μl)	1 μl
Purified PCR (0.15 pmol/μl)	1 μl
Deionized water	3 μl

Place the tubes in a thermal cycler (GeneAmp 9700).

Cycle 25 times	96°C	10 s
	50°C	5 s
	60°C	30 s

Postextension Treatment Unincorporated [F]ddNTPs will comigrate with the fragments of interest. Removal of the 5′ phosphoryl groups by phosphatase treatment alters the migration of the unincorporated [F]ddNTPs, thus prohibiting interference.

To the reaction mixture add 0.5 units of shrimp alkaline phosphatase, incubate at 37°C for 1 h and then at 72°C for 15 min.

Electrophoresis on the ABI310

Sample Preparation To 1 μl of sample add 10 μl deionized formamide in a 200-μl tube. Cap the tube, vortex briefly, and spin down in a centrifuge. Denature the samples at 95°C for 5 min and then place them on ice until you are ready to load them onto the 310 Genetic Analyzer.

ABI310 Setup The ABI310 should be set up with a 36-cm Well to read capillary, primed with POP4 polymer. Control module Fast GS Template E can be used, which has parameters as shown below:

Injection time	10 s
Electrophoresis voltage	15 KV
Collection time	24 min
EP voltage	15 KV
Heat plate temperature	60°C
Syringe pump time	150 s
Preinjection time	120 s

This is sufficient to collect data on minisequencing primers up to 60 bp in length. For smaller primers, it is possible to reduce collection times and thus increase throughput.

Data Analysis

Once collection has finished, data can be imported in to a Genescan Project and analyzed. The sample files, once analyzed, can be imported into Genotyper and macros can be written to automate allele calling.

17

SNP GENOTYPING USING TAG-ARRAY MINISEQUENCING

Snaevar Sigurdsson and Ann-Christine Syvänen
Department of Medical Sciences, Uppsala University, Uppsala, Sweden

■ INTRODUCTION

A promising approach toward high-throughput genotyping of single nucleotide polymorphisms (SNPs) is to use arrays of immobilized oligonucleotides in miniaturized assays (1, 2). Significant advantages of performing the assays in microarray formats are the reduced costs of genotyping because of the simultaneous analysis of many SNPs in each sample and the small reaction volumes employed on the microarrays. There are three major reaction principles that are currently used for genotyping of SNPs: hybridization with allele-specific oligonucleotide (ASO) probes (3), oligonucleotide ligation (4), and DNA polymerase-assisted primer extension (5) (see Ref. 6 for a recent review on genotyping techniques). All of these methods have been utilized in microarray-based assay formats. Simultaneous genotyping of multiple SNPs by ASO hybridization is hampered by the inability of ASO probes to discriminate between SNP genotypes in large, diploid genomes (7, 8). Therefore, enzyme-assisted methods are gaining acceptance as more specific alternatives than ASO hybridization for multiplex SNP typing (9–12).

In the minisequencing reaction a DNA polymerase is used to extend detection primers that anneal immediately adjacent to the sites of the SNPs. The primers are extended with differently labeled nucleotide analogs that are complementary to the different nucleotides at the SNP sites. The method was initially described for a microtiter solid-phase support, in which the PCR product is immobilized (13, 14). Minisequencing was later adapted for the microarray format, where instead the detection primer is covalently attached to the microarray and extended with radioactively or fluorescently labeled dideoxynucleotides (ddNTPs) (5, 11, 12).

Guide to Mutation Detection, Edited by Graham R. Taylor and Ian N. M. Day.
ISBN 0-471-23444-3

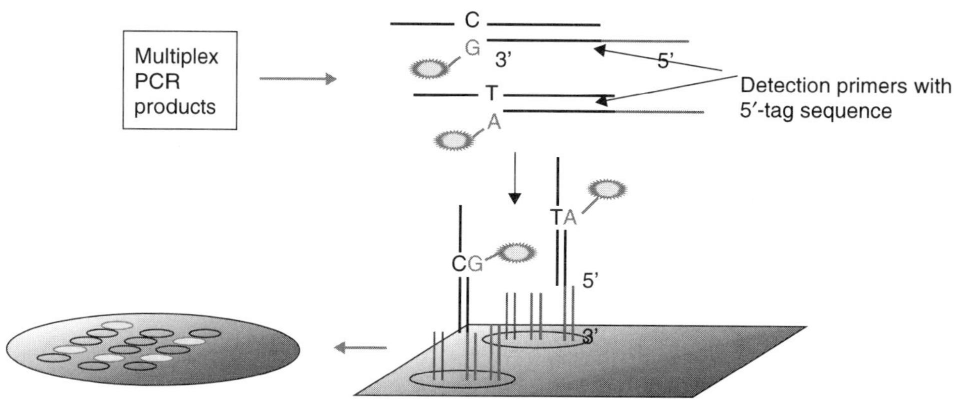

FIGURE 17.1. The principle of tag minisequencing on a microarray.

There is an alternative minisequencing system in which generic oligonucleotides ("cTags") are immobilized on the microarray instead of detection primers. This tag-array minisequencing system involves cyclic minisequencing reactions with fluorescently labeled ddNTPs. The reactions are performed in solution with detection primers with 5′-"tag" sequences. Each SNP has its own specific tag, which is complementary to one of the "cTags" that are attached to the microarray (see Fig. 17.1). When the mix of extended detection primers is applied to the microarray, the tags will hybridize to their corresponding "cTags." From the known locations of the "cTags" on the microarray, the genotypes of the SNPs can be deduced. The concept of using tagged PCR primers was first described for analysis of gene expression in yeast by PCR (15) and was later applied to SNP genotyping by primer extension and capture on fluorescent microparticles (16), high-density oligonucleotide arrays (Affymetrix, GenFlex™ arrays, Ref. 17), and medium-density, custom-made oligonucleotide arrays (18).

For both of these minisequencing microarray systems, one may use either ddNTPs that are labeled with the same fluorophore and used in separate reactions, or ddNTPs labeled with four different fluorophores and used in the same reaction (19). For economic and throughput reasons, the four-color system is preferable and will be used in this exercise.

The aim of this exercise is to genotype 61 mitochondrial SNPs in 20 individual DNA samples and 3 pooled DNA samples by using the four-color tag-array "minisequencing" primer extension method. The method can easily be scaled up for the analysis of 200–300 SNPs per sample. Figure 17.2 outlines the steps of the method, and they are described in the Method section below.

METHOD

PCR

The key to high-throughput assays for SNPs is a simple and efficient method for DNA amplification to achieve the required specificity and sensitivity. In practice, performing multiplex PCR is a bottleneck because of the difficulties of designing more than a few primer pairs that do not interact with each other. To avoid these

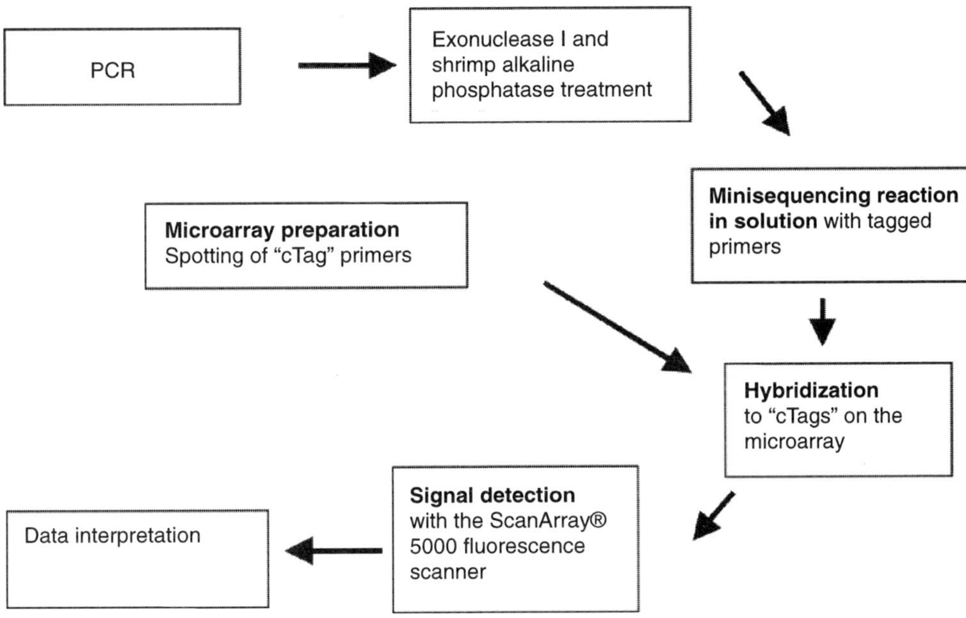

FIGURE 17.2. An outline of the tag minisequencing method.

problems, the PCR primers may be designed to contain common 5′ tail sequences to uniform the kinetics of the primer annealing (20) or to avoid primer dimers using homo-tag PCR (21). Also, complementary 3′ overlap of the primers can be avoided by designing primers with common 3′ terminal dinucleotides (22).

Exonuclease I and Shrimp Alkaline Phosphatase Treatment

Exonuclease I and shrimp alkaline phosphatase are used to remove the excess of deoxynucleotides (dNTPs) and primers from the PCR before the single nucleotide primer extension reaction. Exonuclease I catalyzes the removal of nucleotides from single-stranded DNA in the 3′ to 5′ direction, whereas shrimp alkaline phosphatase removes the 5′ phosphates of dNTPs. This step is necessary to ensure that the dNTPs cannot be incorporated in the primer extension step, which would disrupt the accuracy and specificity of the minisequencing reactions.

Microarray Preparation

The arrays used here are prepared through contact printing of the oligonucleotides. This was done with a ProSys 5510A instrument (Cartesian Technologies Inc, Irvine, CA) with four Stealth Micro Spotting Pins (SMP3, TeleChem International Inc., Sunnyvale, CA) that delivers 1 nl of the 25 μM 5′-amino-modified oligonucleotide solutions to the three-dimensional surface of CodeLink™ Activated slides (Amersham Biosciences, Uppsala, Sweden) (12, 23, 24). This results in oligonucleotide spots with a diameter of 125–150 μm and a center-to-center distance of 200 μM. Minisequencing primers used in solution in the tag-array assay were designed with a 20-base-long 5′-tag sequence. It is important that they are designed with a low self-complementarity and a lack of complementarity to each other (18). The complementary tag sequences that will be immobilized contain a 3′-amino group and 15T residues as spacers located 3′ of the specific sequence.

Minisequencing Reaction

The minisequencing detection primers are designed to anneal adjacent to the variable nucleotide position. The primers are about 18–25 bp and have melting temperatures around 55°C. In a cyclic reaction the minisequencing primers are extended with a fluorescently labeled nucleotide (ddNTP that terminates further elongation) complementary to the nucleotide at the variable site. The distinction between genotypes of the SNPs is based on the high accuracy of nucleotide incorporation by the DNA polymerase. The cyclic reaction helps to extend as many primers as possible to give as high a signal as possible.

Hybridization to Tag Primers on the Microarray

A custom-made aluminum reaction rack is used as an incubation chamber. A reusable silicon grid is placed over the arrays on the slide to form up to 80 separate reaction wells on each slide in the same configuration as the wells of a standard 384-well microtiter plate (11). The capacity of each well is 200 primer spots at the printing density used here.

Signal Detection

The signals are measured with a four-laser confocal fluorescence scanner (ScanArray® 5000, Packard BioScience). Because the four different ddNTPs are labeled with fluorophores with little or no spectral overlap, the nucleotide compositions of the spots can be determined through four consecutive scans at each fluorophore's optimum excitation wavelength (see Table 17.1).

Data Interpretation

The spot intensities are determined from the scanned images with the QuantArray® software supplied with the scanner. The QuantArray software locates the spots on the slide and extracts a numerical value of the signal intensity from them. To assign the genotypes, a ratio between the signal for one allele and the sum of the signals from the two possible alleles is calculated. The ratios will cluster in three groups, where a ratio close to 1 corresponds to a homozygote for one allele, a ratio close to 0 corresponds to a homozygote for the other allele, and a ratio of approximately 0.5 indicates a heterozygote. To visualize the results, the SNPSnapper v3.1beta software,

■TABLE 17.1. Fluorophore and ddNTP Data

Dideoxynucleotide	Fluorophore	Excitation Wavelength (nm)	Excitation Laser
DdATP	Texas Red	595	Yellow HeNe 594
DdCTP	Tamra	560	Green HeNe 543.5
DdGTP	R110	496	Blue Argon 488
ddUTP*	Cy5	649	Red HeNe 632.8

*ddUTP is used instead of ddTTP because ddTTP is not commercially available.

which was specially written for us for this application, will be used. The SNPSnapper program plots the results as scatter plots where the clusters should be clearly separated, and the user can set the limits between the clusters. Each SNP has different genotype limits depending on many factors, so no general rule can be applied for the genotyping procedure. A webpage describing the software can be found at this location: http://www.bioinfo.helsinki.fi/snpsnapper/.

Laboratory Protocol

The minisequencing procedure includes several steps that can all be done in one or two days. After each step the samples can be put in the freezer for storage.

1. PCR 2–3 h
2. Exo/SAP treatment 1.5 h
3. Cyclic minisequencing 1 h
4. Hybridization/washing 2–3 h
5. Scanning 10–40 min
6. Data analysis

SNPs

The 61 SNPs to be analyzed are present in the mitochondrial genome, which in total spans about 16.6 kb. The DNA fragments spanning the SNPs are amplified in four PCR reactions from twenty genomic DNA samples and three pooled DNA samples.

PCR

Reagents

DNA Polymerase

DyNAzyme™ EXT DNA Polymerase, Finnzymes OY, Helsinki, Finland

10× Concentrated DNA Polymerase Buffer

500 mM Tris·HCl pH 9.0, 150 mM $(NH_4)_2SO_4$, 0.1% Triton X-100

dNTP Mixture

2 mM dATP, 2 mM dCTP, 2 mM dGTP, and 2 mM dTTP (Life-Technologies)

DNA Samples: 5 ng/μl

1. Prepare four PCR master mixes (one for each primer pair) by combining all PCR reagents except DNA according to Tables 17.2 and 17.3.
2. Aliquot 48 μl of each mix into separate PCR tubes.
3. Add 2 μl of each DNA sample or H_2O to the tubes.
4. Prepare the PCR mixtures on ice and place them in the PCR machine when the block is 94°C (hot start).

The PCR is done under conditions specified in Table 17.4, and the success of the reactions is evaluated on a 1% agarose gel.

TABLE 17.2. PCR Mixture

	Initial Concentration	Final Concentration	1× (ml)
DNA polymerase buffer	10x	1x	5
DNTPs	2 mM	0.2 mM	5
$MgCl_2$	25 mM	0.8 mM	1.6
Forward primer	20 μM	0.4 μM	1
Reverse primer	20 μM	0.4 μM	1
DNA polymerase	1 U/μl	0.02 U/μl	1
Water			33.4
DNA	5 ng/μl	10 ng	2
Total volume			50 μl

TABLE 17.3. PCR Primers Used

Forward 1	TAG ACG GGC TCA CAT CAC C
Reverse 1	GGG TTG TAC GGT AGA ACT GC
Forward 2	CTA CTC TAC CAT CTT TGC AGG C
Reverse 2	CAG TAA TGT TAG CGG TTA GGC G
Forward 3	GCC GCA GTA CTG ATC ATT C
Reverse 3	CTG CTG CGA ACA GAG TGG
Forward 4	CTC GAA CTG ACA CTG AGC C
Reverse 4	CTG AGC AAG AGG TGG TGA GG

TABLE 17.4. Conditions for the PCR

1	94°C	2 min
2	94°C	20 s
3	60°C	30
4	68°C	8 min
5	To step 2, 9 times	
6	94°C	20 s
7	60°C	30 s
8	68°C	8 min + 15 s/cycle
9	To step 6, 19 times	
10	68°C	30 min
11	8°C	forever

Exonuclease I and Shrimp Alkaline Phosphatase Treatment

Reagents

ExoI

Exonuclease I 10 U/μl (Amersham Biosciences)

SAP

Shrimp alkaline phosphatase 1 U/μl (Amersham Biosciences)

1. For each DNA sample and the negative control, the four PCR fragments spanning the SNPs are combined into a pool.

TABLE 17.5. Exonuclease I and Shrimp Alkaline Phosphatase Mixture

	Stock Concentration	Final Concentration	1× (μl)	30× (μl)
PCR product			(7)	—
$MgCl_2$	25 mM	3.6 mM	1.5	45
Tris·HCl pH 9.5	1 M	48 mM	0.5	15
ExoI	10 U/μl	5 U	0.5	15
SAP	1 U/μl	1 U	1	30
Total volume			3.5	105

2. All samples will be analyzed in single reactions on the array (however, it is recommended to analyze them in duplicates). Therefore, aliquot 7 μl of the pooled PCR products (or negative control) into test tubes, making a total of 24 tubes.
3. Prepare the reaction mixtures according to Table 17.5 by first mixing $MgCl_2$ and Tris·HCl buffer and then adding the exonuclease I and the shrimp alkaline phosphatase. **NB**: Both exonuclease I and shrimp alkaline phosphatase are thermosensitive and the original vials should be kept on ice!
4. Add 3.5 μl of the reaction mixture to each of the tubes containing 7 μl of the pooled PCR products or water control.
5. Incubate the tubes for 60 minutes at 37°C and inactivate the enzymes for 15 minutes at 99°C. This will be done using a thermocycler.

NB: Make sure that all tubes are labeled before they are placed in the thermocycler!

The Minisequencing Reaction

Reagents

Primer-ddNTP mix: 26 nM of each tagged minisequencing primer, 4.1 nM of reaction control template, 5 mM $MgCl_2$, 0,1% Triton-X100, 284 mM Tris·HCl pH 9.5, 0.5 μM labeled ddNTPs (Texas Red-ddATP, Tamra-ddCTP, R110-ddGTP), and 1 μM Cy5-ddUTP (PerkinElmer Life Sciences, Boston, MA, USA). Any solution containing these is light sensitive and should be covered with aluminum foil to avoid bleaching.

DNA Polymerase

Thermosequenase (Amersham Biosciences) diluted to 0.5 U/μl.

1. Combine 90 μl of the primer-ddNTP-mix with 60 μl of the diluted DNA polymerase (0.5 U/μl).
2. Add 5 μl of each of the ddNTP-DNA polymerase mixes to the tubes containing 10.5 μl of the exonuclease I- and shrimp alkaline phosphatase-treated PCR product or negative control.
3. Place the tubes in the thermocycler for the extension reaction to take place (the thermocycling protocol is given in Table 17.6).

TABLE 17.6. The Thermocycling Process

95°C 3 min
95°C 20 s; 55°C 20 s (repeat 33 times)
4°C indefinitely

Hybridization to "cTags" on the Microarray

Reagents

20× SSC

3 M NaCl, 0.3 M sodium citrate pH 7.0

HybKtamra

45 nM hybridization control, oligonucleotide labeled with Tamra.

Microarrays

In-house printed according to Figure 17.3.

1. Place the slide with the tag oligonucleotide arrays in the reaction rack with the rubber grid and the Plexiglas cover and preheat the assembly to 42°C.
2. Mix 206.4 µl of 20× SSC and 3.6 µl of Tamra hybridization control. **NB**: The hybridization control as well as the minisequencing mixtures are light sensitive.
3. Add 7 µl of this mixture to the tubes containing the minisequencing products.
4. Pipette the mixtures (22 µl) into the reaction wells on the slide as indicated in Figure 17.3. **NB**: For interpretation purposes, it is important that the samples are pipetted in a known pattern.
5. Place the rack into a humid chamber* previously heated to 42°C and allow the hybridization reaction to proceed for approximately 2 h at 42°C.

Washing of the Microarrays

Reagents

4× SSC
2× SSC with 0.1% SDS
0.2× SSC

(1× SSC: 0.15 M NaCl, 0.015 M sodium citrate, pH 7.0)

1. Remove the slide from the rack and rinse briefly with room temperature 4× SSC.

*The reaction wells should be kept humid during the hybridization time because drying of the reaction mixture will result in background fluorescence, which complicates the signal detection. A humid environment is accomplished, for example, by placing a wet tissue on the Plexiglas lid and covering with plastic film.

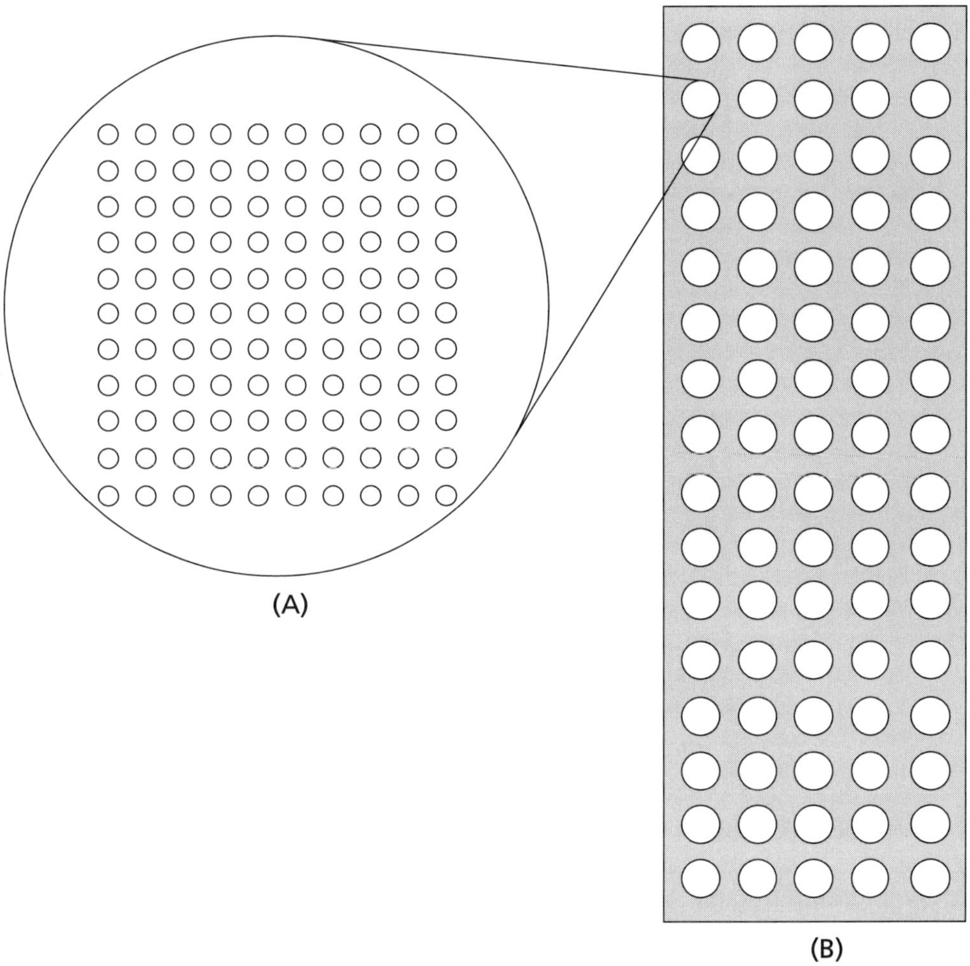

FIGURE 17.3. (A) Oligonucleotide spots, (B) placement of samples on the microarray.

2. Place the slide in a 50-ml Falcon tube and wash twice for 5 min in prewarmed 2× SSC, 0.1% SDS at 42°C using a shaker and avoiding exposure to light.
3. Wash briefly (1 min) twice with 0.2× SSC at room temperature
4. Dry the slide by spinning it for 7 min at 900 rpm.
5. Store the slide in darkness, for instance, loosely wrapped in aluminum foil.

Scanning and Quantification

Follow the manufacturer's guidelines: Our laboratory uses the ScanArray® 5000 and the QuantArray® software package (PerkinElmer Ltd). If allowed by the scanner used, balance the signal intensity from each laser channel so that no signals are saturated and the signals from the four fluorophores are equally strong. Balancing is easy if a reaction control with signals from all four fluorophores is included on the array. An example of a scanning result is given in Figure 17.4.

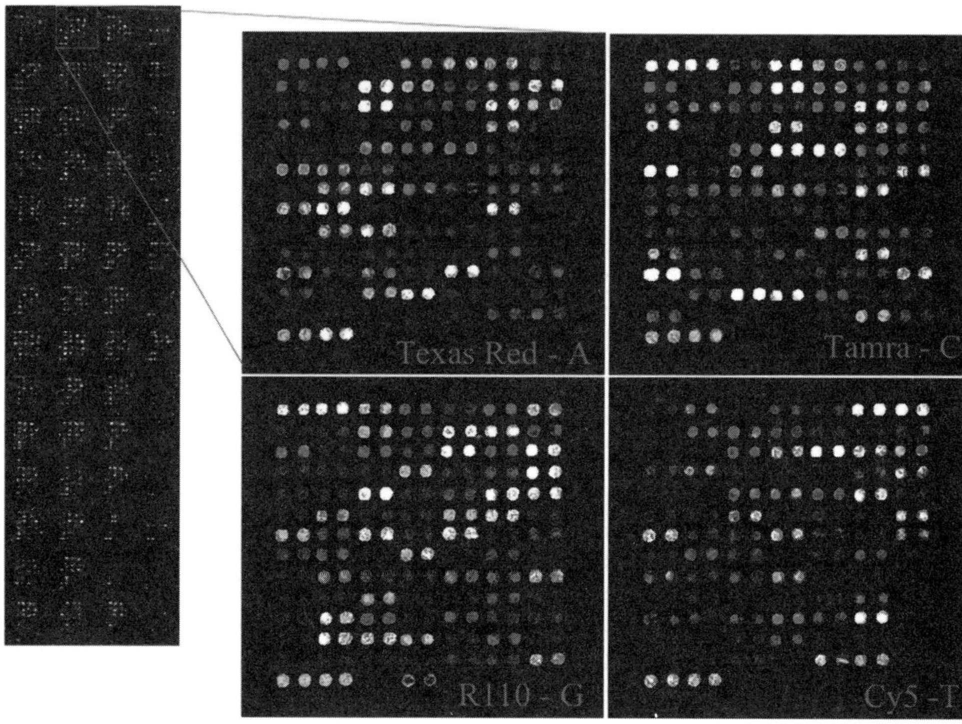

FIGURE 17.4. Example of a scanned microarray image.

Data Analysis and Interpretation

A quantification program such as QuantArray handles the scanning images and quantitates the signals from each spot. The raw data is collected as an Excel sheet. From the signals from each channel subtract the background measured either around the spots or at negative control spots, that is, spotted cTags without corresponding tagged primers. Assign the genotypes of the SNPs in each sample by calculating the ratios between the signals from one of the alleles and the sum of the signals from both the alleles: $Signal_{Allele1}/(Signal_{Allele1} + Signal_{Allele2})$. A scatter plot with this ratio on the X-axis and the sum of the signals from both alleles on the Y-axis is used for assigning the genotypes. This scatter plot should give three distinct clusters, with the homozygote samples clustering at each side and the heterozygotes in the middle. The ratios may vary between SNPs depending on the sequence surrounding it, the type of nucleotide incorporated, and the light intensity of the fluorophores.

REFERENCES

1. Southern EM (1996) DNA chips: Analysing sequence by hybridization to oligonucleotides on a large scale. Trends Genet 12: 110–115.
2. Hacia JG (1999) Resequencing and mutational analysis using oligonucleotide microarrays. Nat Genet 21: 42–47.
3. Hacia JG, Sun B, Hunt N, Edgemon K, Mosbrook D, Robbins C, Fodor SP, Tagle DA, Collins FS (1998) Strategies for mutational analysis of the large multiexon ATM gene using high-density oligonucleotide arrays. Genome Res 8: 1245–1258.

4. Gerry NP, Witowski NE, Day J, Hammer RP, Barany G, Barany F (1999) Universal DNA microarray method for multiplex detection of low abundance point mutations. J Mol Biol 292: 251–262.
5. Pastinen T, Kurg A, Metspalu A, Peltonen L, Syvanen AC (1997) Minisequencing: A specific tool for DNA analysis and diagnostics on oligonucleotide arrays. Genome Res 7: 606–614.
6. Syvanen AC (2001) Accessing genetic variation: Genotyping single nucleotide polymorphisms. Nat Rev Genet 2: 930–942.
7. Wang DG, Fan JB, Siao CJ, Berno A, Young P, Sapolsky R, Ghandour G, Perkins N, Winchester E, Spencer J, Kruglyak L, Stein L, Hsie L, Topaloglou T, Hubbell E, Robinson E, Mittmann M, Morris MS, Shen N, Kilburn D, Rioux J, Nusbaum C, Rozen S, Hudson TJ, Lander ES et al. (1998) Large-scale identification, mapping, and genotyping of single-nucleotide polymorphisms in the human genome. Science 280: 1077–1082.
8. Cho RJ, Mindrinos M, Richards DR, Sapolsky RJ, Anderson M, Drenkard E, Dewdney J, Reuber TL, Stammers M, Federspiel N, Theologis A, Yang WH, Hubbell E, Au M, Chung EY, Lashkari D, Lemieux B, Dean C, Lipshutz RJ, Ausubel FM, Davis RW, Oefner PJ (1999) Genome-wide mapping with biallelic markers in Arabidopsis thaliana. Nat Genet 23: 203–207.
9. Southern E, Mir K, Shchepinov M (1999) Molecular interactions on microarrays. Nat Genet 21: 5–9.
10. Fortina P, Delgrosso K, Sakazume T, Santacroce R, Moutereau S, Su HJ, Graves D, McKenzie S, Surrey S (2000) Simple two-color array-based approach for mutation detection. Eur J Hum Genet 8: 884–894.
11. Pastinen T, Raitio M, Lindroos K, Tainola P, Peltonen L, Syvanen AC (2000) A system for specific, high-throughput genotyping by allele-specific primer extension on microarrays. Genome Res 10: 1031–1042.
12. Raitio M, Lindroos K, Laukkanen M, Pastinen T, Sistonen P, Sajantila A, Syvanen AC (2001) Y-chromosomal SNPs in Finno-Ugric-speaking populations analyzed by minisequencing on microarrays. Genome Res 11: 471–482.
13. Syvanen AC, Aalto-Setala K, Harju L, Kontula K, Soderlund H (1990) A primer-guided nucleotide incorporation assay in the genotyping of apolipoprotein E. Genomics 8: 684–692.
14. Syvanen AC, Sajantila A, Lukka M (1993) Identification of individuals by analysis of biallelic DNA markers, using PCR and solid-phase minisequencing. Am J Hum Genet 52: 46–59.
15. Shoemaker DD, Lashkari DA, Morris D, Mittmann M, Davis RW (1996) Quantitative phenotypic analysis of yeast deletion mutants using a highly parallel molecular barcoding strategy. Nat Genet 14: 450–456.
16. Cai H, White PS, Torney D, Deshpande A, Wang Z, Marrone B, Nolan JP (2000) Flow cytometry-based minisequencing: A new platform for high-throughput single-nucleotide polymorphism scoring. Genomics 66: 135–143.
17. Fan JB, Chen X, Halushka MK, Berno A, Huang X, Ryder T, Lipshutz RJ, Lockhart DJ, Chakravarti A (2000) Parallel genotyping of human SNPs using generic high-density oligonucleotide tag arrays. Genome Res 10: 853–860.
18. Hirschhorn JN, Sklar P, Lindblad-Toh K, Lim YM, Ruiz-Gutierrez M, Bolk S, Langhorst B, Schaffner S, Winchester E, Lander ES (2000) SBE-TAGS: An array-based method for efficient single-nucleotide polymorphism genotyping. Proc Natl Acad Sci USA 97: 12164–12169.
19. Lindroos K, Sigurdsson S, Johansson K, Ronnblom L, Syvanen AC (2002) Multiplex SNP genotyping in pooled DNA samples by a four-colour microarray system. Nucleic Acids Res 30: E70.
20. Shuber AP, Grondin VJ, Klinger KW (1995) A simplified procedure for developing multiplex PCRs. Genome Res 5: 488–493.
21. Brownie J, Shawcross S, Theaker J, Whitcombe D, Ferrie R, Newton C, Little S (1997) The elimination of primer-dimer accumulation in PCR. Nucleic Acids Res 25: 3235–3241.

22. Zangenberg G, Saiki RK, Reynolds R (1999) Multiplex PCR: Optimization guidelines. In *PCR Applications: Protocols for Functional Genomics*. Sninsky JJ (ed.), Academic Press. London, UK. pp 73–94.
23. Guo Z, Guilfoyle RA, Thiel AJ, Wang R, Smith LM (1994) Direct fluorescence analysis of genetic polymorphisms by hybridization with oligonucleotide arrays on glass supports. Nucleic Acids Res 22: 5456–5465.
24. Lindroos K, Liljedahl U, Raitio M, Syvanen AC (2001) Minisequencing on oligonucleotide microarrays: comparison of immobilisation chemistries. Nucleic Acids Res 29: E69.

18

DYNAMIC ALLELE-SPECIFIC HYBRIDIZATION

Darren Cuthbert-Heavens
Mutation Detection Facility, Cancer Research UK, St. James's University Hospital, Leeds, UK

■ INTRODUCTION

Dynamic allele-specific hybridization (DASH) is a medium-throughput genotyping technique that relies on the reduced annealing temperature of mismatched probes to discriminate alleles, which entails dynamic tracking of probe (oligonucleotide) to target (PCR product) hybridization as reaction temperature is steadily increased (1–3). This protocol uses a commercial system manufactured by Thermo-Hybaid Ltd.

■ METHODS

DASH Recommendations

Genomic DNA

Before running your PCR, check the concentration of your DNA by running what you believe to be 50 ng on a 0.7% agarose gel. Compare your genomic DNA sample to a known DNA ladder (e.g., lambda *Hin*dIII). Run the gel in ~1 cm to visualize the DNA as a compact band. A clean, visible band should be seen under these conditions (even for as little as 10 ng). The DNA can be run a further ~10 cm to check for degradation of your genomic DNA. Intact genomic DNA usually runs ~20 kb. Slightly degraded DNA will not seriously harm your PCRs, but very degraded DNA

Guide to Mutation Detection, Edited by Graham R. Taylor and Ian N. M. Day.
ISBN 0-471-23444-3

will. Based on your gel results, dilute your DNA sample accordingly so that you use 10–25 ng per 25 µl PCR.

PCR Primers

To obtain optimal PCR results, the primers (one biotinylated) should ideally be between 21- and 23 bp (it is possible to use 20 bp if the fragment is particularly GC rich.) All primers and the DNA sequence to be amplified should be checked by appropriate software for aberrant intermolecular base pairing before commencing. Most importantly, the amplified DNA fragment should be 50–70 bp, with the polymorphic position located approximately in the center of the amplified fragment, and **this full-length fragment (both alleles) MUST be checked to ensure that it forms no significant secondary structures**. Contiguous intramolecular base pairing regions of 4 nucleotides or longer are particularly problematic as these can lead to hybridization inhibition in the binding domain. These can be tested by using various "DNA folding" software packages such as the web-based program "MFOLD" by Michael Zuker. This program can be found at: http://mfold2.wustl.edu/~mfold/dna/form1.cgi. Use thc default settings, with the exception of the temperature, which should be set to 50°C (default is 30°C). If the region to be amplified contains severe 2° structure (dG less than –3.0), the primer sequences can be modified to destabilize hairpin structures. Alter no more than 2 bases in the primer sequence of a 23-base primer, avoiding bases toward the 3′ end. Note that the prediction of high 2° structure does not preclude use of the sequence for analysis but is merely a prediction of the likelihood of producing a weakened DASH signal. It is possible to use target regions of slightly longer than 70 bp, but the potential for forming secondary structures increases exponentially with fragment length.

For successful DASH, the biotinylated primer must be used in limiting amounts so that it all becomes incorporated into the PCR product. The recommended primer amount for a 25-µl PCR is 4 pmol (20 ng) biotinylated primer plus 20 pmol (100 ng) nonbiotinylated primer. PCR efficiency is *not* severely affected by using these primer levels, but deviations from these amounts are unlikely to function in subsequent DASH analysis.

DASH Hybridization Probe Design

Probes should be between 15 and 21 bp in length, with the polymorphic position in the centre. For the majority of assays, the optimal probe length is 17 bp (probes must be complementary to the strand that will contain the biotin label). Probes matching to both possible SNP variants should be made, one being used as a control for the other. This is most easily performed by using the complementary probing option (see below) and by reprobing a previous experimental plate (see below). Where DASH is used for typing of indels, identical principles to SNPs are used.

The biotinylated PCR primer should not excessively overlap the sequence to which the probe will hybridize because biotinylated primers that are not incorporated during the PCR reaction will bind to the streptavidin-coated plate. These primers, if complementary to the probe sequence over more than 3 bases, result in increased background fluorescence.

PCR Protocol

NB: It is extremely important that the PCR product formed contains no primer dimers. To prevent this we strongly recommend the following:

- Preventing primer dimer demands that hot start protocols be followed. DASH will fail in up to 50% of cases if hot-start protocols are not employed.
- AmpliTaq Gold is highly recommended because of its "hot start" activation properties. This polymerase works best at 3.0 mM $MgCl_2$ (rather than at 1.5 mM). Using AmpliTaq Gold has prevented primer-dimer formation in *all* of the 90 different assays so far tested for DASH analysis.

Always include a no-DNA control. A two-step PCR cycle can be used because of the short length of the DNA amplicon, that is, an extension step is not required. An initial heating step at 95°C for 5–10 min is required to activate the AmpliTaq Gold polymerase.

PCR Product Gel Running Conditions

Running 10–20% of a selection of your PCR products on a 4% low-melting agarose gel is recommended to check for successful amplification (only 50% of a full PCR volume is needed for subsequent DASH analysis). The gel buffer level should reach the top of your gel or ~am over (this is to ensure that the gel doesn't overheat and thereby diffuse your band). A high voltage is recommended to run the gel. The gel should be run 4–8 cm to separate your product from primer dimers. If primer dimers are present they will run at ~50 bp and will be present in your PCR negative control. However, using the above PCR method should prevent primer dimers from occurring.

Primer Dimers

It is extremely important for a successful DASH experiment that the primer dimers are not formed. In the absence of hot start conditions it is possible to form primer dimers even when there is little (if any) complementary sequence between the primers. It is also possible for inverted repeat structures to form between two primers of the same sequence, resulting in hairpin structures. As Sybr Green dye fluoresces in the presence of double-stranded DNA, any double-stranded DNA artifacts present during DASH will contribute to background fluorescence and potentially obscure your genuine "DASH signals."

DASH PCR Amplification Protocol

For DASH, typically a 25-μl amplification reaction is performed for each sample to be tested. Where a sample will be probed with both hybridization oligos, it is recommended that a 50-μl reaction is made, allowing 20 μl for each DASH hybridization and 10 μl for gel analysis. Where a strong DASH signal is obtained for a given set of reagents, DNA bound to streptavidin plates may be probed first with one and then, after washing of the well with NaOH, with the other oligo. When this is the case, a single 25-μl reaction will allow analysis using both probes.

PCR Conditions

The following is a standard protocol, which should work with all SNPs; however, alteration of the primer concentrations by variation of the biotin primer amounts can be performed to optimize the PCR. **Note**: HPLC-purified primers *only* should be used.

PCR Setup

Water	15 μl	☐
$MgCl_2$ (25 mM)	3 μl	☐
10 × PCR buffer	2.5 μl	☐
Biotin-labeled primer (4 pmol/μl)	1 μl	☐
Nonlabeled primer (15 pmol/μl)	1 μl	☐
Taq pol (0.5 U/μl)	1 μl	☐
DNA (10–75 ng/μl)	1 μl	☐
dNTP mix (10 mM each)	0.5 μl	☐
Total	25 μl	

Run the PCR

Suggested PCR cycling: 1 × [95°C: 5–10 min]
38 × [94°C: 15 s, → Annealing temperature: 30 s]

PCR product should be clearly visible as a single band, best visualized with a 4% low-melt agarose. Run 10 μl on a gel. Best results are achieved where a faint band is obtained with low background.

Where no product is seen at the calculated optimal annealing temperature, it is recommended that PCR is performed with an annealing step several degrees lower than the predicted optimum and gradually increased to find the appropriate conditions. Where such optimization is required it is most efficiently performed using a gradient block such as is available for the Hybaid PCR Express.

Occasionally, the PCR may not be efficient when using unequal amounts of each primer (note, however, that only a small amount of product, as little as 120 ng of DNA, may be expected to be formed where 4 pmol of the biotinylated primer is used). Where no PCR product is observed and no DASH signal is obtained, the amount of biotinylated primer may be increased; however, DASH signal may be compromised in such cases.

The Dash Assay

Binding Product to Plate and Hybrid Preparation

1. Add 37.5 μl of 1 × DASH Buffer Gold to each well.
2. Add to each well 12.5 μl of PCR product and leave for 1 h. (The long binding time serves to increase signal. We have had good results with binding times

of 10 min or less with our reference DNA. The amount of PCR reaction product added can be varied to a minimum of 5 and a maximum of 25 μl with buffer added to give a total of 50 μl).

3. Remove solution from well with pipette. (If desired, this solution can be saved and reused at a later date).
4. Add 50 μl 0.1 M NaOH, wait 5 min, rinse with 50 μl fresh 0.1 M NaOH.
5. Add 50 μl DASH Buffer Gold, remove immediately (2×).
6. Add 50 μl of DASH Buffer Gold/Dye mix (prepare 10 ml of DASH Buffer Gold/Dye solution by addition of 1 μl of SYBR Green I Dye to 10 ml of DASH Buffer Gold).
7. Add 2 μl of probe (30 pmol/μl).
8. Place the lid on the microtiter plate.
9. Place the microtiter plate/lid on the 0.2-ml block of the PCR Express Thermal Cycler.
10. Place a compression pad on top of the microtiter plate.
11. Close the lid of the PCR Express and ensure that the pressure selection switch is set to "PLATE" setting.
12. Using PCR block, heat plate to 85°C for 1 s, and then cool to 25°C at a rate of 0.07°C/s.
13. Replace solution with fresh DASH Buffer Gold/Dye mix.

Reading a DASH Plate

1. Ensure that the microtiter plate lid is fitted to the microtiter plate.
2. Eject the 0.2-ml block from the DASH machine, disengage the slide locking mechanism on the lefthand side of the block, and raise the heated lid.
3. Place the microtiter plate/lid on the 0.2-ml block of the DASH machine, close the heated lid down, and lock in place with the slide locking mechanism.
4. Click on play button on the run screen to start the analysis.

■ REFERENCES

1. Howell WM, Jobs M, Gyllensten U, Brookes AJ (1999) Dynamic allele-specific hybridization. A new method for scoring single nucleotide polymorphisms. Nat Biotechnol 17: 87–88.
2. Prince JA, Brookes AJ (2001) Towards high-throughput genotyping of SNPs by dynamic allele-specific hybridization. Expert Rev Mol Diagn 1: 352–358.
3. Prince JA, Feuk L, Howell WM, Jobs M, Emahazion T, Blennow K, Brookes AJ (2001) Robust and accurate single nucleotide polymorphism genotyping by dynamic allele-specific hybridization (DASH): Design criteria and assay validation. Genome Res 11: 152–162.

19

FLUORESCENCE POLARIZATION SNP DETECTION

PerkinElmer Life Sciences, Inc.

Based on the manufacturer's demonstration at the HUGO Mutation Detection Course, Montpellier, September 2002

■ INTRODUCTION

Single nucleotide polymorphisms (SNPs) are the most common genetic variations between individuals of a species and are therefore thought to be responsible for a large part of individual phenotypic variation. Research on human SNPs is expected to facilitate genetic mapping studies that may lead to a better understanding of the genetic basis for complex diseases and individual variation in drug metabolism.

The AcycloPrime-FP process uses a thermostable polymerase to extend by one base an oligonucleotide primer that ends immediately upstream of the SNP position, using one of two fluorescent dye-labeled terminators. The identity of the base added is then determined by the increased fluorescence polarization (FP) of its linked dye. The process uses AcycloTerminators 'and AcycloPol', a novel mutant thermostable polymerase from the Archeon family. It has been found that AcycloPol has a higher affinity and specificity for derivatized AcycloTerminators than various *Taq* mutants have for derivatized 2′8,3′-dideoxynucleotide terminators (1). The structure of a representative dye-labeled AcycloTerminator, dye-acylATP, is shown in Figure 19.A.

FP is an empirical technique that measures the vertical and horizontal components of the fluorescence emission produced after excitation by plane-polarized light. Polarization values (in mP units) are inversely related to the speed of molecular rotation of the fluorescent target. Because molecular rotation is inversely related to the molecular volume, incorporation of a fluorescent AcycloTerminator

Guide to Mutation Detection, Edited by Graham R. Taylor and Ian N. M. Day.
ISBN 0-471-23444-3

TABLE 19.1. FP-Kit Catalog Numbers

R110/TAMRA	1K SNPs	10K SNPs	100K SNPs
G/A	ACP101A	ACP101B	ACP101C
G/C	ACP102A	ACP102B	ACP102C
G/T	ACP103A	ACP103B	ACP103C
C/A	ACP104A	ACP104B	ACP104C
C/T	ACP106A	ACP106B	ACP106C
A/T	ACP109A	ACP109B	ACP109C
Combo*	—	ACP113B	ACP113C

* ACP113B contains 3 × 1K G/A, 3 × 1K C/T, 1 × 1K G/C, 1 × 1K G/T, 1 × 1K C/T, and 1 × 1K A/T. ACP113C contains 3 × 10K G/A, 3 × 10K C/T, 1 × 10K G/C, 1 × 10K G/T, 1 × 10K C/T, and 1 × 10K A/T.
Caution: Research chemicals for research purposes only.

into a primer oligonucleotide increases its polarization. This is used to determine which of the two labeled terminators has been incorporated.

ACYCLOPRIME™-FP SNP DETECTION KIT

The AcycloPrime™-FP SNP detection kit is designed to determine the base at a SNP location in an amplified DNA sample by a modification of template-directed dye-terminator incorporation with fluorescence polarization detection (TDI-FP) (Table 19.1) (2, 3).

The AcycloPrime-FP process is illustrated schematically in Figure 19.1.

Kit Components

Reagents supplied are intended FOR LABORATORY USE. Kits are shipped on dry ice. For maximum stability, all reagents should be stored at –20°C as described below and on the vial labels. If desired for convenience, the AcycloTerminator Mix, 10 × Reaction Buffer, and PCR Clean-Up Reagent Diluent Buffer may be stored at 4°C for up to 1 month.

NOTE: The AcycloTerminator Mix, particularly if it contains R110-acyGTP (ACP101, ACP102, ACP103, and ACP113), is sensitive to light. Under ordinary circumstances, the brown bottle is sufficient to protect the solution from light effects. If the solution is removed from the brown bottle it is shipped in, take precautions to minimize prolonged light exposure. Exposure to normal laboratory lighting during routine assay operations is not detrimental to subsequent stability.

10× Reaction Buffer

Tris · HCl containing $MgSO_4$ and detergent. Store at –20°C on receipt.

AcycloPol™

Contains a mutant thermostable polymerase in buffer with salts, reducing agent, and glycerol. Store at –20°C on receipt.

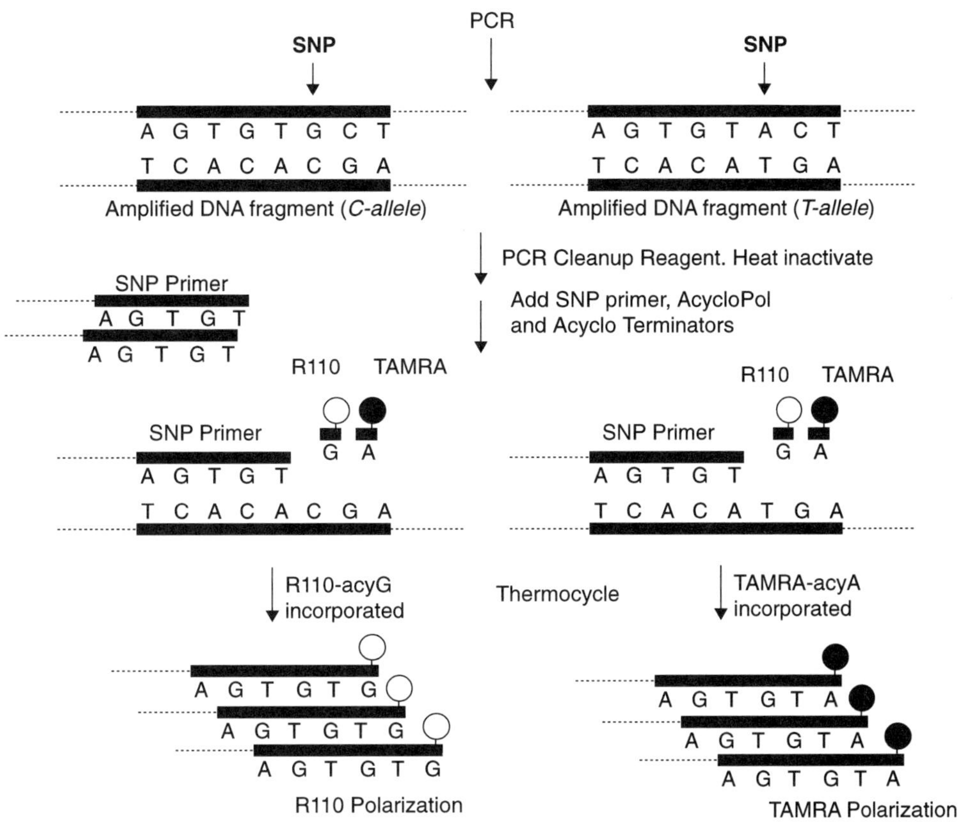

FIGURE 19.1. Schematic of the Acyclo-Prime FP process.

AcycloTerminator™ Mix

Each mix contains one R110-labeled AcycloTerminator, one TAMRA-labeled AcycloTerminator, and two unlabeled terminators. The vial label specifies which labeled AcycloTerminators are in the particular AcycloTerminator Mix. The first base listed is labeled with R110 and the second base with TAMRA, for example, the G/A AcycloTerminator Mix contains R110-acyGTP and TAMRA-acylATP. Different AcycloTerminator Mixes contain different pairs of labeled AcycloTerminators and are used for different SNPs. Verify that you have ordered and received the correct AcycloTerminator Mix for the reactions to be performed.

On Choosing the Correct Kit

Which bases are incorporated depends on which orientation is chosen to prepare the SNP primer. To illustrate, consider the following sequence for two heterozygous chromosomes containing a G/A SNP (bold) on the upper strands. Notice that this is a C/T SNP on the lower strands.

5′-ACTTAGGCCGACCGCTAGCGTAC**A**GAGGCCTTAGACATGCATGCAGT-3′
3′-TGAATCCGGCTGGCGATCGCATG**T**CTCCGGAATCTGTACGTACGTCA-5′
5′-ACTTAGGCCGACCGCTAGCGTAC**G**GAGGCCTTAGACATGCATGCAGT-3′
3′-TGAATCCGGCTGGCGATCGCATG**C**CTCCGGAATCTGTACGTACGTCA-5′

If the SNP primer has the sequence underlined on the upper strands, the next base added will be either an A or a G, so use the G/A kit. If the SNP primer has the sequence underlined on the lower strands, the next base added will be either a C or a T, so use the C/T kit. Note that this means that two different kits are required to perform the assay with both forward and reverse SNP primers unless the SNP is either G/C or A/T.

PCR Clean-Up Reagent, 10×

This is a custom formulation of Exo-SAP-IT™ sold under agreement with USB® Corporation. It is optimized to degrade primers and dNTPs from the amplification steps to prevent interference with primer extension. This material is significantly different from Exo-SAP-IT and is used differently as described below under **Processing of PCR Product**.

1 vial, 0.2 ml (1K SNPs) or
1 bottle, 2.0 ml (10K SNPs) or
1 bottle, 20.0 ml (100K SNPs)

Contains a mixture of exonuclease I and shrimp alkaline phosphatase in buffer with glycerol. Store at –20°C on receipt.

PCR Clean-Up Dilution Buffer

1 bottle, 2.0 ml (1K SNPs) or
1 bottle, 20.0 ml (10K SNPs) or
1 bottle, 200 ml (100K SNPs)

Contains a buffer. Store at –20°C on receipt.

Instruction Manual

Includes performance data, instructions for reagent preparation, and a protocol for primer extension.

Additional Materials and Equipment Required

Thermocycler

Preferred models can utilize a skirted, PCR-compatible microplate.

Microplates, 384 or 96 Wells, Black

Microplates from a number of manufacturers have been used satisfactorily in the assay. For PCR amplification, it is convenient to use a black 384-well PCR plate such as catalog number MSP-3862 from MJ Research. These plates can be used directly in MJ Research and some other thermocyclers, and all of the subsequent assay steps can be performed directly in the well with no sample transfer steps. Similarly designed black 96-well PCR plates (HSP-9666, MJ Research) can be used in the

TABLE 19.2. Victor Filters

	Dye	Excitation	Emission
Recommended filters for the Victor^2	R110	485 (30)	535 (40)
	TAMRA	544 (15)	595 (60)

same way with no modifications to the assay protocol. If evaporation is found to be a problem when using 96-well plates, some users have found it helpful to add 1 µl of extra water to each well. To work well with most FP readers, a skirted plate must be used.

Fluorescence Polarization Reader

The assay has been shown to work well with several instruments including the PerkinElmer Victor^2, the Tecan Ultra, and the Molecular Devices Analyst™. To ensure optimal assay performance, **it is essential to use the appropriate spectral filters, and dichroic mirrors if applicable** (see Table 19.2). These may not be provided as "standard" components with particular FP readers.

G-Factor Calibration

The instrument G-factor should be adjusted to facilitate comparisons between various instruments or different sites and to help with troubleshooting. On the Victor, the two polarization directions are called S and P. With the G-factor set to 1.0, read the S and P intensities for 20 µl of a solution containing 1 µl of any Terminator Mix and 2 µl of 10× Reaction Buffer. It is useful to average the results for several wells to get an accurate measurement. Then use the equation below to calculate the G-factor that will give a value of 50 mP for the unincorporated terminators:

$$G = [(950*S)/(1050*P)]$$

After adding the new G-factor into the Victor software, reading the mP again should give a result near 50 mP. This must be done twice, once for R110 and once for TAMRA.

Reagents for Amplification of Target DNA

If PCR is used, purchase a licensed PCR kit. Polymerase Chain Reaction patents are owned by Roche Molecular Systems, Inc. and Hoffmann-La Roche, Inc. and are licensed to various suppliers of these reagents.

SNP Primer(s)

A SNP primer is a synthetic oligonucleotide complementary to the sequence adjacent to the SNP site. The next base added to the primer is therefore complementary to the base at the SNP location on the target DNA strand. Although limitations are imposed by the specific sequence around the SNP site, the best approach to primer design is to use standard criteria to ensure a melting temperature (T_m) in the range of 60–80°C. It is helpful to select sequences that minimize secondary structure (e.g., hairpin loops) and that have approximately a 50% G + C content. The length should be kept short (usually 20–30 bases). An important control is to design primers complementary to both the sense and antisense strands and to run the assay in both the "forward" and "reverse" directions.

DNA

This is the target DNA containing the SNPs of interest. The DNA quality must be such that it is compatible with the chosen method of amplification, e.g., PCR.

Allele-Calling Software

A series of Excel-based workbooks for data analysis and allele calling are available for download from http://lifesciences.perkinelmer.com/products/snp.asp. For the Victor, files allowing automatic data transfer into the workbook are: SNP macro Victor 96 well, SNP macro Victor 384 well, or SNP macro Victor 384 4 × 96 well. The latter file enables four different genotypes to be analyzed in a single 384-well plate. For manual data input from the Victor, use Victor 96 well or Victor 384 well. For other instruments, use GENERIC 96 well or GENERIC 384 well. For the Molecular Devices Analyst or Tecan Ultra, similar software is available on request. To work efficiently with the workbooks, the instrument protocol should be set up so that R110 polarization is read before TAMRA polarization.

Other Equipment

Pipettors and/or pipettes that can accurately and precisely deliver required volumes. Low-retention pipette tips are recommended.

Laboratory vortex mixer

Test tube rack

Polypropylene tubes

■ REAGENT PREPARATION AND ASSAY PROTOCOL

As described in detail below, the 20 µl final AcycloPrime-FP reaction volume consists of:

5 µl of amplified DNA sample

2 µl of PCR Clean-Up Reagent, 1×

13 µl of AcycloPrime Mix

Each analysis uses a single microplate well, with each step performed sequentially by adding the next set of reagents. To avoid sample transfer steps, the microplate must be compatible with both the thermocycler and the FP reader.

Amplification of Target DNA

The sequence containing the SNP of interest must be amplified, both to increase its quantity (for sensitivity) and to reduce the complexity of the target (for specificity).

TABLE 19.3. PCR Conditions

PCR Conditions	Recommended	Maximum Limit
Primers	100–200 nM	1 μM
dNTPs	100–150 μM	200 μM
Genomic DNA	5–10 ng/5 μl	20 ng/5 μl
Amplicon size	80–200 bp	1000 bp
Cycles	35	45

Although other amplification methods can be used, these instructions assume amplification by AmpliTaq Gold® PCR in a total volume of 5 μl. Even if larger amplification reactions are performed, only 5 μl of the solution should be used in subsequent steps.

Most assay problems result from issues in the PCR reaction. The amount of PCR reaction product desired for subsequent steps is 50–500 fmol in 5 μl. For a 200-bp amplicon, this will be 6–60 ng of product in a 5-μl PCR reaction. The best results will be obtained if the amount of product in each sample is consistent.

If too little PCR product is produced, insufficient signal will be produced to see the polarization; if too much PCR product is produced, misincorporation errors can sometimes result. If multiple bands are produced, the amount of specific target sequence available for primer extension is decreased and there is a possibility that nonspecific sites may be represented that can bind the SNP primer. We recommend the PCR conditions shown in Table 19.3 as robust and reliable for most assays.

Typical PCR cycling conditions are as follows: an initial denaturation and enzyme activation step at 95°°C for 10 min, followed by 25–45 cycles consisting of 94°C for 15 s (denaturation), 55°C for 30 s (annealing of primer), and 72°C for 60 s (extension). After a final primer extension step at 72°C for 10 min, the reactions are cooled to 4°C until further use.

Processing of PCR Product

Primers and dNTPs used in the PCR reaction must be eliminated because they will interfere with the AcycloPrime-FP reaction. This is accomplished by enzymatic digestion with the PCR Clean-Up Reagent, 10×, and the PCR Clean-Up Dilution Buffer. Dilute the PCR Clean-Up Reagent 10-fold with the PCR Clean-Up Dilution Buffer provided in the kit to prepare PCR Clean-Up Reagent, 1×.

NOTE: The dilution step should be performed just before use. If it is necessary to dilute sooner, keep the diluted reagent cold until use to avoid deterioration of the enzymes. The diluted reagent is stable for 1–2 days at 4°C, but loses up to 30% of its activity in 4 h at room temperature.

Add 2 μl PCR Clean-Up Reagent, 1×, to each 5-μl amplification reaction. Incubate at 37°C for 60 min.

Inactivate the enzymes in the PCR Clean-Up Reagent by heating at 80°C for 15 min.

AcycloPrime-FP Mix

For storage, SNP primers should be dissolved at 100 μM in TE Buffer (10 mM Tris_HCl, pH 7.6 containing 0.1 mM EDTA). Before use in the AcycloPrime-FP

TABLE 19.4. Reaction Mixes

Reagent	μl/Single Reaction	μl/96 Wells	μl/384 Wells	μl/1K Reactions	μl/10K Reactions
AcycloPol	0.05	6	0.022	0.055	0.525
10× Reaction buffer	2	240	0.88	2.2	21
AcycloTerminator mix	1	120	0.44	1.1	10.5
SNP primer (10 μM)	0.5	60	0.22	0.55	5.25
Water	9.45	1134	4.158	10.395	99.225
Total volume	13 μl	1560 μl	5.72 ml	14.3 ml	136.5 ml

reaction, dilute the 100 μM SNP primer 1/10 with water to give a concentration of 10 μM. If a different primer concentration is used, adjust the quantities of primer and water in the AcycloPrime Mix appropriately.

For each SNP primer used, prepare the appropriate amount of AcycloPrime-FP Mix as described in Table 19.4 (some excess allowed for losses in liquid transfer):

AcycloPrime-FP Protocol

1. Add 13 μl of AcycloPrime-FP Mix to each well containing 7 μl of amplified and processed target DNA.
2. After an initial denaturation at 95°C for 2 min, perform 10–30 thermal cycles* consisting of 95°C for 15 s and 55°C for 30 s.
3. Read the fluorescence polarization in an appropriate instrument.

***NOTE:** The AcycloPrime-FP reaction is very stable. This property can be used to establish the optimum conditions for a particular assay by, for example, reading the result after 10 cycles and then performing additional cycles before reading the plate again, until a sufficient difference between the negative controls and the samples is achieved. Many users find that 25 cycles is an optimum number for many assays. When finished, bring to room temperature and read. If not reading immediately, hold at 4°C. If desired, finished plates may be stored at 4°C *in the dark* for at least several weeks without significant loss of signal.

Calculations

The FP value is calculated in the instrument software by the following formula:

$$\text{mP} = 1000 \times [I_{vv} - I_{vh}]/[I_{vv} + I_{vh}]$$

where I_{vv} is the emission intensity measured when the excitation and emission polarizers are parallel and I_{vh} is the emission intensity measured when the emission and excitation polarizers are oriented perpendicular to each other.

Allele scoring: As shown in the next section, a plot of the polarization of R110 vs. the polarization of TAMRA (Fig. 19.2) should show four data clusters representing the two homozygotes, the heterozygotes, and the negative controls. The separation between the clusters should be at least 50 mP to determine the genotypes.

Note that the genotype on the target strand is the complement of the labeled terminator incorporated into the SNP primer.

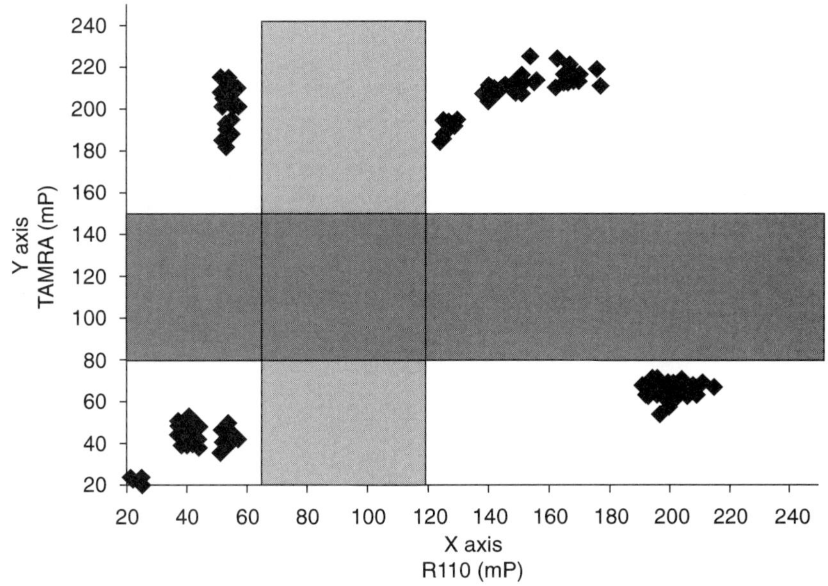

FIGURE 19.2. Plot of the polarization of R110 vs. TAMRA.

Safety Considerations

10× Reaction Buffer contains tris(hydroxymethyl)aminomethane, which can be irritating to eyes and skin. AcycloPol and the PCR Clean-Up Reagent, 10×, contain glycerol, which can be irritating to eyes and skin. Target organ: kidney.

■ TROUBLESHOOTING

Amplification Problems

Many of the problems encountered in performing AcycloPrime-FP occur in the sample DNA amplification steps. It is important to follow accepted guidelines for the selection of PCR primers. Best results are obtained with short amplicons of 200 bp or less. This minimizes the likelihood of a cross-reacting sequence in the PCR product and improves the PCR yield. It is usually sufficient to begin the amplification with 5 ng of high-quality genomic DNA. If the DNA is of lower quality, it may be necessary to start with 10 ng or more to achieve the necessary PCR yield.

The PCR conditions recommended in section on **Processing of PCR Product** will limit the total productivity of the PCR reaction to a level that is unlikely to permit misincorporation. Because limiting conditions will reduce the PCR efficiency, it is recommended that at least 35 PCR cycles be performed to ensure that sufficient product is made.

Normally it is not necessary to analyze the products of a PCR reaction. For troubleshooting purposes, it is helpful to demonstrate that the production of the right amount of product (50–500 fmol/5 μl PCR reaction) is present by running a gel. About 0.5 ng of product will give a detectable band. There also should be only a single predominant band.

TABLE 19.5. Synthetic Oligo Template and Primer Sequences

Oligo	Sequence (5′ → 3′)
A Template	ATTGGATTATTTGTAACTC**A**AAGGATAAGTGCATAAGGGG
C Template	ATTGGATTATTTGTAACTC**C**AAGGATAAGTGCATAAGGGG
G Template	ATTGGATTATTTGTAACTC**G**AAGGATAAGTGCATAAGGGG
T Template	ATTGGATTATTTGTAACTC**T**AAGGATAAGTGCATAAGGGG
SNP Primer	CCCCTTATGCACTTATCCTT
Reverse SNP Primer	ATTGGATTATTTGTAACTC

AcycloPrime-FP

It is desirable to run several controls in each reaction. Replacing DNA in some reactions with water provides negative controls. Other desirable controls include known homozygous and heterozygous samples for the SNP of interest. It is also possible to use synthetic oligonucleotide templates with a corresponding SNP primer to demonstrate appropriate behavior of the AcycloPrime-FP reaction.

For particular types of sequences, it may be necessary to modify the recommended primer extension conditions. For example, if the SNP sequence being studied is particularly high in AT, it may not be possible to design a SNP primer with a T_m of 60°°C or greater. In such a case, it may be necessary to lower the annealing temperature to a value 5 or more degrees below the T_m. Because of the high representation of the target sequence in the PCR reaction product, the reduced hybridization specificity caused by decreasing the annealing temperature does not cause a corresponding loss of specificity in the overall reaction.

A set of 40-mer synthetic oligonucleotide templates derived from reference SNP rs656 has been used in our laboratories, and their sequences are shown in Table 19.5. By using the complements of the chosen template sequences, it is possible to create dsDNA targets and/or evaluate the reverse primer reaction. Many alternative sequences should work just as well as these.

FP is a calculated value that includes the signals from both the free and the incorporated labeled AcycloTerminators. FP will reach its largest value when all of the labeled AcycloTerminator is incorporated into SNP primer. Unincorporated labeled AcycloTerminators contribute to the background signal. Therefore, it is desirable to push the reaction to completion.

However, when the correctly base-paired AcycloTerminator is exhausted, pushing the reaction too much can lead to the nonspecific incorporation of whatever labeled AcycloTerminators remain. Misincorporation typically results in poor separation of data clusters and can result in all samples appearing to be heterozygotes.

Reading plates after various numbers of AcycloPrime-FP cycles allows stopping the process before misincorporation becomes a problem. It has occasionally been found useful to double the AcycloTerminator Mix in the primer extension reaction for a difficult SNP. If the suggested PCR conditions are used, the likelihood of misincorporation is reduced because the amount PCR product produced is automatically limited to the useful concentration range.

We do not encourage the practice of adding unlabeled terminators to effectively dilute or compete with the labeled AcycloTerminators in the reaction as has been suggested by others. In general, we urge the optimization of the process by controlling the extent of amplification in the PCR by appropriate assay composition and in the AcycloPrime-FP reactions through optimizing the number of thermal cycles.

■ LICENSING

The AcycloPol mutant DNA polymerase is sold under license from New England BioLabs®, Inc. and is covered by U.S. Patents 5,352,778, 5,500,363, and 5,756,334. Purchase of this kit does not include a license to perform PCR, but purchase of a product from an approved vendor includes such a user license. The methods of use for the PCR Clean-Up Reagent are covered under U.S. Patents 5,741,676 and 5,756,285 issued to USB as well as other patents pending.

■ SUPPLIER

Worldwide Headquarters
PerkinElmer Life Sciences
549 Albany Street
Boston, MA 02118-2512
USA
800-551-2121
techsupport@perkinelmer.com

European Headquarters
PerkinElmer Life Sciences
Imperlastraat 8
BE-1930 Zaventem
Belgium
techsupport.europe@perkinelmer.com

■ REFERENCES

1. Gardner AF, Jack WE (2002) Acyclic and dideoxy terminator preferences denote divergent sugar recognition by archaeon and Taq DNA polymerases. Nucleic Acids Res 30: 605–613.
2. Chen X, Levine L, Kwok P-Y (1999) Fluorescence polarization in homogeneous nucleic acid analysis. Genome Res 9: 492–498.
3. Hsu TM, Chen X, Duan S, Miller RD, Kwok P-Y (2001) Universal SNP genotyping assay with fluorescence polarization detection. BioTechniques 31: 560–570.

20

PYROSEQUENCING

Pyrosequencing AB

Based on the manufacturer's demonstration at the HUGO Mutation Detection Course, Montpellier, September 2002

■ INTRODUCTION

To perform DNA analysis with Pyrosequencing™ technology, PCR products must be processed to yield single-stranded DNA with annealed sequencing primer. The simple and robust sample preparation process, outlined below, produces high-quality DNA from crude PCR reactions without prior purification (1). The PCR product, with one strand 5Åå-biotinylated, is captured on streptavidin-coated beads. Subsequently, the nonbiotinylated DNA strand is removed by alkali treatment and washed away from the biotinylated strand attached to the bead. The washing step also serves to neutralize the pH. Finally, the beads with attached DNA are transferred to buffer containing the sequencing primer and the primer is annealed to the single-stranded template by heating and cooling. The DNA is now ready for analysis.

■ TEMPLATE PREPARATION WITH STREPTAVIDIN SEPHAROSE™ HP (Fig. 1)

Additional Materials Needed

- Mixer/shaker for microtiter plates (room temperature)
- Heating block (80°C)
- Vacuum manifold basic kit (e.g., Millipore no. MAVM 096 0R)
- Vacuum pump (e.g., Millipore Vacuum/Pressure Pump, 220 V/50 Hz, no. XX55 220 50)

Guide to Mutation Detection, Edited by Graham R. Taylor and Ian N. M. Day.
ISBN 0-471-23444-3

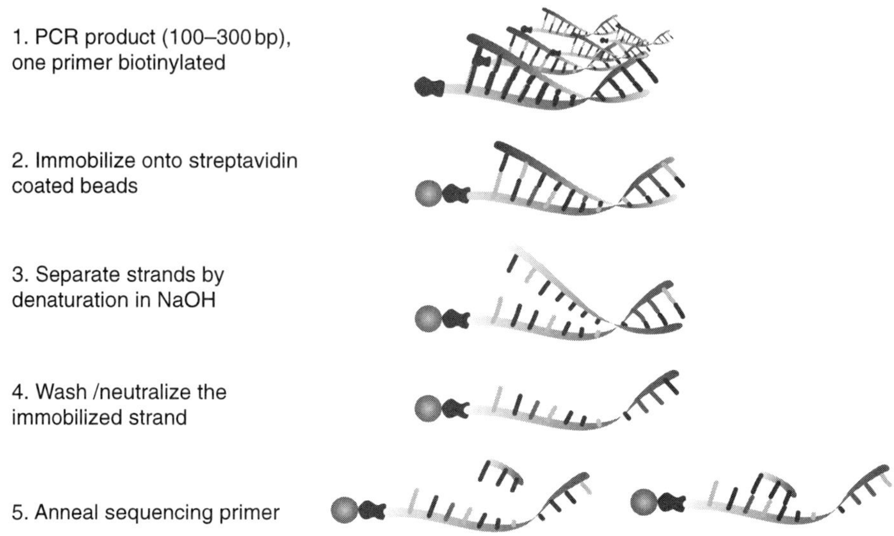

FIGURE 20.1. Template preparation for the pyrosequencing method.

- PSQ 96 Plate Low
- PSQ 96 Reagent Cartridge (40-0022)
- PSQ 96 Sample Prep Thermoplate
- 96-Well PCR plate
- Strip lids or sealing tape

Additional Reagents Needed

- Biotinylated PCR template
- Streptavidin Sepharose HP (5 ml, no. 17-5113-01, Amersham Biosciences, Uppsala, Sweden)
- PSQ 96 Sample Preparation Kit 10×96 (40-0007)
- Binding buffer
- Denaturation solution
- Washing buffer
- Annealing buffer
- 96-Well filter plates
- Sequencing primer
- High-purity water (MilliQ 18.2 MΩ × cm or equivalent)

Capture of PCR Product to Beads

Biotinylated PCR products are immobilized on streptavidin-coated Sepharose beads (streptavidin Sepharose HP, Amersham Biosciences). Parallel immobilization of several samples can be performed in a sealable 96-well PCR plate.

All steps are performed at room temperature unless otherwise stated.

- Allow all solutions to reach room temperature before starting.
- Shake the bottle with streptavidin Sepharose beads until a homogenous solution is obtained.
- Transfer the total required amount of streptavidin Sepharose beads (4 μl per sample) to a tube.
- Add binding buffer to a volume equivalent to the total volume of the PCR product to be used (10–50 μl per sample). Mix.
- In a 96-well PCR plate, mix 10–50 μl of a well-optimized PCR reaction (at least 50% yield of a reaction based on 0.2 μM PCR primers) with an equal volume (1:1) of the binding buffer-streptavidin Sepharose bead mix.
- Using a mixer/shaker, incubate at room temperature for 10 min, agitating constantly to keep the beads dispersed.

Tip: During the immobilization incubation, prepare the sequencing primer-annealing buffer solutions needed for primer annealing.

Strand Separation

- Transfer the immobilized templates to a 96-well filter plate by pipetting.
- Apply vacuum to remove all liquid.
- Release the vacuum and add 50 μl denaturation solution.
- Incubate 1 min. Apply vacuum to remove the denaturation solution.
- Wash the beads twice with 150 μl washing buffer. Make sure that all wash solution is removed before adding new solution.

Tip 1: When using only a few wells, covering the rest of the plate can accelerate the filtration.

Tip 2: Excess wash liquid may remain underneath the wells after the wash step. Remove the liquid by gently tapping the filter plate on a piece of lint-free tissue.

Primer Annealing

Choose one of the procedures below:

Procedure A

- Resuspend the beads with the immobilized template by adding 45 μl annealing buffer per well and pipetting up and down 4–5 times.
- Transfer 35 μl from the filter plate to a PSQ 96 Plate Low.
- Add 5 μl of sequencing primer (3 μM in annealing buffer)

Procedure B

- Dilute the sequencing primer to 0.40 μM in annealing buffer.
- Add 50 μl per well and resuspend the beads with the immobilized template by pipetting up and down 4–5 times.
- Transfer 40 μlL from the filter plate to a PSQ 96 Plate Low.

Tip: If handling a full 96-well plate, only resuspend the beads in one column at a time and transfer them to a PSQ 96 Plate Low before moving on to the next row.

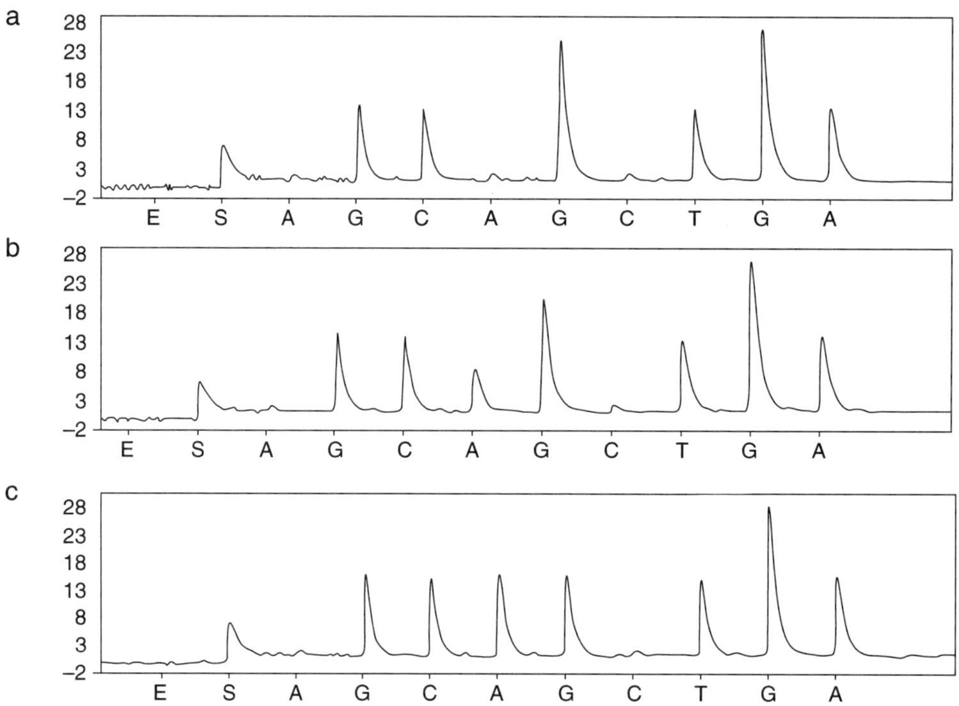

FIGURE 20.2. Typical pyrosequencing results. The DNA sequence is GCA/GGTGGA. (a) Homozygous GG; (b) heterozygous AG; (c) homozygous AA.

To recover all beads, it is advisable to tilt the filter plate slightly during resuspension and transfer.

- Heat the plate with the samples at 60°C for 5 minutes with the PSQ 96 Sample Prep Thermoplate. Let the samples cool to room temperature.
- After reaching room temperature, continue with the sequencing reaction (Fig. 20.2).

PREPARATION OF MULTIPLEX TEMPLATES WITH STREPTAVIDIN SEPHAROSE HP

Capture of PCR Product to Streptavidin Sepharose Beads

- Allow all solutions to reach room temperature before starting.
- Shake the bottle with streptavidin Sepharose beads until a homogenous solution is obtained.
- Transfer the total required amount of streptavidin Sepharose beads (4 μl per well) to a tube.
- Add binding buffer to a volume equivalent to the **total** volume of the PCR products to be used (10–50 μl per sample). Mix.

Depending on the type of multiplex, choose one of the procedures below.

Pooled Simplex PCR Products

- In a 96-well PCR plate, pool 10–30 µl of each PCR product and mix the pooled PCR products with an equal volume (1:1) of the binding buffer-streptavidin Sepharose bead mix.

Multiplexed PCR Products or One PCR Product with Several SNP Positions

- In a 96-well PCR plate, mix 10–50 µl PCR product with an equal volume (1:1) of the binding buffer-streptavidin Sepharose bead mix.
- Using a mixer/shaker, incubate at room temperature for 10 min, agitating constantly to keep the beads dispersed.

Tip: During the immobilization incubation, prepare the sequencing primer-annealing buffer solutions needed for primer annealing.

Strand Separation

- Transfer the immobilized templates to a 96-well filter plate by pipetting.
- Apply vacuum to remove all liquid.
- Release the vacuum and add 50 µl denaturation solution.
- Incubate 1 min. Apply vacuum to remove the denaturation solution.
- Wash the beads twice with 150 µl washing buffer. Make sure that all wash solution in the first wash step is removed before adding new solution.

Tip 1: When using only a few wells, covering the rest of the plate can accelerate the filtration.

Tip 2: Excess wash liquid may remain underneath the wells after the wash step. Remove the liquid by gently tapping the filter plate on a piece of lint-free tissue.

Primer Annealing

Choose one of the procedures below:

Procedure A

- Resuspend the beads with the immobilized template by adding 45 µl annealing buffer per well and pipetting up and down 4–5 times.
- Transfer 35 µl from the filter plate to a PSQ 96 Plate Low.
- Add 5 µl of sequencing primer mix (3 µM of each primer in annealing buffer)

Procedure B

- Dilute the sequencing primers to 0.40 µM of each primer in annealing buffer.
- Add 50 µl per well and resuspend the beads with the immobilized template by pipetting up and down 4–5 times.
- Transfer 40 µl from the filter plate to a PSQ 96 Plate Low.

Tip: If handling a full 96-well plate, only resuspend the beads in one column at a time and transfer them to a PSQ 96 Plate Low before moving on to the next row.

To recover all beads, it is advisable to tilt the filter plate slightly during resuspension and transfer.

- Heat the plate with the samples at 60°C for 5 min with the PSQ 96 Sample Prep Thermoplate. Let the samples cool to room temperature.
- After reaching room temperature, continue with the sequencing reaction.

DISPENSATION ORDER

If the beginning of a sequence is constant and known for a certain SQA application, the obtained read length may be improved by applying a sequence-directed dispensation order for these bases before the following cyclic dispensation scheme. These known bases may then also be entered as reference peaks in the software, to act as an internal control in the base-calling analysis and further improve the base-calling performance.

Single-Stranded DNA Binding Protein

SSB (single-stranded DNA binding protein) binds to the single-stranded DNA and prevents structures like hairpin loops, duplexes, etc. from forming. SSB, which improves the general sequence quality and resolves most DNA secondary structures, is thus an excellent tool, especially when long read lengths are desired. However, please note that the signal intensities for some assays may decrease when SSB is added to the reaction. To maintain efficient sequencing primer annealing in the presence of SSB it is advisable to increase the T_m of the sequencing primer by adding some extra nucleotides at the 5′-end.

Add 3 µl diluted SSB to the sequencing reaction after the annealing step. For detailed instructions, please refer to SSB Pyrosequencing Grade Kit, Instructions for Use.

If the standard amount of SSB fails to completely resolve the secondary structures of the template, try to dilute the SSB 1:2 (1+1) in SSB dilution buffer. Add 3 µl of this dilution to the sequencing reaction after the annealing step.

REFERENCE

1. Ahmadian A, Gharizadeh B, Gustafsson AC, Sterky F, Nyren P, Uhlen M, Lundeberg J (2000) Single-nucleotide polymorphism analysis by pyrosequencing. Anal Biochem 280: 103–110.

21

SINGLE NUCLEOTIDE POLYMORPHISMS (SNPs) BY ELECTRONIC HYBRIDIZATION

Nanogen

Based on the manufacturer's demonstration at the HUGO Mutation Detection Course, Montpellier, September 2002

■ INTRODUCTION

The following application allows the determination of SNP genotypes based on microarray electronic transport and hybridization. SNP genotypes are becoming increasingly important in disease diagnosis, pharmacogenomics, functional genomics, and forensics.

■ METHODS

Materials Required

Oligos and amplicons for Amplicon-Down protocol
Amplicon sample using biotin-bound primer
Reporter oligo labeled with appropriate fluorophore (CY3®® or CY™ 5)
Stabilizer oligo (unlabeled)

Solutions and Reagents

MilliQ H_2O (dH_2O) with a resistivity > 18 MΩ
High-salt buffer—50 mM $NaPO_4$, pH 7.0, 500 mM NaCl
Low-salt buffer—50 mM $NaPO_4$, pH 7.0

Guide to Mutation Detection, Edited by Graham R. Taylor and Ian N. M. Day.
ISBN 0-471-23444-3

100 mM histidine
50 mM histidine
Standard amplification reaction reagents

Equipment

96-Well plates
p-200 Pipette
p-200 Pipette tips
Thermocycler
Heating block
Desalting columns (one for each amplified sample)

Note: Nanogen recommends using the Bio-Rad-G6 Biospin™ columns, Qiagen's QIAquick™ PCR Purification Kit, Amicon® MultiScreen PCR, Microcon®-PCR, or YM-50 Microcon® Centrifugal filter units; other desalting products may be used, but Nanogen cannot guarantee the results obtained on the NanoChip Instrument. See Appendix A for modified desalting protocols.

Solution Preparation

Prepare the following solutions:

50 mM histidine:	Prepare 500 ml for the loader and place in buffer bottle in the loader buffer drawer.
100 mM histidine:	Prepare 5 ml for sample preparation.
High-salt buffer:	Prepare 250 ml 50 mM $NaPO_4$, pH 7.0, 500 mM NaCl for the reader and place in Bottle B in the reader buffer drawer. Identify as 50/500.
Low-salt buffer:	Prepare 250 ml 50 mM $NaPO_4$, pH 7.0 for the reader and place in (50 mM $NaPO_4$, pH 7.0) Bottle C in the reader buffer drawer. Identify as 50 mM $NaPO_4$.
dH_2O:	Place 500 ml of dH_2O in water bottle for the loader.

Note: Store the 50 mM histidine at 4°C in the dark after using for the day; all other solutions may be kept at room temperature. Check the conductivity of the 50 mM histidine before each use (should not exceed 100 μS). Replace the 50 mM histidine once a week and replace other solutions once a month.

SNP Protocols

There are two types of SNP protocols that can be run on the NanoChip Molecular Biology Workstation: Amplicon-Down and Capture-Down. Each protocol contains three components: the amplicon of interest, the stabilizer oligo, and the reporter oligo.

The Amplicon-Down SNP protocol addresses a biotinylated amplicon to the NanoChip and, at the same time, hybridizes a nonbiotinylated stabilizer oligo to the amplicon. This protocol requires less time on the loader than the Capture-Down protocol.

The Capture-Down SNP protocol uses the stabilizer strand as the biotinylated oligo instead of the amplicon. Hence, the stabilizer strand is now a capture oligo as well as a stabilizer. This is useful if the amplicon is not biotinylated. A biotinylated capture oligo is addressed to the chip before amplicon addressing. The biotinylated capture oligo should be designed exactly the same as the stabilizer oligo, except it should terminate with a biotin at the 3′ end. The amplicon should not contain a biotin for the Capture-Down protocol. The reporters will be designed the same as for the Amplicon-Down protocol.

The Amplicon-Down oligo and amplicon design are discussed in detail below. These designs can also be applied for Capture-Down protocols, keeping in mind that the amplicon does not require a biotin but the stabilizer oligo does.

Amplicon and Oligo Design for SNP Applications

Introduction

The SNP protocol contains three components: the amplicon of interest, the stabilizer oligo, and the reporter oligo.

Base-stacking energies, the sharing of electrons between nucleotides along a DNA strand, play an important role in designing each component, especially the stabilizer and reporter oligos. The design of the stabilizer and reporter oligos takes advantage of the target DNA-directed juxtaposition of the 3′ terminal base of the stabilizer and the 5′ terminal base of the reporter. If a reporter sequence is short (<15 bases), direct hybridization will be more stable in the presence of base-stacking interactions across the junction of these two molecules. Each base pair combination contains varying degrees of base-stacking energy. The most stable base-stacking interaction is 5′G:C3′ and the least stable is 5T:A3′.

General Guidelines

- Use commonly available DNA sequence analysis programs to check for secondary structures and homodimers for designing the oligos as well as for the annealing temperatures of amplicon primers.
- Select the highest base-stacking interaction sequence for the capture/stabilizer and reporter oligos, and use the opposite strand for the biotinylated amplification primer.
- Use standard DNA sequence analysis software to establish the %GC content and to determine the melting temperatures for the oligos.
- Purify the oligos with HPLC (oligos synthesized on a 100-nmol scale should provide enough material to analyze at least 100 cartridges).
- Resuspend all oligos in dH_2O (if vendor normally ships oligos in this buffer, request that they are shipped lyophilized or in dH_2O).
- Stabilizer should be diluted to 20 μM and reporter to 50 μM.
- Store all oligo stocks at –70°C.

SNP Components

Amplicon Sample(s)

Ideally, the location of the amplification primers should be chosen to produce an amplicon of less than 150–200 bp in length. In some cases, the system has effectively produced results on amplicons up to 1000 bp in size. However, the fluorescent signals in longer amplicons may be reduced as length is increased. Center the SNP location within the amplicon.

For amplicon-down SNPs, attach a biotin to the 5′ end of one of the amplification primers. The biotin is necessary for attachment to the streptavidin in the permeation layer of the microchip array. The biotinylated amplicons should be complementary to the stabilizer and reporter oligos.

Stabilizer Oligos

Stabilizer oligo sequences contain 25–30 nucleotides following the 3′-end nucleotide of the reporter. The terminal nucleotide at the 5′-end is the base just before the SNP. Length should be varied to obtain a melting temperature (based on %GC) of 70°C.

Before preparing the stabilizer oligo, check for secondary structures (switch to the opposite strand if the stacking interactions are suitable).

Reporter Oligos

Reporter oligos are designed to have the SNP as the terminal base at the 3′-end of the oligo and the CY3/CY5 fluorophore on the end opposite the base-stacking interaction. For this description, the CY3/CY5 fluorophores are phosphoamidites, which are placed on the 5′-end of the oligo.

Note: If an application requires fluorophores conjugated to the 3′-end, these guidelines still apply, but the relative positions of the capture/stabilizer and reporter oligos would need to be switched.

Analyze the location of the reporter sequence to determine the optimal base-stacking interaction of the reporter oligo with the stabilizer oligo. Avoid base-stacking pairs (5′→3′) of A:T, T:T and T:C (with A:T being the least desirable combination). If these pairings are encountered, switch strands and orient the reporter oligo in the opposite direction.

The reporter oligo contains the SNP as the terminal base on the 3′-end with nucleotides added in the 5′ direction until the melting temperature (based on %GC) reaches 35–37°C (approximately 10–13 bases). WT and mutant reporters must be labeled with different fluorophores (WT CY3 and mutant CY5). If multiple loci are intended for a single cartridge, match melting temperatures as closely as possible. Keep the WT and mutant temperatures as close as possible—the user may have to add extra bases on the mutant to accomplish this.

Note: Working stocks of the reporter oligo should be stored in dark tubes, because of the light sensitivity of the fluorophore, and should be prepared in small volumes (~50 μl) to avoid damage of multiple freeze/thaw cycles. Keep reporter oligos on ice when using.

Amplicon Sample Preparation

For instructions on preparing and desalting amplified samples, please refer to Appendix A in the Applications guide.

Running SNP Protocols on the NanoChip Molecular Biology Workstation

The Amplicon-Down and Capture-Down protocols each consist of three steps: sample/stabilizer loading, reporter hybridization, and thermal discrimination. The two protocols only differ in the sample/stabilizer loading step. The sample/stabilizer loading step will be discussed separately for each protocol, but the reporter hybridization and thermal discrimination steps described will apply for both protocols.

SNP Amplicon-Down Amplicon/Stabilizer Oligo Address

1. Turn on the loader (see Basic Operation—the Loader).
2. Select an appropriate protocol in the loader protocols for a SNP assay. Edit the protocol as necessary (see Section IV—Creating and Editing Loader MAP files).
3. If a new protocol needs to be written, see Section IV—Creating and Editing Loader MAP files. For Amplicon-Down SNP analysis, the user will need to write a target/amplicon map corresponding to the amplicon/stabilizer mixtures in the 96-well plates.

A background control should be run on at least three test sites per cartridge. The background control should be a nonspecific biotinylated oligo.

Note: It is also very important to run a heterozygote control on each cartridge. A known heterozygote is required to normalize red and green lasers and the difference in hybridization efficiency between the green- and red dye-labeled reporter oligos.

4. Combine the desalted amplicon and the stabilizer oligo with 100 mM histidine and dH_2O to an appropriate volume, depending on the number of cartridges being used. Final amplicon concentration is 5–40 nM. Final stabilizer oligo concentration is 250 nM. Final concentration for histidine is 50 mM.
5. Denature the mixture at 95°C for 5 min.
6. Spin down the mixture in a microcentrifuge.
7. Load the sample mixture into the 96-well plate. Mixtures from several samples can be loaded into the 96-well plate and addressed to the NanoChip in one loader session. Load the plate into the loader.
8. Insert the cartridge(s) into the loader. Close door
9. Run a conductivity test from the Loader Runtime Toolbar and then run the selected protocol.
10. Remove the cartridge(s) from the loader when the protocol is complete.
11. Proceed to the Reporter Hybridization and Thermal Discrimination steps.

SNP Capture-Down Stabilizer Oligo and Amplicon Address

1. Turn on the loader (see Basic Operation—the Loader).
2. Select an appropriate protocol in the loader protocols for a SNP assay. Edit the protocol as necessary (see Section IV—Creating and Editing Loader MAP files).
3. If a new protocol needs to be written, see Section IV—Creating and Editing Loader MAP files. For Capture-Down SNP analysis, the user will need to write two maps within the MAP file:
   - **a.** A capture map to address the biotinylated capture oligos (this map should be run first)
   - **b.** A target map for amplicon addressing

Note: Do not address more than ten pads simultaneously. If an individual capture oligo is to be used over a larger portion of the array, several wells and addressing steps of the capture oligo will be needed for the capture map.

A background control should be run on at least three test sites per cartridge. The background control should be a nonspecific biotinylated oligo.

Note: It is also very important to run a heterozygote control on each cartridge. A known heterozygote is required to normalize red and green lasers and the difference in hybridization efficiency between the green- and red dye-labeled reporter oligos.

1. Combine the desalted amplicon with 100 mM histidine and dH_2O to an appropriate volume, depending on the number of cartridges being used. Final amplicon concentration is 5–40 nM. Final concentration for histidine is 50 mM.
2. Denature the mixture at 95°C for 5 min.
3. Spin down the mixture in a microcentrifuge.
4. Dilute the capture (stabilizer) oligo as follows:

Combine the stabilizer oligo with 100 mM histidine and dH_2O to an appropriate volume, depending on the number of cartridges being used. The final concentration for the stabilizer oligo is 250 nM. The final concentration for the histidine is 50 mM.

1. Load the stabilizer oligos and denatured amplicons into the 96-well plate, using the wells designated in the MAP file. Load the plate into the loader. Close loader door.
2. Insert the cartridge(s) into the loader.
3. Run a conductivity test from the Loader Runtime Toolbar and then run the selected protocol.
4. Remove the cartridge(s) from the loader when the protocol is complete.

Proceed to the Reporter Hybridization and Thermal Discrimination steps.

SNP Reporter Hybridization

1. Wash the cartridge with the high-salt buffer several times to remove the histidine buffer. Flush each chamber with the high-salt buffer to remove any remaining histidine buffer. To remove existing buffer, push air through the tubing connected to each chamber.

Note: Any histidine buffer remaining in the cartridge may cause poor reporter hybridization.

2. Dilute both WT and mutant reporters to 500 nM in the high-salt buffer to a final volume of 50 µl.
3. Pipette the reporter solution over the array.
4. Incubate the cartridge for 3 min at room temperature.
5. Wash the cartridge 4 times in high-salt buffer.
6. Add 65 µl of high-salt buffer over the cartridge array. Make sure that the buffer is over the array by pushing air through the tubing if necessary.

Note: Buffer must be present over the array for the automatic focus feature on the reader to operate properly.

7. Image the cartridge using a FLUORESCENCE process to determine hybridization efficiency, correct PMT voltage gain, and accumulation time (if necessary, adjust the PMT voltage gain and accumulation time and scan the cartridge again—if changes to the default settings are necessary, make sure to change the values for all the imaging stages in the script of the NAF file).
8. Continue to SNP Thermal Discrimination section.

SNP Thermal Discrimination

Discrimination between a matched and a mismatched reporter is achieved by using temperature to denature the capture/target/reporter complex. The mismatched reporter will denature at a lower temperature than the matched reporter. Therefore, after thermal discrimination is achieved, the wild-type allele complex will contain a CY3 reporter, the mutant allele complex will contain a CY5 reporter, and the heterozygote complex will contain an equal amount of both reporters.

The optimal temperature for thermal discrimination is the temperature at which a significant amount of the mismatch reporter denatures but the match reporter does not. For each locus, the optimal temperature will vary and must be optimized.

To determine the optimal temperature for a new locus, run known homozygous and heterozygous samples if possible. The optimal temperature is approximately 2° below the melting temperature of the least stable of the two reporter oligos. The optimal temperature should be determined by using a temperature ramping protocol, which starts at a temperature 2°C below the melting temperature of the least stable reporter and is increased in 1° increments for 4°C. For each temperature increment, ramp the cartridge to the desired temperature, wash in low-salt buffer, return the cartridge to room temperature, and image with both lasers. Once the optimal temperature is determined, the assay can be run at a single temperature using the following steps as a guideline:

1. Wash the cartridge with low-salt buffer on the reader using a Fill Reagent C step. Repeat the wash.
2. Take an initial image of the cartridge with both lasers.
3. Ramp the cartridge to optimal temperature.
4. Wash the cartridge 2 times with low-salt buffer (using a Fill Reagent C step).
5. Return the cartridge temperature to ambient temperature (24°C).
6. Image the cartridge again with both lasers.

Note: To achieve cleaner discrimination repeat steps 3–6, or increase the temperature 1°C from the optimal temperature and then repeat steps 3–6.

If analyzing multiple loci, remove the remaining reporters by washing in low-salt buffer for 2 minutes at 50°C. Although multiplexing is possible, combinations should be compared to monoplex results to determine whether cross-hybridization is taking place.

22

MANUAL SSCP AND HETERODUPLEX ANALYSIS GELS

Mark D. Robinson
Regional DNA Laboratory, St. James's University Hospital, Leeds, UK

■ INTRODUCTION

Single-strand conformational polymorphism (SSCP) (1) detects mutations by the effect of base changes on DNA secondary structure. These conformational changes affect fragment mobility under nondenaturing conditions. SSCP results are dependent on the type of polymer used, buffer type and strength, and temperature. The mobility of double-stranded DNA is reduced by mismatched base pairs, and gel systems similar to SSCP can be used to reveal mutations by heteroduplex analysis (HA) (2, 3). Because the same gel conditions are used, it is possible to analyze single-stranded (SSCP) and double-stranded (HA) DNA on the same gel (4).

Equipment

Cooling water bath (e.g., Grant)
12- to 15-cm polyacrylamide gel system (e.g., Protean, Bio-Rad)
Photographic trays for silver staining
Heated gel dryer (e.g., Bio-Rad)

■ SOLUTIONS

40% Acrylamide Stock

Use Sigma premix (49:1), dissolve as instructed and deionize with 0.5 g of Amberlite MB1 ion-exchange resin (Bio-Rad).

Guide to Mutation Detection, Edited by Graham R. Taylor and Ian N. M. Day.
ISBN 0-471-23444-3

10× TBE

Use Boehringer ready-made solution, have a stock of 1× TBE refrigerated.

0.89 M Tris	107.8 g/liter
0.89 M boric acid	55.0 g/liter
20 mM EDTA	8.2 g/liter OR 40 ml 0.5 M EDTA/liter

10% Ammonium Persulfate

Ammonium persulfate   1 g/10 ml (make up fresh)

Formamide Loading Buffer

Formamide	10 ml
0.5 M EDTA	200 µl
Xylene cyanol	1.5 mg
Bromophenol blue	3 mg

TEMED (*N,N,N*[1],*N*[1]-tetramethylethylenediamine)

50% Glycerol

Glycerol   50 g

Dissolve the glycerol in distilled water and make up to 100 ml.

Silver Staining Solution A

10% ethanol	100 ml
Glacial acetic acid	5 ml

Make up to 1.0 liter with distilled water.

Prepare on day of use. Store at room temperature, can be used up to 5 times.

Silver Staining Solution B

Prepare a 10× stock solution.

Silver nitrate   5.0 g

Dissolve in 500 ml distilled water and store at room temperature in brown bottles in the dark.

The 1× (0.1%) solution (500 ml) can be stored in clear bottles away from strong light and can be reused for staining up to 5 gels.

Silver Staining Solution C

Sodium hydroxide	6.0 g
37% Formaldehyde solution	1.6 ml

Dissolve the sodium hydroxide in 400 ml distilled water and then add the formaldehyde immediately before use.

Make fresh each time and use once only

Silver Staining Solution D

Prepare a 10× stock solution
Sodium carbonate 75 g/l
Make fresh each time and use once only

Flat Capillary Gel Loading Tips

(e.g., Stratatips, available from Stratagene, Cat #410007)

■ GEL PREPARATION

8% SSCP Gel

To make 30 ml of gel:

1. Switch on the cooling water bath set to 8°C
2. Place 6 ml of deionized 40% 49:1 stock acrylamide solution into a small beaker.
3. Add 3 ml of 10× TBE buffer followed by 18 ml distilled water and 3 ml 50% glycerol.
4. Mix well, then add 25 µl of TEMED followed by 200 µl of ammonium persulfate. Mix well.
5. Draw the mixture carefully into a 50-ml disposable syringe without introducing air bubbles.
6. Assemble the glass plates in the casting stand according to the Bio-Rad instruction manual.
7. Insert an appropriate comb, taking care not to trap any air bubbles.
8. Leave the gel for 40 min and monitor for leakages. Care should be taken when inserting the comb into the gel to avoid gel solution spurting out into your eyes.
9. When the gel has set, allow to precool by placing in the gel tank and adding TBE stored in the refrigerator.

The composition for other gels is shown below.

	Volume of gel required		
Reagent	30 ml	40 ml	60 ml
40% (49:1) deionized acrylamide solution	6 ml	8 ml	12 ml
10× TBE electrophoresis buffer	3 ml	4 ml	6 ml
Deionized water	18 ml	24 ml	36 ml
50% glycerol	3 ml	4 ml	6 ml
TEMED	25 µl	35 µl	50 µl
10% Ammonium persulfate	200 µl	265 µl	400 µl

8% Heteroduplex Gels

The compositions are identical to the SSCP gels except that water is substituted for the 50% glycerol.

SAMPLES, PREPARATION, AND ELECTROPHORESIS

Samples for combined SSCP/HA are usually PCR products amplified under standard laboratory conditions. No special preparatory steps are necessary, the only requirement being that the amplification should be strong and specific. A minimum of 10 µl of product is required for combined SSCP/HA electrophoresis. The PCR products should ideally be less than 300 bp, although good results can be obtained with fragments up to 500 bp. Large fragments can be analyzed by cutting at internal restriction sites with appropriate restriction endonucleases. It can be advantageous to run restricted and unrestricted DNA simultaneously, provided the complete fragment is below 500 bp, because the sensitivity of mutation detection can be compromised close to the restriction site.

ENDONUCLEASE DIGESTION

When restriction digestion is required, place 60% of the reaction mixture in a separate microcentrifugation tube and digest according to the supplier's instructions.

1. Add an equal volume of formamide loading buffer with sample and mix well.
2. Preload the gel with 1–2 µl of the sample. Electrophoresis MUST NOT be commenced at this stage.
3. Place the remainder of the samples on a heated block or thermal cycler set at 95°C for 3 min and snap cool in a bath of crushed ice to prevent reannealing of the single-stranded DNA.
4. Reload the wells with 6–10 µl of each sample using thin gel loading tips (the optimum volume will depend on the size of the wells and the amplification efficiency). Load a DNA size ladder into one lane of the gel to facilitate orientation.
5. Carry out the electrophoresis at 180–500 V (depending on fragment size and run time). Double-stranded DNA and xylene cyanol comigrate, but there can be considerable sequence-dependent variability. Optimum electrophoresis times are best determined empirically.

SILVER STAINING OF POLYACRYLAMIDE GELS

Slide the spacers from between the glass plates and gently prise the plates apart; avoid the use of metal implements that may scratch or chip the glass. Ensure that the gel is attached to the lower plate before removing the upper plate. Place the glass plate with gel attached into a staining tray. Very carefully lift the bottom edge of the gel and fold over a 10-cm length; repeat this action until the gel is completely rolled up. The gel can now be easily dislodged from the plate into the staining tray and stained according to the following protocol:

1. Fix the gel for 8 min in Solution A, decant.
2. Impregnate with silver by immersing gel in solution B for 15 min, rinse twice with distilled H_2O.

3. Reduce silver by immersing gel in freshly made Solution C for 15–20 min, decant.
4. Stop the silver development by addition of solution D, immerse for 10 min, decant.
5. Rinse gels in distilled H_2O and place between two sheets of cellophane membrane backing (Bio-Rad).
6. Dry under vacuum for 2 h at 80°C.

Guide to Interpretation

1. Single-stranded DNA (ssDNA) has a much lower mobility than double-stranded DNA (dsDNA) in the type of native acrylamide gels used for SSCP/HA. Obviously, the further a band migrates the greater the potential for resolving small mobility shifts, but the limiting factor in any combined SSCP/HA gel is the need to retain dsDNA on the gel. As a rough guide, on an 8% (49:1) gel, xylene cyanol migrates at approximately 180 bp.
2. The silver staining technique tends to stain ssDNA, which forms the SSCPs, a brown/orange shade. The dsDNA, which can form heteroduplexes, is stained dark gray/black.
3. SSCPs from a homozygote typically give rise to two distinct migrating bands corresponding to the two complementary DNA strands. Often more than two SSCP bands occur, some of which may stain more weakly than others. Providing the PCR amplification has been clean and specific, these bands are most likely to be due to alternative stable configurations.
4. Rarely there is no discrete SSCP band formed by one or both ssDNA strands. If this occurs a faint, in-track, smear is usually observed. This corresponds to the fragment adopting a whole series of stable configurations, each of slightly different mobility. The presence of a mutation often destroys this delicate balance, creating a SSCP band of typical appearance.
5. Occasionally a fragment produces a single SSCP band. This occurrence is due to the strand and its complement having coincidentally similar mobilities and should not compromise sensitivity.
6. When a genuine sequence difference gives rise to a SSCP the intensity of staining of each individual band will be reduced relative to other samples of the same concentration, as the same quantity of DNA is distributed over a greater gel volume. This reduction in intensity can be used to justify the exclusion of artifacts that may have originated from spurious amplification.
7. Excess primers reannealing to the ssDNA after gel loading can cause a third set of bands intermediate in mobility to the ssDNA and dsDNA. Their pattern of migration usually shadows that of the corresponding SSCPs, although they tend to stain, like dsDNA, gray/black. They can serve as a useful adjunct to interpretation.
8. Heteroduplexes usually manifest as a band, or bands, associated with the dsDNA but with a lower mobility. They typically stain less intensely than the homoduplexes.
9. Sometimes the homoduplexes themselves show mobility shifts. These shifts are usually caused by an insertion or deletion mutation that alters the molecular weight. This being the case, there are usually clearly visible heteroduplexes.

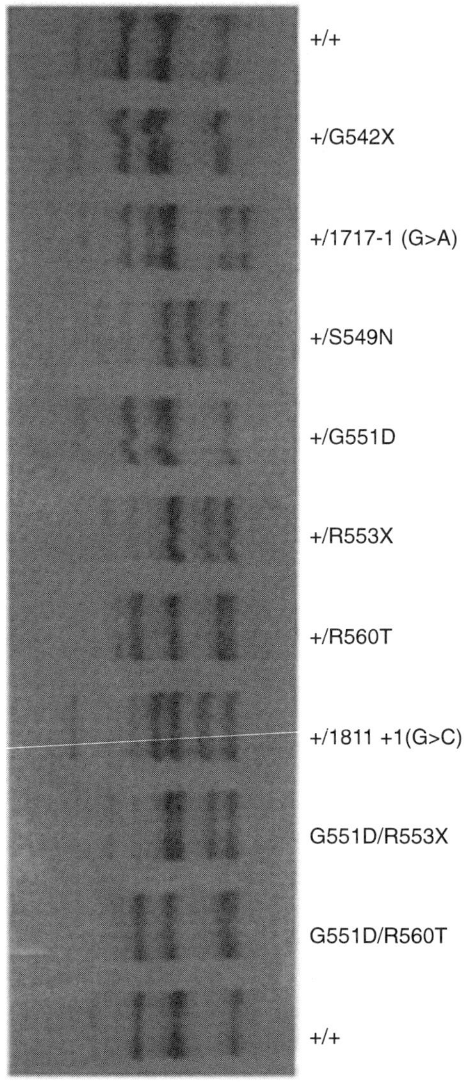

FIGURE 22.1. Silver-stained SSCP gel of a series of mutations in *CFTR* exon 11 flanked by wild-type amplicons.

Some single base substitutions can also give a marked homoduplex mobility shift (see Fig. 22.1). This effect is probably due to the base change creating or destroying sequence-dependent curvature, for the bending of DNA is known to affect gel mobility.

Troubleshooting Guide

1. Formaldehyde, which is used in silver staining solution C, is usually supplied as a 37% solution (formalin) because pure formaldehyde is a gas at room temperature. Remember to take this into account when calculating volumes because the formaldehyde concentration is critical.
2. The presence of only small quantities of contaminating salt in dH_2O can cause $AgNO_3$ solutions to change to a milky white appearance. If this occurs,

the standard of staining will be poor. The most common reason for this is the exhaustion of the ion-exchange cartridge used for desalting the tap water supply; replacing the cartridge almost always cures the problem.

3. Holding gels in place while pouring silver staining solution away may cause the gels to become marked. This appears to be due to the talc used in some brands of latex gloves. Microtouch gloves seem to be satisfactory. The problem can be avoided by holding the gel in place with an old piece of X-ray film.
4. If a gel in silver staining solution B (0.1% $AgNO_3$) comes into contact with a strong salt or alkaline solution, it may give a milky white precipitate. Attempting to continue under these circumstances is futile. The gel can often be saved by washing twice with dH_2O and then immersing in a 2.5% ammonia solution followed by two further rinses in dH_2O. Silver staining can then be restarted with solution A.
5. Silver-stained gels can be heat sealed between layers of plastic for temporary storage. Care must be taken while handling gels stored in this way because they are delicate and are especially susceptible to pressure damage.
6. Leakage from the upper buffer chamber can be a problem with the Protean system. Dampening the seal on the upper chamber can help.
7. If solution D is made up incorrectly, any bands present on the gel may fade or disappear completely during the drying step.
8. Gels containing more than 8.5% acrylamide are prone to cracking as they are dried down. To minimize this problem, they should be dried at room temperature.
9. Occasionally a white powdery deposit may build up on the surface of the dried gels. This can be removed by spraying with any household furniture polish and rubbing with a soft cloth.
10. Replicates of gels may be markedly different in appearance. This can usually be traced to gel and environmental variability. Take great care in measuring reagents before pouring gels. Because buffer and gel temperatures are such important parameters it is best to use a strict routine when setting up SSCP/HA experiments. Pour the gels at a fixed time of the day; similarly, place the gel and electrophoresis buffer in the cold room at the same time and load the gels at the same time of day for every experiment. Your results will then become much more reproducible, although a certain degree of variability should be expected.

REFERENCES

1. Orita M, Iwahana H, Kanazawa H, Hayashi K, Sekiya T (1989) Detection of polymorphisms of human DNA by gel electrophoresis as single-strand conformation polymorphisms. Proc Natl Acad Sci USA 86: 2766–2770.
2. Anglani F, Picci L, Camporese C, Zacchello F (1990) Heteroduplex formation in polymerase chain reaction. Am J Hum Genet 47: 169–170.
3. Keen J, Lester D, Inglehearn C, Curtis A, Bhattacharya S (1991) Rapid detection of single base mismatches as heteroduplexes on Hydrolink gels. Trends Genet 7: 5.
4. Nataraj AJ, Olivos-Glander I, Kusukawa N, Highsmith WE, Jr (1999) Single-strand conformation polymorphism and heteroduplex analysis for gel-based mutation detection. Electrophoresis 20: 1177–1185.

23

FLUORESCENT SSCP: SLAB GELS AND CAPILLARY ELECTROPHORESIS

Lucy Jenkins
Regional DNA Laboratory, St. James's University Hospital, Leeds, UK

Ruth Charlton
Molecular Genetics Laboratory, Institute of Child Health, London, UK

■ INTRODUCTION

SSCP using fluorescent labeling either by the use of fluorescent primers or by post-PCR labeling is a popular mutation scanning tool. The advantages over conventional SSCP include multiplexing and automated sample handling and data collection. Unlike manual systems, most high-voltage automated systems are unsafe to run at temperatures below 15°C; therefore, more than one run condition may be required for maximum sensitivity.

■ FLUORESCENT SSCP ON THE ABI 377 SLAB GEL SYSTEM

Equipment

ABI 377/3100 Sequencer
Cooling water bath (ABI 377)

Guide to Mutation Detection, Edited by Graham R. Taylor and Ian N. M. Day.
ISBN 0-471-23444-3

Solutions (ABI 377)

40% Acrylamide Stock

Use Sigma premix (49:1 or 37:1), dissolve as instructed, and deionize with 0.5 g of Amberlite MB 1 ion-exchange resin.

MDE Gel

Mutation Detection Gel Solution, FMC Corporation. Supplied as 2× concentrate.

Ammonium Persulfate

10% Ammonium persulfate (make fresh)

Other Reagents

10× TBE (Invitrogen).
Glycerol (Sigma)
TEMED (*N,N,N′,N′*-tetramethylethylenediamine) (Sigma)
Deionized formamide (loading buffer) (Applied Biosystems)

Gel Preparation

Assemble glass plates in frame as described in Chapter 2 of the ABI 377 DNA Sequencer Users Manual.

6.5% SSCP Acrylamide Gel

	40 ml (1 gel)	200 ml
40% Acrylamide stock (37:1 or 49:1)	6.5 ml	32.5 ml
Amberlite ion exchange resin	1.0 g	5.0 g
Glycerol	1.25 g	6.25 g
dH_2O	25.5 ml	127.5 ml

1. Vacuum filter 4 ml (20 ml if making 200 ml) of 10× TBE.
2. Mix the acrylamide solution well and add to vacuum filter to remove resin, and allow to degas for 5–10 min.

MDE SSCP Gel

	40 ml (1 gel)	200 ml
2× MDE (as supplied)	10 ml	50 ml
10× TBE	4 ml	20 ml
Glycerol	2.0 g	10 g
dH_2O	up to 40 ml	up to 200 ml

Mix the MDE solution well and vacuum filter, allow to degas for 10 min. MDE gel solution must be degassed before each use.

Casting the Gel

Assemble the gel plates and spacers in the casting frame as described in the ABI Prism 377 users manual, Chapter 2.

Take 25 µl of TEMED and 200 µl of freshly made 10% APS to 30 ml gel solution, gently swirl to mix without creating air bubbles.

Pour the gel, using a syringe with screw-fitting adapter; introduce the gel solution via the casting frame into the base of the gel. Gently tap the plates at the gel front to prevent bubbles forming as the solution spreads up the plates by capillary action.

Make sure there is surplus gel solution at the top of the plates and insert a 34-well 377 castle (square tooth) comb, taking care not to introduce air bubbles. Use three bulldog clips to grip the upper edge of plates around comb. Allow the gel to set for 2 h. For MDE gels, wrap upper edge with plastic wrap and leave casting frame attached to avoid air contacting the gel.

SSCP Sample Preparation

PCR product diluted 1:1 in deionized formamide loading buffer	1.0 µl
Deionized formamide loading buffer*	3.0 µl
GS Tamra 500 size standard	1.0 µl
Total	5.0 µl

1. Heat reactions to 95°C for 2–3 min on a PCR block.
2. Place on ice.
3. Load 1.5–2.0 µl onto gel.

Setting up the 377 Computer for an SSCP Run

The 377 is connected to an Apple computer that, when switched on automatically, opens the ABI 377 collection software. This software controls all run parameters.

Run Control Panel

Push Button Icons

Available for three main functions:

Plate check, prerun, and run
There are also buttons to pause the machine and cancel a run.

SSCP Modules (user preset)

For each run the appropriate modules are selected in advance. The module is selected according to function (plate check, prerun, or run), gel length, filter set, and run voltage. For example:

Plate Check C Chiller (this refers to filter set C, which detects the dyes Fam, Tet, Hex, and Tamra)

GS PR 36C-1200 Chiller (GeneScan PreRun module, 36-cm gel, Filter Set C 1200 volts)

GS Run 36C-1200 Chiller (GeneScan Run module, 36-cm gel, Filter Set C 1200 volts)

* If loading multiple PCRs per lane, this volume should be reduced to maintain a final volume of 5 µl.

Chiller modules are used for SSCP analysis, these disable the 377 pump and allow an external cooling water bath to control the temperature of the run.

Gel's Matrix File (user preset)

For SSCP applications select FamTetHexTam-nat.

This matrix is appropriate for the dyes listed when the machine is run on filter set C. It is also made to be run under the native conditions required for SSCP.

Sample Sheet

It is important that the sample sheet precisely matches the order in which the samples are loaded on the gel—including empty lanes for any wells that have been left out.

When the correct sample sheet is in place on the run control panel, two additional columns need to be amended:

Matrix file: Select the matrix file being used for the run for each sample.

Shortcut: (1) select matrix for sample 1, (2) click on column title to select the whole column, (3) use fill down option to alter whole column (appleD or Edit menu).

Auto analyze: Automatic analysis after the run is not required, so deselect this box for each sample (fill down option can be used). This then clears the last three columns, which are also related to analysis.

Collect Time

A 16-h overnight run is necessary to collect data up to 500 bp for SSCP analysis with 49:1 acrylamide at 20°C. Run times will increase as the temperature is reduced.

Bear in mind that MDE gels are significantly slower running (it is necessary to increase the power setting to collect 500-bp fragments in 16 h).

Longer run time is not advisable because of the available memory on the computer and taking up 377 time for the next morning.

Well to Read Length

36 cm is selected according to length of gel being run.

Lanes

34—for castle comb (recommended for SSCP gels)

Operator

Your name (useful to know for archiving purposes)

To Make a New Sample Sheet

> **File** menu

> **New**—Select Genescan Sample from the box displayed.

A blank sample sheet is now displayed.

	Sample Name	Project Name	Color	Std	Pres	Sample Info.	Comments
1	John Doe		v		X	PCR 99/1234	
			v		X	PCR 99/1234	
			v		X	PCR 99/1234	
			v	υ	X	PCR 99/1234	
2			v				

The sample sheet is filled in as shown for each sample being loaded on the gel (up to 34).

Sample name: Usually patient name or worksheet reference

Project name: This can be left blank.

Color v: A colored square indicating each dye available for use

Std: In each sample lane (1–34), the color used for the size standards is marked by clicking the box for the appropriate color (ALWAYS RED).

Pres: This is used to indicate which color dyes are being used for the run. If all colors are in use, select all four boxes as shown. If the PCR is labeled with FAM only, select Blue for the PCR products and Red for size standards.

If in doubt, all colors can be selected with no adverse effect.

Sample Info: Worksheet for run or information required can be filled in here (this will appear on GeneScan printouts.

Comments: Leave blank.

Once completed, save the sample sheet:

> **File** menu

> **Save as.** Name the sample sheet according to the worksheet (i.e., 99/1234).

> Close sample sheet.

The new sample sheet will then appear for selection in the sample sheet dropdown on the run control panel.

To Amend a Sample Sheet

A selected sample sheet is shown on the run control panel; this cannot be altered directly.

Remove sample sheet from run control panel by selecting < none > in the sample sheet dropdown.

Now the original sample sheet has to be opened:

> **File** menu

> **OPEN.** Select correct sample sheet and make the required changes.

Save sample sheet.

Reselect altered sample sheet in the dropdown on the run control panel.

Setting up the 377 for an SSCP Run

Equipment Start-Up

Before switching on the 377 or water bath, connect cooling water bath:

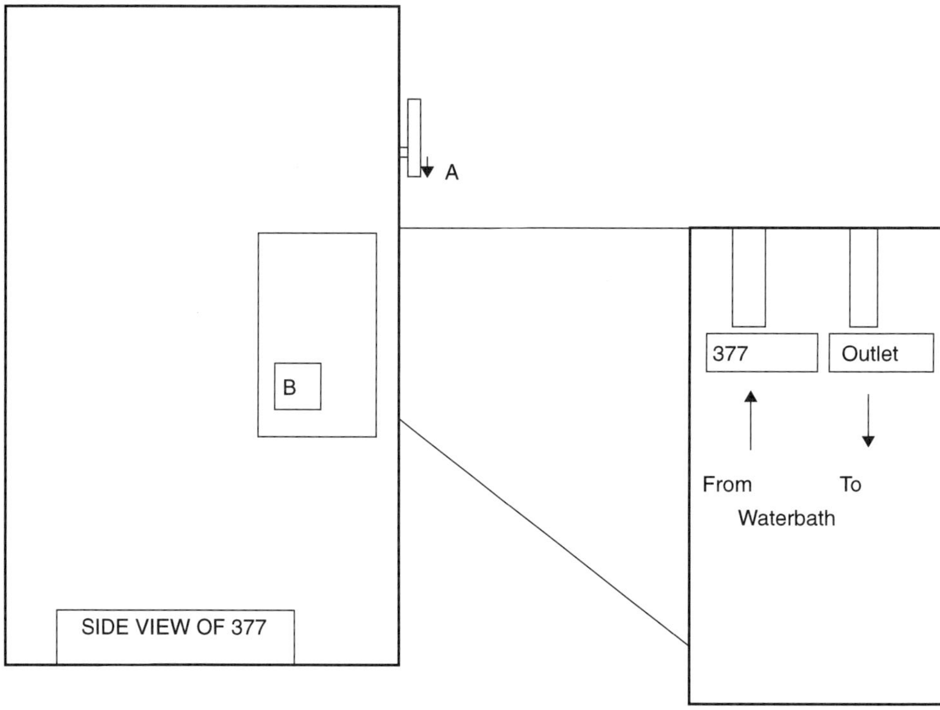

FIGURE 23.1. Side elevation of 377, front is on the left.

1. Remove coolant bottle from behind door **B** (on the side of the machine) by unscrewing it—some cooling fluid will probably leak out. Remove the tubes attached to the inlet and outlet by pulling firmly. Place all three to one side.
2. The cooling water bath inlet and outlet tubes are connected to each other when not in use. Separate these. Attach the outlet tube to the 377 inlet (tube labeled in red); attach the inlet tube to the 377 outlet (tube labeled in green).
3. Turn valve **A** through 180° so that the raised arrow on the handle points downwards. This allows the coolant to be pumped externally, with the 377 internal pump turned off.
4. Both 377 and water bath can now be switched on. By holding down the SET button on the water bath the required temperature can be selected (15–35°C). Selecting cooler temperatures is not recommended because condensation inside the 377 may occur.

Monitoring Temperature

The Status box shows the internal temperature of the machine along with other run parameters (see Figs. 23.1 and 23.2).

The ABI 377, when switched on, takes a few moments to warm up.

The three display lights are: green (machine running), amber (machine idle/run completed), and red (error warning).

The modules described above are then used in turn (Plate Check, PreRun, and Run) to set up the 377. During setup it is useful to monitor both the status and log of the machine:

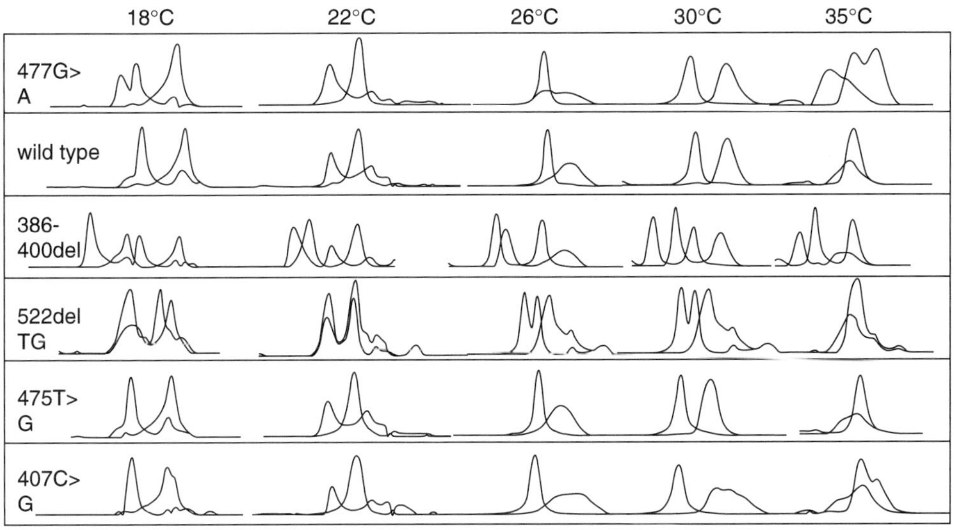

FIGURE 23.2. SSCP mobility variants of VHL amplimers over a range of temperatures.

Status of 377

The status box is useful for checking the status of the 377 during setup and also for monitoring the progress of a run:

> **Window** menu
> **Status**
Values shown are:

Instrument state: Running/Idle/Paused
Laser power: On/Off
Electrophoresis power: On/Off
Instrument door: Open/Closed

Plate Check

To check that the plates are clean, the gel in its frame is placed in the machine and held in place by turning the clips at all four corners—*It is important not to force clips. If they do not turn easily, remove the gel and adjust it within the frame.*

On computer:

> **Plate Check Module** = **Plate Check C Chiller**
> **Plate Check**

A scan window is displayed and the laser scans across the gel to provide a baseline image for each dye. These should be completely flat.

If the trace is not completely flat, clean both sides of the of gel with distilled water and repeat the plate check. This process should be repeated until you are satisfied the plates are clean.

Prerun

1. Remove comb from top of gel.
2. Using a pipette, fill the wells with a dilute solution of agarose gel loading buffer. During electrophoresis this runs in to mark out each well, which makes the gel easier to load.
3. Attach top buffer tank and plug in electrode.
4. Attach heat plate and plug in cooling water tubes plus thermistor.
5. Fill both buffer tanks with 1× TBE (approximately 1.5 liters).

On the Computer

> Check Prerun module is correct.

> Click Prerun button.

Check Status box.

Allow to prerun for 5–10 min, then Pause.

Electrophoresis stops, but the gel is maintained at a cool temperature.

The wells should be flushed with buffer before loading denatured samples

Starting the Run

Once samples are loaded:

1. Place lid on top buffer tank to prevent evaporation.
2. Start run on computer:
3. **Check Run Module is correct.**
4. Click **Run** button.
5. A message is displayed that requests a name for the gel file: Enter worksheet number.
6. OK

The run will now proceed, but it is necessary to reduce the electrophoresis power—this helps to maintain the cool temperature.

> Window Menu > Manual Control

A manual control box is now displayed. Click on the parameter box and select Electrophoresis power.

Now enter a value: 60 W for 49:1 or 37:1 acrylamide gels

80 W for MDE gels

Confirm this change using the Status box (see status of 377 above). The target value should now show the revised setting.

The progress of the SSCP run can be viewed by selecting Gel Image (window menu).

After the set time the run will automatically stop. Data are automatically processed, and a final gel image is presented for analysis.

The run can be stopped early by pressing the Cancel button.

Select **Stop and Analyze** data to complete the run.

To view the gel:

> Apple HD > ABI Prism 377 > Runs

Select the correct run folder according to date and time, rename this folder by clicking on the file name. Use the GeneScan Analysis Package to process your gel file.

FLUORESCENT SSCP ON THE ABI 3100 CAPILLARY SYSTEM

Introduction

Fluorescent SSCP (single-stranded conformation polymorphism) analysis can be performed on the ABI 3100. The sensitivity has been estimated to be approximately 93% when performed under three temperature conditions. Fragment analysis is carried out on PCR fragments 200–500 bp in length. To maximize sensitivity, both primers should be labeled with different fluorescent dyes (FAM, HEX, or NED).

Equipment and Reagents

Capillary array 36 cm (Applied Biosystems part number 4315931)

96-Well tray (part number AB-0600)

96-Well septa (Applied Biosystems part number 4315933)

10× EDTA buffer (Applied Biosystems part number 402824)

Genescan polymer (7%, 401885 Applied Biosystems)

Deionized formamide (Applied Biosystems or equivalent)

ROX500 size standard (Applied Biosystems or equivalent)

SSCP Polymer

35.5 g Genescan polymer (401885, Applied Biosystems 7%)

5 g glycerol (Sigma G-5516)

5 g 10× buffer with EDTA (402824 Applied Biosystems)

Deionized water to 50 ml (54.5 g)

Filter with a Nalgene 0.22-μm filter and store at 4°°C, labeled and dated. Stable for 1 month.

Method

1. Check that PCR has worked on agarose gel.
2. Prepare samples for loading.

Preparation of Samples

Dilute samples in dH_2O—as a guide

Weak band dilute 1/10 to 1/15

Average band dilute 1/20 to 1/30

Bright band dilute 1/50 to 1/70

Prepare 96-well plate as on your sample sheet with the following in each well:

GS500 ROX	0.5 µl
HDF	10 µl
Diluted PCR	1 µl

1. Fill any unused wells within the run (set of 16) with 10 µl HDF.
2. Place the septa onto the plate and centrifuge to bring samples to the bottom of the wells.
3. Heat to 95°C for 1 to 1.5 min and then place immediately onto wet ice.
4. Dab away excess ice/liquid and place plate into 3100 holder, making sure white lid clicks into place. Ensure that the plate well positions match that of the plate well holder.
5. Sample plate is now ready. Check polymer levels and, if required, fill the syringe.
6. Prepare the sample sheet.

Preparation of Sample Sheet

Take the next numbered 96-well SSCP sheet from the worksheet tray in the main office.

Fill in the sample details starting from A1 corresponding to the sample plate.

Position, PCR no., and sample no. e.g. A1 = 01_1_01, B1 = 02_1_02, A2 = 09_2_01.

H	G	F	E	D	C	B	A
08_1_8	07_1_7	06_1_6	05_1_5	04_1_4	03_1_3	02_1_2 10_2_2	01_1_1 09_2_1

1. Click on the 3100 Data collection icon if not already open.
2. Select the sample sheet icon (2nd from left)
3. Type in the plate number as on the sample sheet, e.g., SSCP_1
4. Select GeneScan and confirm plate is 96 well; add any comments and click finish.
5. The SSCP sample sheet will now open.

Fill in accordingly

- Sample name = label each of the wells that contains a sample according to sample sheet.
- Dye set = D
- bioLIMS = 3100 project
- Run module = You can set up to five run modules
- Analysis = no sizestd.gsp

Click "finish", closing the sample sheet and placing it in the pending plate records box.

The run can now be started.

Push the tray button on the outside of the machine (making sure button does not stick). The autosampler will move to front of machine. When it has stopped. open the doors.

TABLE 23.1. Example SSCP Run Modules

	18°C_15kV_40	22°C_15kV_40	30°C_15kV_20
Run temperature	18	22	30
Cap fill volume	184	184	184
Prerun voltage	15	15	15
Prerun time	180	180	180
Injection voltage	3	3	3
Injection time	10	10	10
Run voltage	15	15	15
Data delay time	1	1	1
Run time	1800	1800	1200

Place the sample plate onto the autosampler using A or B position (B only in 3100 Avant).

Check that the buffer/water septas are laying flat with the plate.

Close the doors; autosampler will return automatically, making area A or B yellow on the plate view screen.

Click on the pending plate record and then the yellow plate, linking the plate that will turn green. The sample sheet will now be in the linked plate records.

Select the start button on the toolbar (green arrow 4th from left).

You can view your data while running in array or capillary view.

Collecting and Transferring the Data

1. In the run view tab, completed appears in the status column when the run is finished.
2. Extend the run ID section to reveal the run ID number; take note of this along with their temperatures.
3. Open the shortcut to extracted runs folder on the main screen and identify your runs. The runs should be transferred into the 3100 Working SSCP folder.
4. Create a new folder, named according to your plate name e.g., SSCP_1
5. Within this, create other new folders for each of the run temperatures. e.g., 18°C, 22°C, 30°C.
6. Select your runs from the extracted runs folder and drag across to the appropriate folders.
7. When complete, drag the same runs to the copied SSCP folder, also in the extracted runs window.
8. Close extracted runs and other windows, leaving only the 3100 data collection window open.

Analyzing Data

Genescan

1. Double click on the GeneScan icon
2. Click file and then new, then select project; this opens untitled analysis control.
3. Click project and add sample files; this brings up selection window.
4. Select your data.

5. Open the runs individually within a temperature and click "add all" to select.
6. Click "finish."
7. Double click on a sample and note the range of peaks; decide your parameters, e.g., 4000–8000 data points
8. Open parameters for that sample by clicking the arrow in the parameters column.
9. Select "define new."
10. Analysis range = this range 4000–8000 data points (whatever range fits)
11. *You should not need to change anything else*
12. Data processing = None
13. Peak detection = 50
14. Size call range = Full range
15. Size calling method = 2nd-order least squares
16. Save the setting as your name and close; you can overwrite your previous save, fill down.
17. Set the size standard by clicking the arrow of the sample in the size standard column.
18. Select "define new."
19. Select 3–5 peaks spanning the range and size these on the table, dividing data point by 10 to give an arbitrary size value.
20. Select "file" then "save as"; save as your name and close (can overwrite previous save).
21. Fill down.
22. Select colors you wish to analyze and click "analyze."
23. When finished select "file" then "save project as." Save within the appropriate temperature folder, naming the file according to temperature of run, e.g., 18°C. Close project to finish and then exit GeneScan.

Genotyper

1. Open Genotyper. Untitled window opens automatically.
2. Select "**file**", "**import**", then "from **GeneScan** files".
3. Select your file from the appropriate folders, then open the analyzed temperature folder. Select runs required and click "finish".
4. Select the colors required and click the plot icon (3rd down right side) to view the samples separately.
5. Zoom in on the peaks by dragging the cursor around the peak (shows as a square) and use controls to zoom in and out.

   Control R = Zooms into range selected

   Control H = Zooms out to show whole plot

   Control + = Zooms in whole plot

   Control – = Zooms out of whole plot
6. Click on the peaks to size.
7. Once complete, save genotyper file within the correct temperature folder as PCR no. and temperature of run, e.g., 123_18°C.
8. Data can be scored and checked directly from the computer screen.

24

DENATURING GRADIENT GEL ELECTROPHORESIS

Gert de Voss
Ingeny International BV, Goes, The Netherlands

■ INTRODUCTION

Denaturing gradient gel electrophoresis (1, 2) and related methods provide one of the most sensitive mutation scanning methods yet developed. The technique is robust and does not require elaborate or expensive equipment.

■ METHOD

The **INGENYphorU-2** cassette consists of the following:

- Electrophoresis cassette
- One U-shaped spacer
- One comb (20 well, 32 well, or 48 well)
- Acrylic glass pressure unit
- One glass plate with large notches
- One glass plate with small notches
- Seven nylon screws

The **INGENYphorU-2x2** cassette consists of the following:

- Electrophoresis cassette
- Two U-shaped spacers

Guide to Mutation Detection, Edited by Graham R. Taylor and Ian N. M. Day.
ISBN 0-471-23444-3

- Two combs (20 well, 32 well, or 48 well)
- Two acrylic glass pressure units
- Two glass plates with large notches
- Two glass plates with small notches
- Fourteen nylon screws

Assembly of the Electrophoresis Cassette

The electrophoresis cassette is assembled as follows:

1. Place the INGENYphorU-2 front faced on the table (Fig. 24.1).
2. Insert the nylon screws a few turns.
3. Place the plexi-glass™ pressure unit in the cassette, **open side up and the circular protuberances to the back** (Figs. 24.2 and 24.3). The circular protuberances will press against the glass plate.

Note: Space between protuberances is essential for gel to buffer contact.

Preparing the Cassette for Perpendicular Electrophoresis

1. Place the glass plate with the large notches (notches up) flat on the table (Fig. 24.4).
2. Place the Teflon spacer on this glass plate. Make sure the small enforcement at the bottom of the spacer is fitted against the bottom of the glass plate (Fig. 4).
3. Place the glass plate with the small notches on the spacer, notches to the bottom. Make sure the tops of both glass plates are leveled.
4. Insert the glass plate sandwich carefully into the electrophoresis cassette. The Teflon spacer must slide smoothly in the slits on each side of the cassette. **Do**

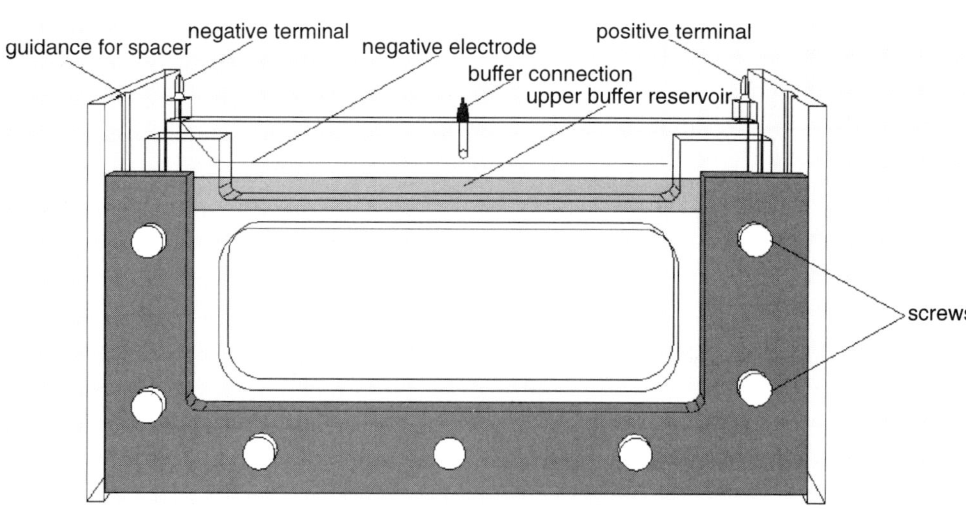

FIGURE 24.1. Schematic diagram of the INGENYphorU electrophoresis cassette.

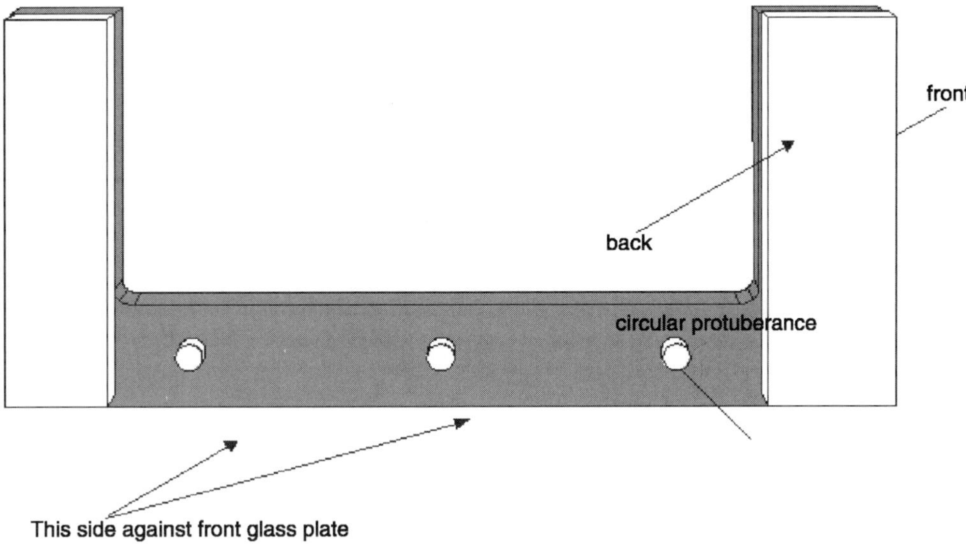

FIGURE 24.2. Plexi-glass pressure unit.

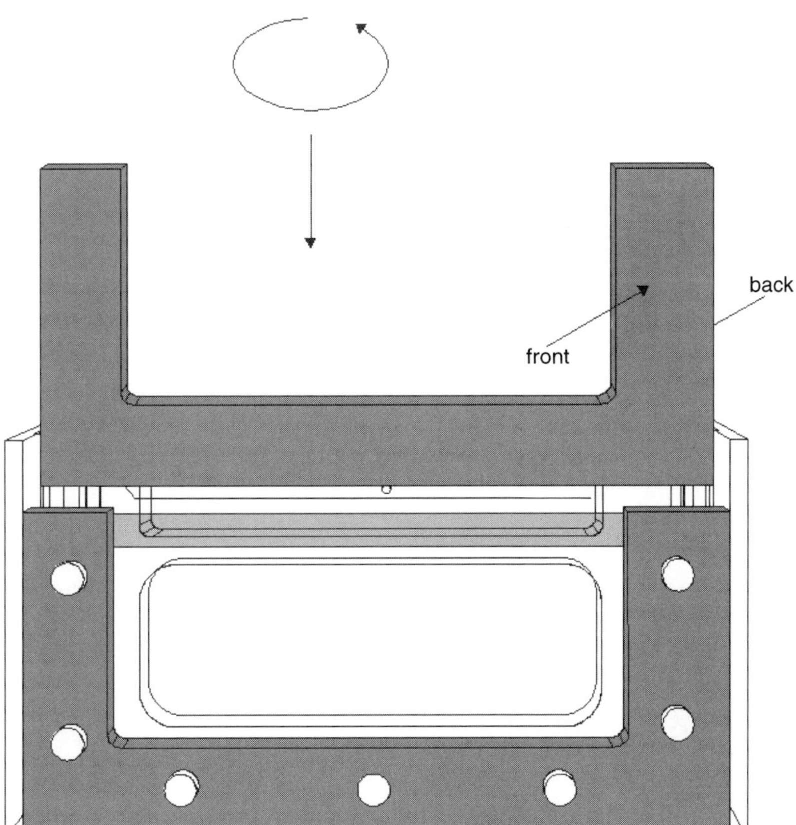

FIGURE 24.3. Orientation of pressure unit in electrophoresis cassette.

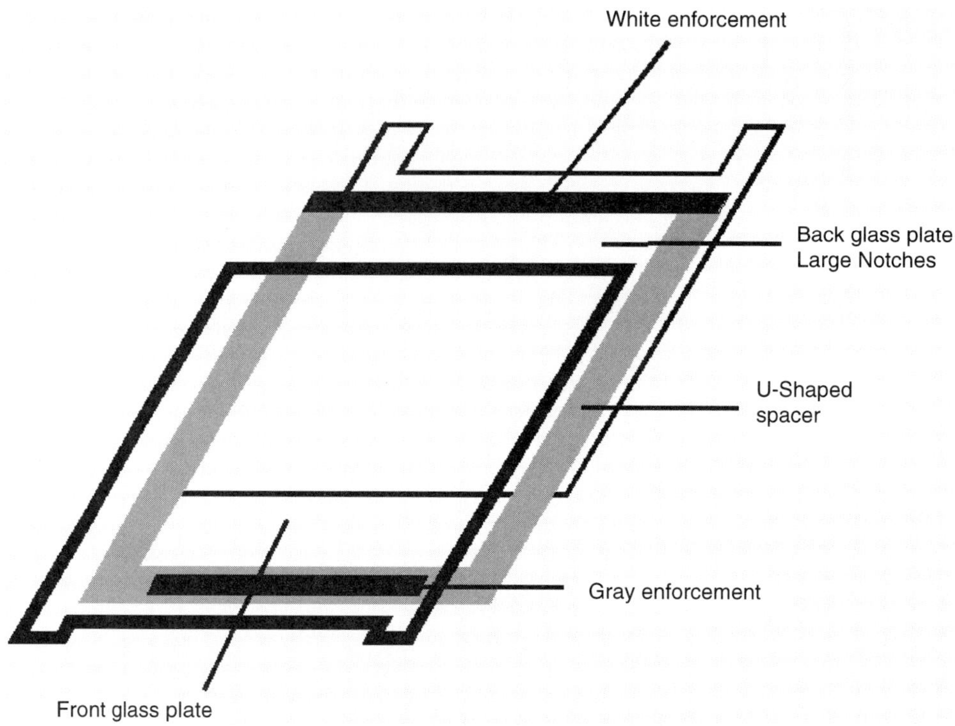

FIGURE 24.4. Assembly of the glass plate sandwich.

not force. If the sandwich does not go in smoothly, check whether the screws were loosened enough.

5. The glass plates should be inserted to touch the bottom of the cassette. The spacer must be pulled up as far as possible without lifting the glass plates. Tighten the three bottom screws.
6. Insert the comb, screws to the front. Make sure both the Teflon comb and the 1-mm-thick strip are behind the front glass plate (Fig. 24.5).
7. Tighten all 7 screws; the cassette is now ready for gel casting.

Casting a Gel in the INGENYphorU-2

In the INGENYphorU-2 both neutral and denaturing gels can be produced. For perpendicular gradient gels an adapter kit is available from Ingeny International BV.

Ingeny International BV	**Tel. +31 113 222920**
Amundsenweg 71	**FAX +31 113 222923**
4462 GP Goes	
The Netherlands	

For a detailed description of neutral and gradient gel casting we refer to **"Two-Dimensional DNA Typing, a Laboratory Manual"** available from Ingeny International BV.

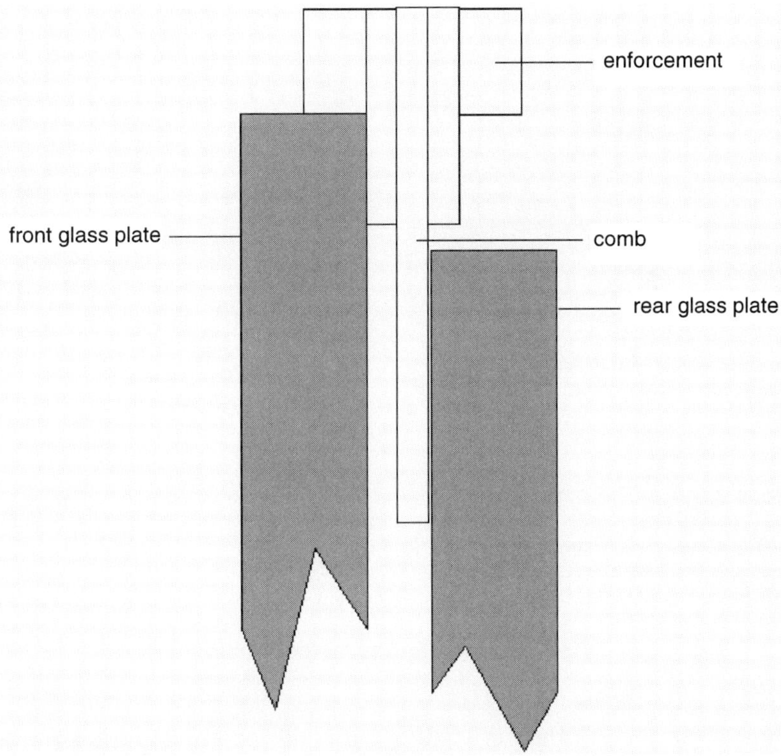

FIGURE 24.5. Comb between glass plates, cross section.

Casting Neutral Gels

1. With a pipette, the acrylamide solution can be poured between the glass plates. Use the opening between the comb and the spacer on either side of the comb. The total amount of acrylamide needed for one gel in the INGENYphorU-2 is 55 ml; 55 µl of TEMED and 375 µl of APS 20% are added just before gel casting. Avoid air bubbles under the comb by tilting the cassette slightly if necessary.
2. Complete polymerization of the gel will take approximately 1 h at room temperature.

Casting Denaturing Gradient Gels

1. With Ingeny's gradient maker, a magnetic stirrer, and a peristaltic pump (flow approximately 10 ml/min) a denaturing gradient gel with a linear gradient can be produced in the INGENYphorU-2 (Fig. 24.6).
2. Two different acrylamide solutions are used. Solution A has a high concentration of *urea* and *formamide* (UF), whereas solution B has a low concentration of UF. For one gel, 27.5 ml of each solution is needed (for second-dimension gels in 2-dimensional DNA-typing use 25 ml of each solution). Add 150 µl of 20% APS and 12 µl of TEMED to solution A as well as solution B.

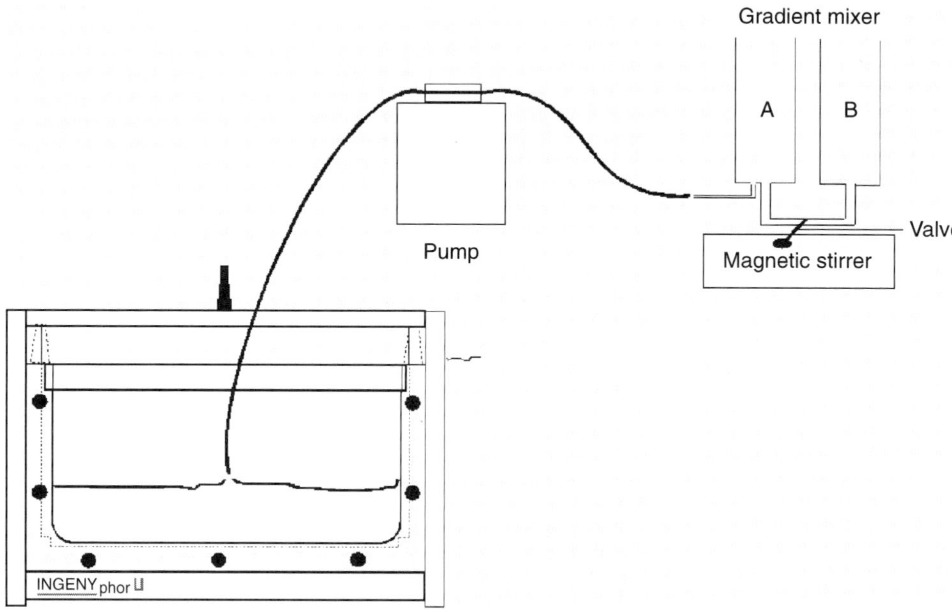

FIGURE 24.6. Setup for gel casting.

3. Connect the tube to the peristaltic pump and the gradient maker. Pour solution A in the left leg of the gradient maker (the leg connected to the tube). and open the valve a little until the connection between the two legs is just free of air; close the valve. Pour solution B in the right leg (Fig. 24.6).
4. Connect the pouring device (or the needle of a syringe) to the tubing of the peristaltic pump and put it between the glass plates, left or right of the comb. Start the stirrer (the solution in the left leg should be stirred) and open the connection. Start the pump to start the gel casting.

Note: APS (ammonium persulfate) and TEMED (*N,N,N′,N′*-tetramethylethylenediamine) are the propagator and initiator in the polymerization reaction of acrylamide and bisacrylamide into polyacrylamide. The propagator (APS) can be added to the gel solution without implications. After addition of TEMED the polymerization reaction starts.

5. Let the complete gel polymerize for approximately 1 h.
6. The cassette is ready for electrophoresis.

Running a Gel

1. All screws have to be loosened until they just touch the plexi-glass pressure unit.
2. Place the electrophoresis cassette in a buffer tank (a microprocessor-controlled buffer reservoir with integrated heater and pump is available from Ingeny International). Connect the buffer flow to the blue connector on the electrophoresis cassette (Fig. 24.7). If not using INGENY's buffer reservoir, the buffer flow for circulation in the upper buffer reservoir of the elec-

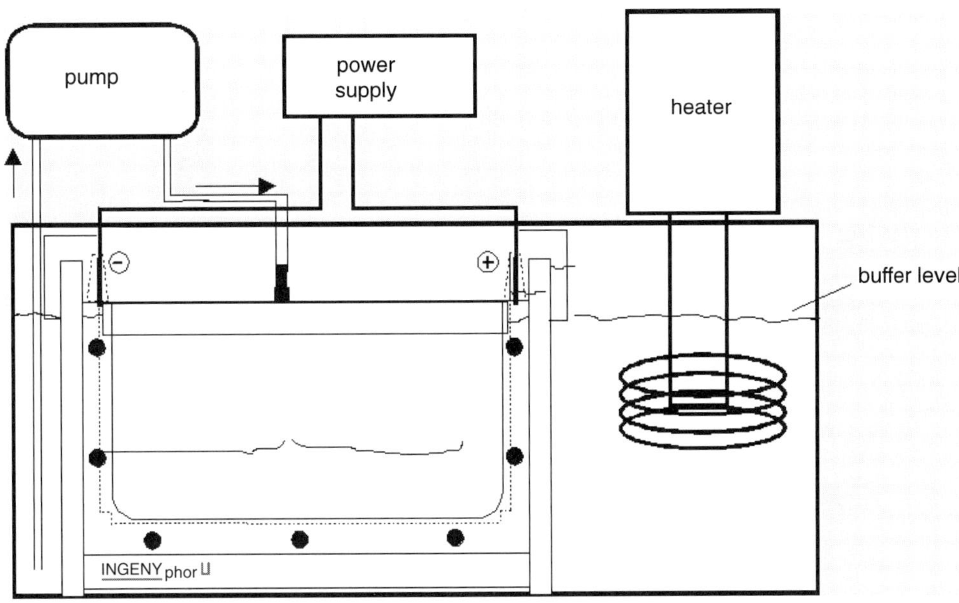

FIGURE 24.7. Electrophoresis unit placed in a heated buffer tank with buffer flow and electrical wires connected.

trophoresis cassette must be made adjustable, for example, by using a clamp around the tubing (Fig. 24.7). Ingeny's buffer system has a valve on the outside of the system to control the buffer flow in the upper buffer reservoir.

> > NEVER RUN AT TEMPERATURES OVER 70°C < <
> > DAMAGE TO THE INSTRUMENT MAY OCCUR< <

3. Fill the upper buffer reservoir with buffer and close the valve at the side of the buffer system or close the tube connected to the blue connector. The comb can now be removed slowly *(never remove the comb without filling the upper buffer reservoir first)*. Carefully push the U-shaped spacer all the way down. Little air bubbles under the gel can be immobilized in the left and right corners of the gel by holding the cassette at an angle (Fig. 24.8).
4. Retighten the upper 2 screws on each side of the cassette.
5. Now the power supply has to be connected. The black electrical lead should be connected to the negative terminal of the power supply and the red lead to the positive terminal.
6. Make sure that the upper buffer reservoir is filled with buffer. Before loading the samples, close the buffer flow to the upper buffer reservoir to prevent the samples from being flushed out of the wells. **Note**: Electrical current will only be possible with both platinum electrodes submerged in buffer.

To minimize diffusion of the samples while loading a voltage of 10–15 V is recommended (a 12-V loading voltage is available in the Ingeny buffer system without the use of an external power supply). With the use of very narrow pipette tips the samples can be loaded just above the bottom of the wells.

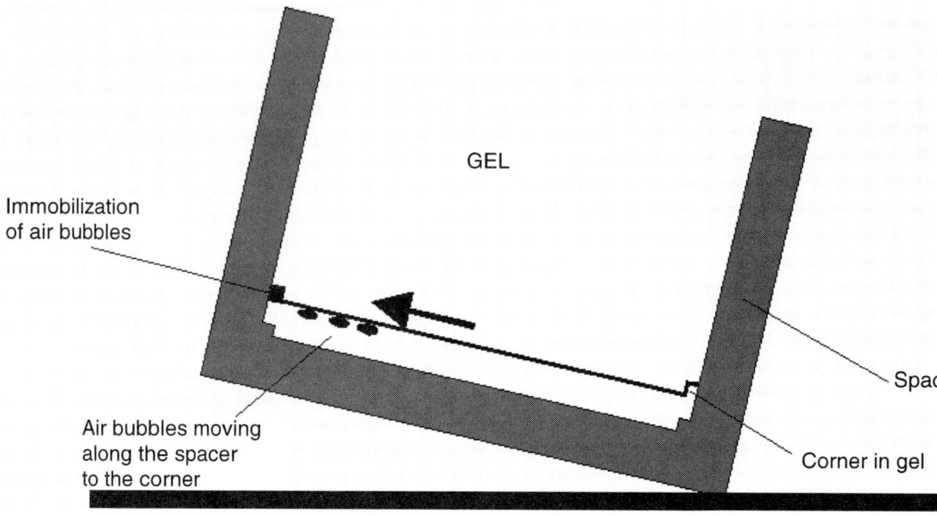

FIGURE 24.8. Immobilizing of air bubbles in corners of gel.

7. Set the power supply for the desired voltage and time. Run the electrophoresis for approximately 5 min before opening the buffer flow to the upper buffer reservoir. The reservoir should overflow slowly through the ducts in the back of the cassette (INGENYphorU-2) or the side of the cassette (INGENYphorU2x2).

Gel Handling After Electrophoresis

1. After completing electrophoresis, switch off the power supply and disconnect the electrical leads.
2. Switch off the heater and circulation pump and disconnect the tubes.
3. Lift the electrophoresis cassette from the buffer tank and loosen all screws. The glass plate sandwich can be removed from the cassette.
4. With a spatula, one of the glass plates can be lifted from the gel. The gel left on the other glass plate can subsequently be processed, for example, staining with ethidium bromide or another suitable dye (e.g., Sybr Green from Molecular Probes). ***Always run, stain, and rinse the gel in the same buffer solution to prevent shrinking or swelling of the gel.***
5. The electrophoresis cassette should be rinsed thoroughly with distilled water after use to remove all buffer solution. Left salts might shorten the life of the cassette.

>>>>> NEVER USE ACETONE, CHLOROFORM, ALCOHOL, OR
>>>>> OTHER DISSOLVENTS TO CLEAN THE CASSETTE. <<<<
>>>>> THIS WILL DAMAGE THE INSTRUMENT. <<<<<

Step-by-step Manual for the INGENYphorU Buffer System

1. Remove all tape from your new buffer system and carefully check for any visible damage.

2. Place the buffer system on a flat horizontal surface.
3. Make sure the valve at the side of the instrument is closed (lower valve of the two).
4. Fill the tank with buffer. ***Never operate a system without buffer.*** For demonstrations use distilled water only.
5. Plug in the system and switch on the power (green switch). The pump should start now and the cooling fan at the back of the system should be blowing. Be careful not to block the air stream from the fan.
6. Set the desired temperature by pressing "SET" and "△" or "▽" simultaneously. Heating the buffer from 20°C to 60°C will take approximately 45 min.
7. Close the lid to prevent evaporation.
8. The buffer will only be heated if the gray level switch is submerged.
9. Once the buffer has reached its temperature, the electrophoresis cassette may be placed in the tank.
10. Connect the tube of the buffer system to the blue connector of the electrophoresis cassette. With the large orange valve at the side of the buffer system the buffer flow in this tube may be controlled.
11. Connect both leads at the inside of the buffer system to the connectors on the cassette. Connect the wires on the outside of the buffer system to an external electrophoresis power supply.
12. Before loading, close the buffer flow to the upper buffer reservoir with the orange valve at the side of the buffer system to prevent the samples from being flushed away. During sample loading a 12-V voltage is available by switching the red/yellow switch to LV; LV will be lit.
13. Close the lid, switch the red/yellow switch to HV (HV should be lit now) to switch from low voltage to high voltage. Set the voltage and run time on your power supply. **Note: For safety reasons high voltage (HV) is only available with the lid closed.**
14. After completing the experiment switch of the buffer system.
15. The buffer will last approximately 4 to 5 days when running once a day at 200 V for 8 h per experiment (6400–8000 volt-hours). The buffer may be changed with the small valve at the side of the system. The system should be switched off during draining.

■ BRCA1 AND BRCA2 DGGE

Amplification

- 5 µl 10× nucleotides (2 mM of each)
- 5 µl 10× reaction buffer
- 2 µl primer mix (25 pmol of each)
- 100–200 ng DNA
- 0.5 U polymerase
- Add sterile water to 50 µl.

Stepdown program

94°C 1 min, 52°C 1 min, 72°C 1 min (5×)
94°C 1 min, 50°C 1 min, 72°C 1 min (5×)

94°C 1 min, 48°C 1 min, 72°C 1 min (25×)
extra 72°C step for 3 min.
Heteroduplex step
94°C 3 min
50°C 10 min

DGGE

BRCA1 gradient 20–65% UF at 58°C, 16 h, 120 V
BRCA2 gradient 30–65% UF at 55°C, 16 h, 100 V

Making Gradient Gels

For preparation of gradient gels, make use of stock solutions:

1. 9% PAA, 0%UF
   = 25 ml 20× TAE (end concentration 0.5× TAE)
   300 ml 30% acrylamide-bisacrylamide (37.5:1)
   Add distilled water up to 1 liter
2. 9%PAA, 100%UF
   = 25 ml 20× TAE (end concentration 0.5× TAE)
   300 ml 30% acrylamide-bisacrylamide (37.5:1)
   420 g urea
   400 ml formamide (deionized)
   Add distilled water up to 1 liter

By mixing these two stock solutions you can make your desired gradient.

With Ingeny's gradient maker (Fig. 6), a magnetic stirrer, and a peristaltic pump (flow approximately 5 ml/min), a denaturing gradient gel with a linear gradient can be produced in the INGENYphorU-2.

Two different acrylamide solutions are used. For example, for BRCA1 DGGE, a 20%–65%UF gradient is used. Solution A has a high concentration of UF, whereas solution B has a low concentration of UF.

For one gel, 24 ml of each solution is needed. Add 50 µl of 20% APS and 5 µl of TEMED to solution A as well as solution B.

Pour solution A in the left leg of the gradient maker (the leg connected to the tube) and open the valve a little until the connection between the two legs is just free of air; close the valve. Pour solution B in the right leg. Connect the pouring device (or the needle of a syringe) to the tubing of the peristaltic pump and put it between the glass plates, left or right of the comb. Start the stirrer (the solution in the left leg should be stirred) and open the connection. Start the pump to start the gel casting. This will take about 10 min. After gel casting, rinse the gradient maker and tubing with water.

Let the gel polymerize for 1 h. Discard the liquid that is formed on top of the gel. Rinse with water.

For formation of slots, a stacking gel is poured on top of the gradient gel. This is a fast-polymerizing gel.

Prepare 5 ml 9%PAA, 0%UF, and add 100 µl of 20% APS and 10 µl of TEMED. Pour this solution immediately on top of the gradient gel. Make sure that no air bubbles are trapped under the comb. After 5 min, the gel is polymerized. Remove the comb and press down the spacer. Put the cassette in the tank and load your sample. Use 15–20 µl PCR product (with loading dye) for each lane.

Silver Staining of Polyacrylamide Gels

Solutions

Silver Staining Solution A

Ethanol	100 ml
Glacial acetic acid	5 ml

Make up to 1.l with distilled water. Store at room temperature; can be reused up to 5 times

Silver Staining Solution B

Prepare a 10× stock solution.

Silver nitrate	5.0 g

Dissolve in 500 ml deionized water and store at room temperature in foil-wrapped bottles in the dark.

The 1× solution (400 ml) can be stored in clear bottles away from strong light and can be reused for staining up to 5 gels.

Silver Staining Solution C (make fresh)

Sodium hydroxide	6.0 g
37% Formaldehyde solution	1.6 ml

Dissolve the sodium hydroxide in 400 ml of distilled water. Add the formaldehyde immediately before use.

Silver Staining Solution D

Prepare a 10× stock solution of 7.5% sodium carbonate.

Sodium carbonate	37.5 g

Dissolve the sodium carbonate in water and make up to 500 ml.

Method

Pour 400 ml of solution A onto the gel; leave on a slow orbital shaker for 3 min. The gel should naturally unfurl during this stage.

Pour off the solution A and add 400 ml of solution B. Gently shake for 15 min, then pour off solution B. Solution B should be poured back into the stock bottle and reused until staining efficiency declines. Rinse the gel quickly with two changes of distilled water.

To the gel add 400 ml of freshly prepared solution C and allow the gel to develop, shaking gently in a fume cupboard, for 20 min. A small amount of metallic silver should be formed in the solution as a black precipitate when the solution is added to the gel. Bands will slowly appear in the gel during this stage.

Pour off solution C and add 400 ml of solution D; leave to fix for 15 min.

Dry the gel.

■ REFERENCES

1. Fischer SG, Lerman LS (1983) DNA fragments differing by single base-pair substitutions are separated in denaturing gradient gels: Correspondence with melting theory. Proc Natl Acad Sci USA 80: 1579–1583.
2. Lerman LS, Silverstein K, Grinfeld E (1986) Searching for gene defects by denaturing gradient gel electrophoresis. Cold Spring Harb Symp Quant Biol 51: 285–297.

25

PROTEIN TRUNCATION TEST

Mark D. Robinson
Regional DNA Laboratory, St. James's University Hospital, Leeds, UK

■ INTRODUCTION

The protein truncation test (PTT) allows mutation detection at the protein level rather than the DNA level (1). PTT specifically uncovers only those disease-causing mutations causing premature translation termination. This technique is invaluable in diseases in which there are many such mutations; for example, 95% of mutations in the APC gene are truncating mutations (see Fig. 25.1) (2)There are two main advantages over other mutation detection systems.

1. A single PTT reaction allows the analysis of large gene fragments (2–3 kb).
2. PTT detects mutations of pathological interest—those that lead to truncated proteins; phenotypically silent mutations are not detected.

To summarize, in this method the coding sequence is screened for the presence of translating terminating mutations using de novo protein synthesis from amplified DNA.

The PTT assay has four steps:

1. Isolation of genomic DNA or RNA from blood or tissue
2. Amplification of target gene coding sequence with PCR from genomic DNA or reverse transcription-PCR (RT-PCR) from RNA
3. In vitro transcription and translation of these PCR products into protein. This stage can now be carried out very rapidly with the TNT quick kit from

Guide to Mutation Detection, Edited by Graham R. Taylor and Ian N. M. Day.
ISBN 0-471-23444-3

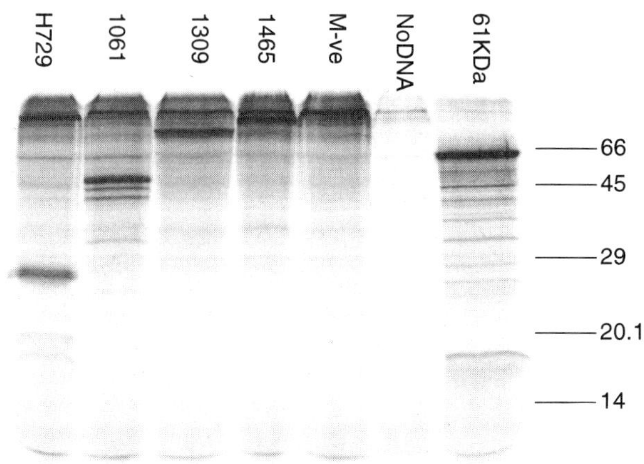

FIGURE 25.1. PTT products resolved on a 12% polyacrylamide gel. The amplicon was a fragment of *APC* exon 15, lane 1 is from a tumor cell line showing a truncated product. Lanes 2–4 are truncations caused by frameshift mutations at codons 1061, 1309, and 1465, respectively. Lane 5 is a wild-type control.

Promega. The kit contains all the components to perform in vitro transcription and translation in a one-tube reaction.

4. SDS-PAGE analysis of the synthesized protein. The proteins can be detected by either radioactive or nonradioactive methods. The most common method is to incorporate radiolabeled amino acids during protein synthesis with ^{35}S-methionine, ^{35}S-cysteine, or ^{3}H-leucine. The proteins are then separated by SDS-PAGE and detected by autoradiography. Nonradioactive detection methods are increasingly popular, and proteins are labeled by direct incorporation of biotin-charged lysines. After transfer to a membrane, the proteins are then detected by a streptavidin conjugate and visualized with a chemiluminescent substrate. A truncated protein is smaller in molecular weight than the wild-type protein.

The key feature of PTT is a specially designed forward primer that contains four specific regions:

1. A T7 promoter sequence at the 5′ end that directs the production of RNA with the corresponding RNA polymerase
2. A spacer sequence of 3–6 bp
3. A eukaryotic translation initiation sequence (Kozak sequence)
4. A gene-specific sequence (17–20 bp) in frame with the ATG codon from the Kozak sequence.

PCR products with this type of forward primer can direct in vitro protein synthesis in cell-free extracts. To facilitate cloning, a restriction site sequence can be engineered into the 5′ end of the primer.

If a gene contains several kilobases of coding sequence it is usually necessary to analyze the region in 1- to 2-kb segments. To ensure that mutations in the 5′ and 3′ end of the fragments are not missed, adjacent fragments should have an overlap of 350–500 bp.

Once a truncated protein is found the mutation is identified and confirmed by sequencing. The size of the truncated protein localizes the site of the mutation fairly precisely so only a small part of the gene needs to be sequenced to identify the exact mutation.

Mutation detection by PTT may be limited in the following ways:

1. Failure to amplify the mutant allele (problem with all PCR)
2. Failure to detect very small in-frame deletions and insertions as the mobility shifts are too small.
3. Missense mutations are undetected.
4. If RNA is being used as the starting material, differences in transcription efficiency or RNA stability could affect the results.

STORAGE OF REAGENTS

Promega in vitro TNT-coupled transcription-translation mix: Store in 38-μl aliquots (enough for 10 PTT reactions) at –80°C. Do not freeze/thaw more than twice.

61-kDa luciferase control: Store at –80°C. Use at 1/8 concentration.

^{35}S Promix (Amersham): Store at –80°C. Prepare and store 5 aliquots to avoid extensive hydrolysis on thawing. This product contains approximately 70% L-[^{35}S]methionine and 30% L-[^{35}S]cysteine.

10× SDS-PAGE running buffer: Store at 4°C.

40% BDH/Merck 29:1 cross-linking ratio acrylamide: Store at 4°C.

1× SDS Laemmli sample buffer. Store 0.5-ml aliquots frozen at –20°C.

Separating gel 4× buffer/stacking gel 4× buffer: Store in the dark at room temperature.

SOLUTIONS

Separating Gel 4 × Buffer

Tris	18.17 g
10% SDS	4 ml

Adjust to pH 8.8 with Conc. HCl and add double-distilled water to a final volume of 100 ml.

Stacking Gel 4 × Buffer

Tris	6.06 g
10% SDS	4 ml

Adjust to pH 6.8 with conc. HCl and add double-distilled water to a final volume of 100 ml.

SDS-PAGE 10× Running Buffer

Tris	30 g
Glycine	144 g
10% SDS	100 ml

Make up to a final volume of 1 liter with double-distilled water.

Protein Gel Coomassie Stain

Methanol	30%
Glacial acetic acid	10%
Brilliant Blue	0.2%
ddH_2O	60%

SDS-PAGE Fix

Methanol	10%
Glacial acetic acid	20%
ddH_2O	70%

Protein Gel Destain

Methanol	25%
Glacial acetic acid	10%
ddH_2O	65%

1× SDS Sample Buffer

(0.0625 M Tris·HCl, pH 6.75, containing 2% SDS, 5% mercaptoethanol, 10% glycerol and 0.001% Bromophenol Blue):

Combine:

Tris	1.51 g
Glycerol	20 ml
ddH_2O	35 ml

Adjust to pH 6.75 with conc. HCl then add:

SDS	4 g
2-Mercaptoethanol	10 ml
Bromophenol Blue	0.002 g

Dilute to a final volume of 200 ml with double-distilled water.

■ FORMULATION OF SDS POLYACRYLAMIDE GELS

Volume for Different % Acrylamide Gels

	8.5%	10%	12.5%	15%	17.5%
Water	3.9 ml	3.62 ml	3.15 ml	2.68 ml	2.21 ml
40% Acrylamide (29:1)	1.6 ml	1.88 ml	2.35 ml	2.82 ml	3.29 ml
4× Separating gel buffer	1.9 ml	1.9 ml	1.9 ml	1.9 ml	1.9 ml
10% APS	112 µl	112 µl	112 µl	112 µlL	112 µl
TEMED	5 µl	5 µl	5 µl	5 µl	5 µl

Formulation of Stacking Gel (5% Acrylamide)

Component	Volume
Water	1.075 ml
Stacking gel buffer	445 µl
Acrylamide solution (40%)	225 µl
10% APS	28 µl
TEMED	5 µl

■ POLYMERASE CHAIN REACTION FOR PTT

1. Perform a 20-µl PCR with 60mer forward primers containing a T7 promoter and ribosome binding site and normal 20mer reverse primers. We generally use KlenTaq (Clontech Advantage) with a hot start and 2-min annealing and extension times. Ensure that the correct magnesium concentration is present in the PCR buffer (1.5 mM is usually OK). Promega buffer works well.
2. Electrophorese PCR products (5 µl) on a 0.8% agarose gel to check the size, specificity, and intensity of the PCR product.

■ PTT REACTIONS

The TNT quick kit from Promega contains a master mix of all the reagents necessary to perform in vitro coupled transcription and translation apart from methionine. Methionine is then added that is either radiolabeled or biotinylated, which allows the proteins to be detected.

1. Defrost a vial of ^{35}S met Promix (Amersham) at room temperature.
2. Set up PTT reactions, including a mutation –ve, a mutation + ve, and a 61-kDa control diluted 1/8 (provided with Promega's kit). For one reaction:

Promix ^{35}S	0.2 µl
TNT Quick (Promega)	3.8 µl
PCR product	1 µl
Total volume	**5 µl**

Mix solutions thoroughly by pipetting up and down, but avoid bubbles (they are hard to remove).

Incubate the reactions on a PCR block at 30°C for 60–90 min. Electrophorese immediately or store at –20°C.

■ ELECTROPHORESIS

Electrophoresis is carried out on a Bio-Rad minigel system:

Spacers and comb should be 0.5 mm thick.

1. Clean comb, spacers, and glass plates with EtOH.
2. Cast separating SDS-PAGE gel(s) of desired concentration (e.g., 15%). Acrylamide should be pipetted in so as to leave ~1 cm below the bottom of the comb for the stacking gel (mark this level beforehand).
3. Carefully layer water over the acrylamide so as to form a level base. Leave to polymerize.

4. Remove **all** water on top of the separating gel with a tissue and 3-MM paper and cast the 5% stacking gel. Insert the 10-well comb (one edge first).
5. Carefully remove the gel comb (best done under running water) and rinse the wells out with distilled water. Remove water from the wells using the corner of a tissue.
6. Prepare a liter of 1× SDS-PAGE running buffer and pour it into the assembled gel rig until both buffer chambers are almost full.
7. Prepare samples: mix 1 µl PTT product with 10 µδ 1× SDS sample buffer. Denature samples (95°C) for 2–3 min and load immediately with a duck-billed pipette tip.
8. Electrophorese samples at 200 V until the dye front has reached the end of the gel (~1 h).

GEL FIXING

Gels can either be fixed directly with SDS-PAGE fix or Coomassie stained/destained visualize total protein. Gels should be fixed for ~1/2 h.

Coomassie staining is useful for visualizing nonradioactive MW markers; the Sigma markers currently in use have MW of 205, 116, 97.4, 66, 45, 29. and 20.1 kDa (load 2.5 µl).

RADIOACTIVE ENHANCEMENT

If required, remove all of the fixative and soak the gel in AMPLIFY (Amersham) on a shaker in darkness for 20–30 min.

AUTORADIOGRAPHY

Work in accordance with radiological safety rules.

1. Transfer the fixed SDS-PAGE onto a sheet of Saran wrap.
2. Remove the stacking gel and discard into the ^{35}S incineration bin.
3. Blot the separating gel onto a small square of 3M blotting paper, cover with Saran wrap, and transfer the gel and Saran wrap to a gel drier.
4. Dry down the gel completely, and expose the gel directly to Hyper-film (Amersham). Autoradiography time is usually 12–24 h.

REFERENCES

1. Roest PA, Roberts RG, Sugino S, van Ommen GJ, den Dunnen JT (1993) Protein truncation test (PTT) for rapid detection of translation-terminating mutations. Hum Mol Genet 2: 1719–1721.
2. van der Luijt R, Khan PM, Vasen H, van Leeuwen C, Tops C, Roest P, den Dunnen J, Fodde R (1994) Rapid detection of translation-terminating mutations at the adenomatous polyposis coli (APC) gene by direct protein truncation test. Genomics 20: 1–4.

26

DHPLC

Claire F. Taylor
Cancer Research UK, Mutation Detection Facility, Leeds, UK

■ INTRODUCTION

Denaturing HPLC (DHPLC) (also known as temperature-modulated heteroduplex analysis) exploits the differential melting properties of homo- and heteroduplex DNA to detect mutations (1, 2).

A DNA sample is injected onto a chromatography column and binds to the column matrix via interactions mediated by the ion-pairing reagent triethylammonium acetate (TEAA). Acetonitrile disrupts this interaction and causes DNA to be eluted from the column. The $A_{260\,nm}$ of the column eluate is measured, and so elution of samples is seen as a peak of absorbance in a plot of absorbance versus time. At a nondenaturing temperature, the concentration of acetonitrile required to remove any given fragment depends on the size and sequence of that fragment. Any DNA fragment, therefore, elutes at a characteristic point in a linear gradient of acetonitrile, which corresponds to a characteristic retention time on a plot of absorbance versus time.

When the temperature of the column is increased, elution takes place at progressively lower acetonitrile concentrations as the DNA fragment starts to denature. Heteroduplex (mismatched) DNA has melting characteristics different from homoduplex DNA and under conditions of partial denaturation has a reduced retention time on the column (Fig. 26.1). At a temperature approximately 1°C below the melting temperature of the fragment, samples containing heteroduplex DNA will show an elution profile significantly different from those containing homoduplex DNA alone (Fig. 26.2). Typically, a single peak from a wild-type sample changes to multiple peaks in samples that contain sequence variations .

Guide to Mutation Detection, Edited by Graham R. Taylor and Ian N. M. Day.
ISBN 0-471-23444-3

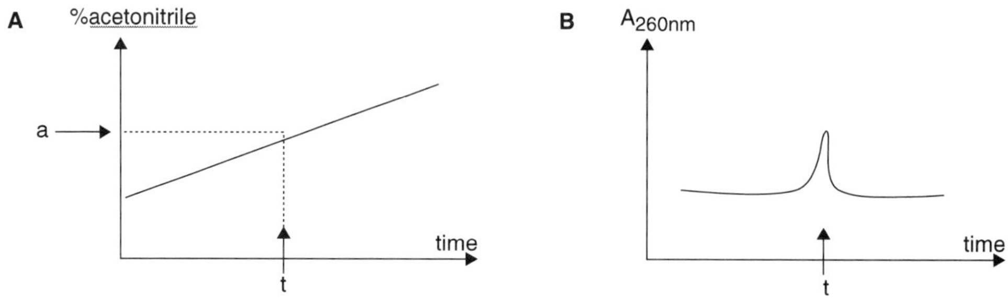

FIGURE 26.1. (A) Linear acetonitrile gradient formed during each analysis. The starting concentration of acetonitrile is low enough for the DNA sample to bind to the column. The concentration of acetonitrile is increased with time and eventually becomes high enough to disrupt the interaction of DNA with the column. This concentration, shown as *a*, corresponds to time *t*. (B) Plot of absorbance in the column eluate vs. time. The DNA sample elutes at a characteristic retention time *t*, which is determined by the acetonitrile concentration, and appears as a peak of absorbance.

To generate an experimental melting curve, aliquots of a wild-type sample are run at temperatures increasing in 2°C increments in the range 50–70°C (it may not be necessary to test all temperatures). A graph of retention time vs. column temperature is plotted. Most mutations are detected at 1°C below the T_m of the fragment. In practice, it may be necessary to test more than one temperature. A biphasic melting curve or a fragment that melts slowly over 10°C or more, can indicate a need for analysis at multiple temperatures.

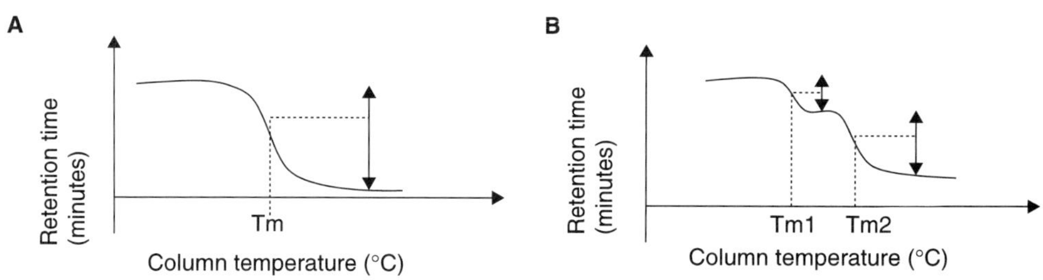

FIGURE 26.2. (A) Typical experimental melting curve showing the decrease in retention time as the column temperature rises. Mutation detection will usually be optimal at 1°C below T_m, although there may be exceptions to this. (B) Biphasic melting curve. Two T_m values and two mutation detection temperatures can be predicted. Analysis should be carried out at both temperatures.

Several studies have shown that DHPLC is a highly sensitive and specific technique (Table 26.1). Some studies detected 100% of mutations tested (3–5), whereas others found that there were mutations that could not be detected under the conditions used (6–8). Although it is clear that analysis at multiple temperatures can increase the detection rate (5, 8, 9), there are still mutations that remain undetected after analysis at all temperatures. The true sensitivity of DHPLC must therefore be <100%.

■ THEORETICAL PREDICTIONS OF MELTING TEMPERATURE

Two programs, Wavemaker (Transgenomic) and Melt (available at http://insertion.stanford.edu/melt.html) can be used to make a theoretical prediction of melting

TABLE 26.1. Sensitivity and Specificity of DHPLC

Study	Number of fragments	Number of mutations tested	% Mutations detected	% False positive results
8	16	67	95	0
7	23	42	95	0
5	42	103	100	n/d
9	33	37	100	0
4	9	15	100	0
6	2	22	92	0
3	2	48	100	0

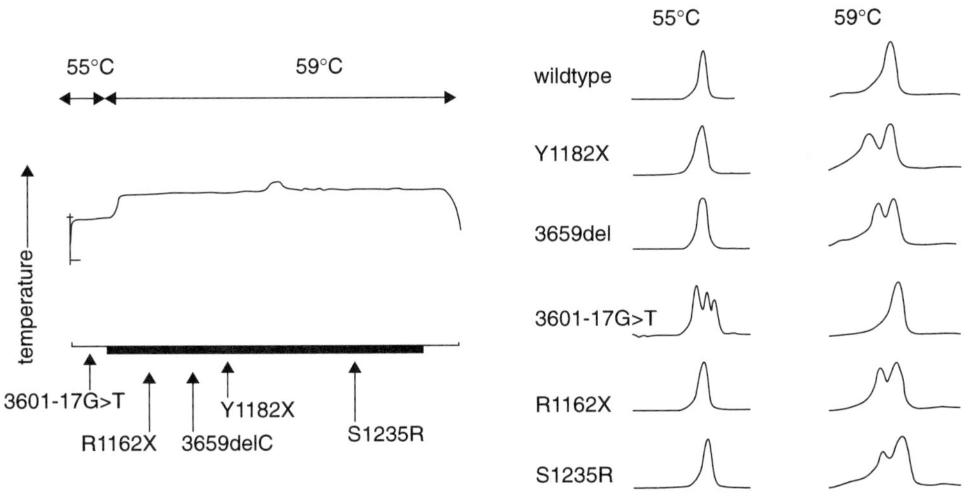

FIGURE 26.3. Melting profile of *ABCC7* (*CFTR*) exon 19.

temperature from the DNA sequence of the fragment under analysis. Using predictions from Melt, Jones et al. (5) recommend analysis at both the temperature predicted by Melt and 2°C higher for maximum mutation detection efficiency. However, there are clearly exceptions to this (8). The prediction of multiple melting domains within a single fragment is one of the most useful features of the theoretical methods; multiple domains may not always be apparent from the experimental melting curve (Fig. 26.3).

SAMPLE PREPARATION FOR DHPLC

Samples suitable for mutation detection by DHPLC are PCR products in the range of 200–500 base pairs. Fragments larger than 500 base pairs can be run, but the sensitivity will be reduced. Fragments of less than about 200 base pairs do not give satisfactory results, as the fragments melt over too narrow a temperature range.

Sample Quality

PCR should be optimized: Poor-quality PCR products will produce poor-quality results. PCR products do not require purification before HPLC; unincorporated nucleotides and primers will elute significantly ahead of the sample peak, whereas template DNA and high-MW contaminants elute after the sample peak. Primer dimers, smears, and contaminating products of a size similar to the sample will interfere with analysis. These contaminants should be eliminated by optimization of the PCR. Gel purification cannot be used to clean up samples for DHPLC, as the reagents and residual agarose damage the column. Any other clean up method should be checked with the HPLC system manufacturer before use.

Sample Quantity

The PCR product should be sufficiently concentrated that 2 µl run on an agarose gel produces a clearly visible band—roughly equivalent to a concentration of at least 20 ng/µl. Usually 3–10 µl (approximately 50–200 ng) of sample would be injected onto the column. Dilute samples produce poor-quality results because of a poor signal-to-noise ratio.

PCR Considerations

The DNASep column is very sensitive to many common PCR reagents, so care must be taken when preparing samples (see Table 26.2).

Heteroduplexing

Mutations are only detected in the form of heteroduplexes. These are easily formed by heating the sample, in the presence of a wild-type DNA spike if necessary, to 95°C, followed by slow cooling to 65°C (Fig. 26.4).

In principle, homozygous mutations can be detected on the basis of a shift in retention time of >0.1 min. In practice, this is not reliable, as the retention time shifts are often within the range of normal variation. To detect homozygous mutations, heteroduplexes must be generated by mixing equal quantities of mutant and wild-type samples (Fig. 26.5).

TABLE 26.2. Reagents and the DNASep Column

Item	Recommended	Comments and Alternatives
Enzyme	Amplitaq Gold	
Buffer	Amplitaq buffer II	Buffer must not contain BSA or formamide. Detergents (e.g., Triton X-100) and up to 10% DMSO do not appear to interfere with analysis or to damage the column.
Primers	Unlabeled	Fluorescent labels will not damage the column but are not recommended. Digoxigenin/biotin-labeled primers have not been tested.
Mineral oil	DO NOT USE MINERAL OIL. A heated lid PCR machine must be used.	Mineral oil will irreversibly damage the column. There are no acceptable clean-up methods.

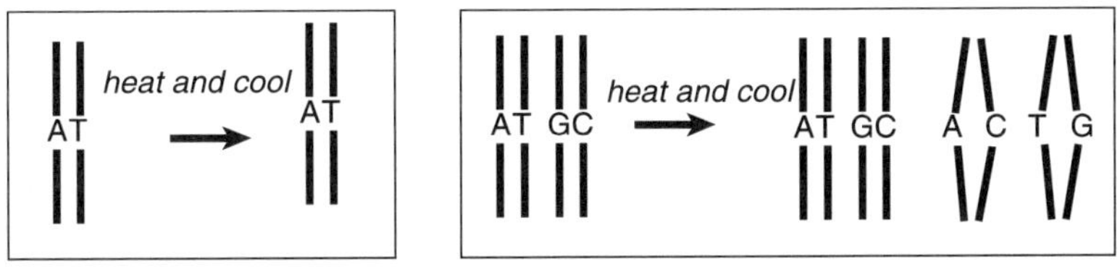

After heating and cooling of a single DNA species, only a single homoduplex molecule is present.
When two DNA molecules, differing by as little as a single mismatch, are heated and cooled together, two homoduplex and two heteroduplex species are formed.
Chromatographic separation of the homo- and heteroduplexes is the basis of DHPLC.

FIGURE 26.4. Formation of heteroduplexes.

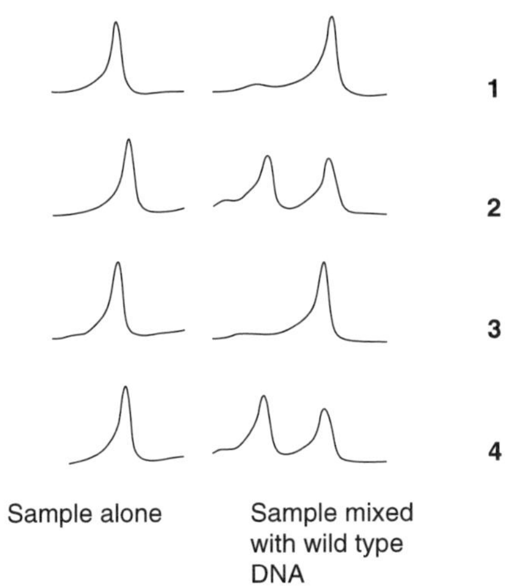

Detection of homozygus mutations.

Four unknown samples were analyzed by DHPLC. Where the sample alone was run, all four samples show identical peak shapes. Each sample was then mixed with 50% of wild type DNA and re-heteroduplexed. Under these conditions, a mutant profile is clearly seen in samples 2 and 4. Sequencing confirmed the presence of a homozygous mutation in both samples 2 and 4.

FIGURE 26.5. Homozygous and heterozygous elution profiles.

Tumor Samples

Samples derived from tumors will often deviate from a 1-to-1 ratio of mutant to wild type. The amount by which this ratio is skewed depends on a number of factors including:

- The relative amounts of tumor and normal tissue in the sample
- Loss of heterozygosity; presence of large deletions or other mechanisms changing gene dosage
- Presence of multiple mutant species in the sample

Ideally, these types of samples should be run as tumor only and as a 1:1 mix of tumor-wild type.

It is not yet clear how large a deviation from 1:1 will be detectable. Kuklin et al. (10) have demonstrated detection at 40:1 wild type: mutant, although this was with the mutation standard control, which is not a typical sample. Liu et al. (6) suggest that a more realistic estimate is likely to be in the range 5:1 to 10:1.

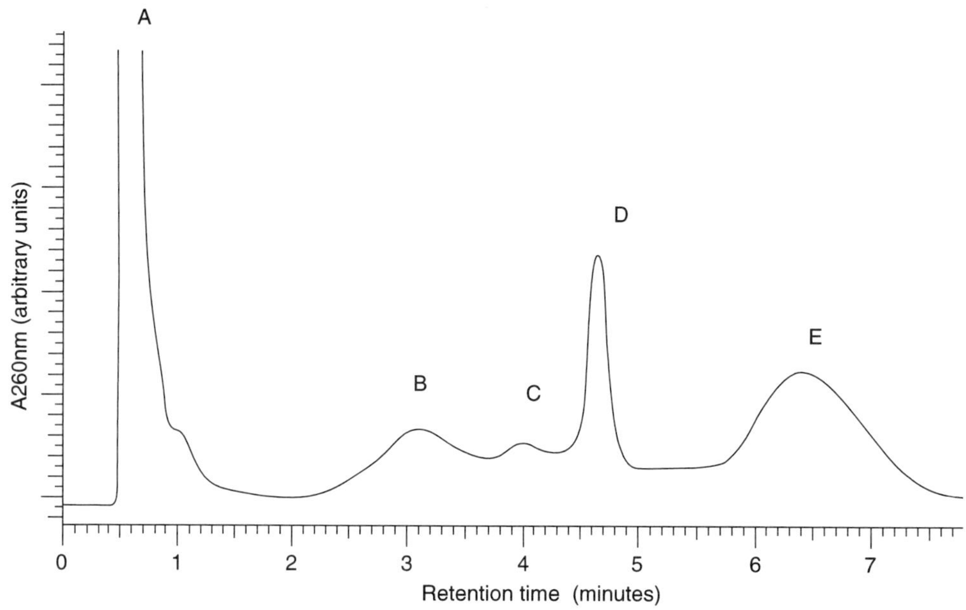

FIGURE 26.6. A typical wild-type chromatogram. A: injection peak: contains unincorporated nucleotides and primers that do not bind to the column. B: variable peak: size of this peak (around 2.5–3 min) is variable. It contains primer dimers and has no relevance for mutation detection. Large amounts of primer dimer in the sample increase the size of the this peak and may obscure the sample peak. C: shoulder on sample peak: size and shape vary between different samples. May be caused either by Taq errors during PCR or by nontemplate A addition. D: sample peak: the majority of wild-type samples appear as a single peak. E: acetonitrile wash peak: acetonitrile is used to elute any remaining bound material from the column after each injection. Acetonitrile absorbs at 260 nm and so appears as a broad peak at 6–7 min. Any high-molecular-weight contaminants of the sample appear as spikes on this peak.

■ INTERPRETATION OF DHPLC DATA

Figure 26.6 illustrates the features of a typical wild-type DHPLC elution profile.

Identification of Mutations

Heteroduplexes have retention properties on the column different from homoduplexes. Most often, the presence of heteroduplexes, and hence of a mutation in the sample, is detected as a change in the number of peaks. The single-peak pattern of the wild type is changed to a two-, three-, or four-peak pattern. Two-peak patterns account for the majority of mutations (see Fig. 26.7).

Wherever possible, a wild-type control should be run and compared with each sample. It is possible for mutations to appear only as a slight broadening of the single peak, or as a subtle change to a shoulder on the peak.

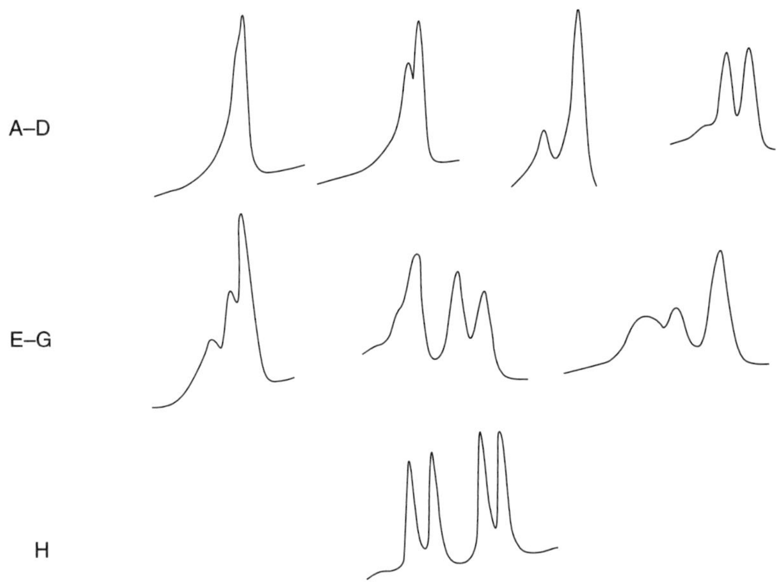

FIGURE 26.7. Examples of elution profiles seen with mutation-wild-type heteroduplexes. It is not yet clear how reproducible and characteristic the mutant patterns produced by any individual mutation are (7, 9). Therefore, once the presence of a mutation has been established by DHPLC, the identity of the mutation should be determined by DNA sequencing.

■ PRACTICAL CONSIDERATIONS FOR DHPLC

The equipment is specialized; users need to develop a good understanding of how the parts function and be prepared to carry out routine maintenance procedures.

A constant supply of buffer to the column needs to be maintained to keep the column in good condition. If the equipment is not kept in constant operation, the system must be fully shut down.

The lifespan of the column is limited: Regular quality control checks must be carried out to ensure that column performance is satisfactory.

Selection of temperatures and analysis of results is subjective.

These factors suggest that, for optimal performance, DHPLC is suited to operation by a core group of specialists within a laboratory or institution.

■ TROUBLESHOOTING GUIDE

Feature	Cause
No sample peak in the chromatogram (Fig. 26.8)	Injection failure, PCR failure
Large spikes in the middle of the chromatogram (Fig. 26.9)	Injection of air onto the column; check the system for leaks
Small spikes on the acetonitrile wash-off peak (Fig. 26.10)	High-molecular-weight products retained on the column until washing
Sample peak eluting on the acetonitrile wash-off peak (Fig. 26.11)	Wrong buffer gradient; increase concentration of buffer B.
Large broad peak eluting at about 2–3 min (Fig. 26.12)	Primer dimer overload, other contaminants

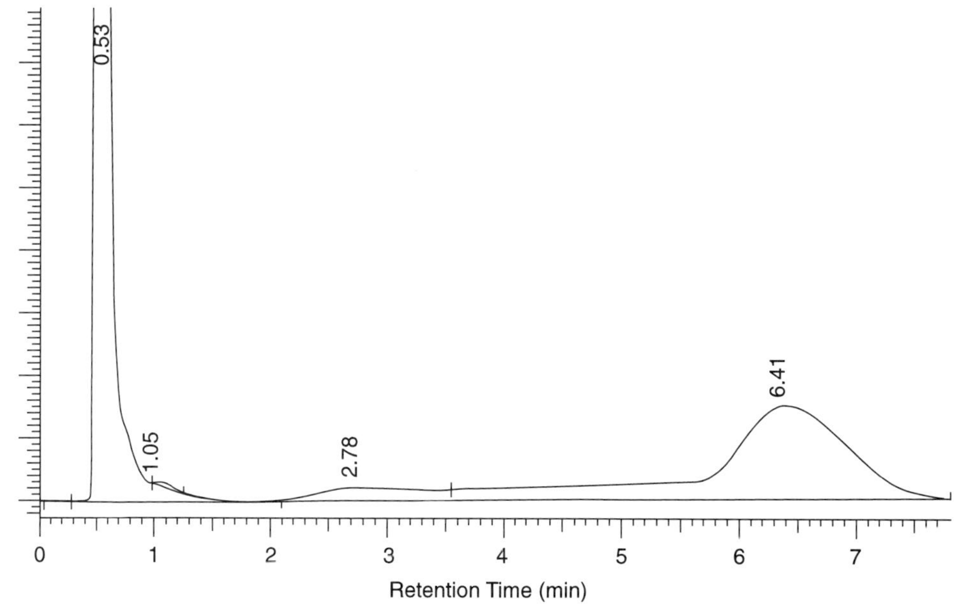

FIGURE 26.8. No sample peak in the chromatogram.

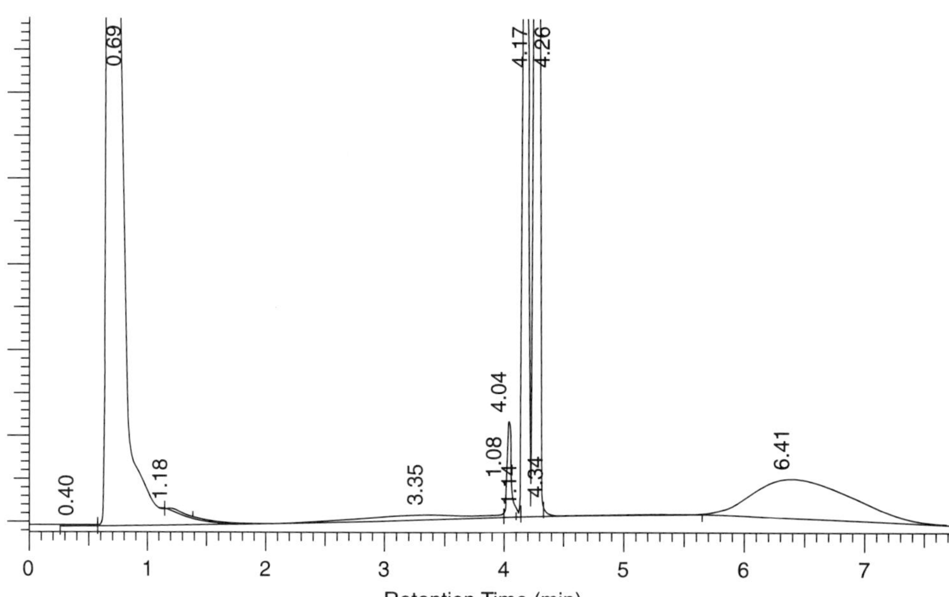

FIGURE 26.9. Large spikes in the middle of the chromatogram.

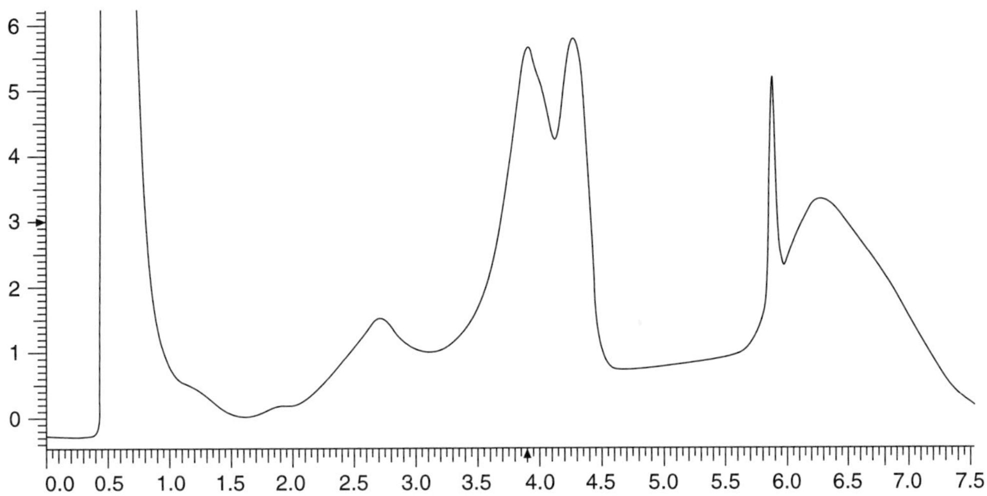

FIGURE 26.10. Small spikes on the acetonitrile wash-off peak.

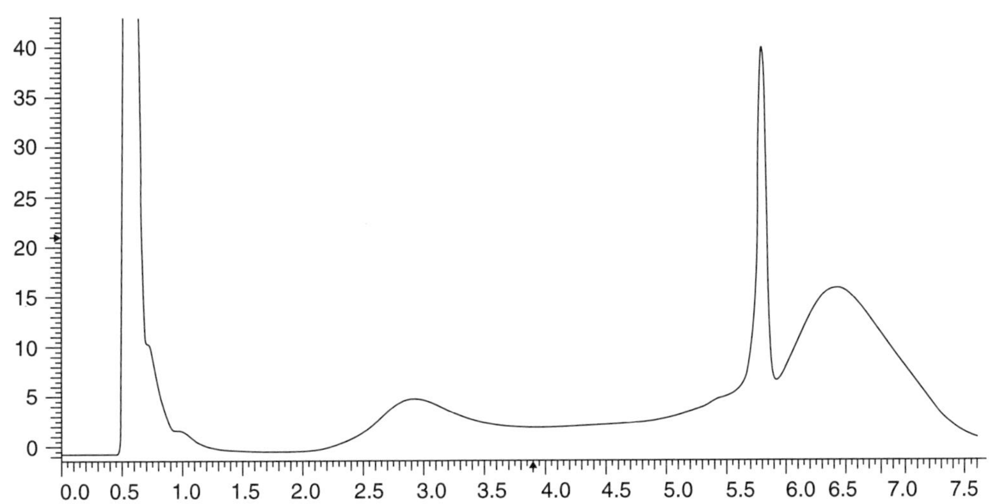

FIGURE 26.11. Sample peak eluting on the acetonitrile wash-off peak.

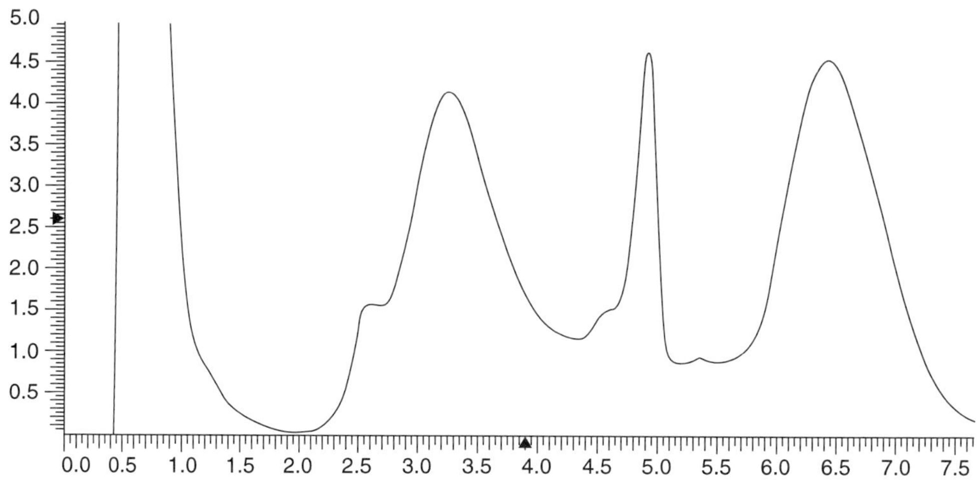

FIGURE 26.12. Large, broad peak eluting at about 2–3 min.

REFERENCES

1. Underhill PA, Jin L, Zemans R, Oefner PJ, Cavalli-Sforza LL (1996) A pre-Columbian Y chromosome-specific transition and its implications for human evolutionary history. Proc Natl Acad Sci USA 93: 196–200.
2. Underhill PA, Jin L, Lin AA, Mehdi SQ, Jenkins T, Vollrath D, Davis RW, Cavalli-Sforza LL, Oefner PJ (1997) Detection of numerous Y chromosome biallelic polymorphisms by denaturing high-performance liquid chromatography. Genome Res 7: 996–1005.
3. O'Donovan MC, Oefner PJ, Roberts SC, Austin J, Hoogendoorn B, Guy C, Speight G, Upadhyaya M, Sommer SS, McGuffin P (1998) Blind analysis of denaturing high-performance liquid chromatography as a tool for mutation detection. Genomics 52: 44–49.
4. Gross E, Arnold N, Goette J, Schwarz-Boeger U, Kiechle M (1999) A comparison of BRCA1 mutation analysis by direct sequencing, SSCP and DHPLC. Hum Genet 105: 72–78.
5. Jones AC, Austin J, Hansen N, Hoogendoorn B, Oefner PJ, Cheadle JP, O'Donovan MC (1999) Optimal temperature selection for mutation detection by denaturing HPLC and comparison to single-stranded conformation polymorphism and heteroduplex analysis. Clin Chem 45: 1133–1140.
6. Liu W, Smith DI, Rechtzigel KJ, Thibodeau SN, James CD (1998) Denaturing high performance liquid chromatography (DHPLC) used in the detection of germline and somatic mutations. Nucleic Acids Res 26: 1396–1400.
7. Dobson-Stone C, Cox RD, Lonie L, Southam L, Fraser M, Wise C, Bernier F, Hodgson S, Porter DE, Simpson AH, Monaco AP (2000) Comparison of fluorescent single-strand conformation polymorphism analysis and denaturing high-performance liquid chromatography for detection of EXT1 and EXT2 mutations in hereditary multiple exostoses. Eur J Hum Genet 8: 24–32.
8. Ellis LA, Taylor CF, Taylor GR (2000) A comparison of fluorescent SSCP and denaturing HPLC for high throughput mutation scanning. Hum Mutat 15: 556–564.
9. Arnold N, Gross E, Schwarz-Boeger U, Pfisterer J, Jonat W, Kiechle M (1999) A highly sensitive, fast, and economical technique for mutation analysis in hereditary breast and ovarian cancers. Hum Mutat 14: 333–339.
10. Kuklin A, Munson K, Gjerde D, Haefele R, Taylor P (1997) Detection of single-nucleotide polymorphisms with the WAVE DNA fragment analysis system. Genet Test 1: 201–206.

27

TEMPERATURE-GRADIENT CAPILLARY ELECTROPHORESIS

Spectrumedix Corp.

Based on the manufacturer's demonstration at the HUGO Mutation Detection Course, Montpellier, September 2002

■ INTRODUCTION

With more DNA sequence information available, genome-wide single nucleotide polymorphism (SNP) discovery and scanning becomes possible. SNPs are of great interest in genetic illness diagnosis and as genetic markers in linkage and association studies. Although many methods have been reported for mutation detection, low-cost, high-throughput, and high-detection-rate methods are still in demand. We have developed a fast and reliable mutation detection system based on temporal temperature-gradient capillary electrophoresis (TGCE) in a massively parallel format. At a temperature close to the T_m of the mutant amplicon, the randomly reannealed combinations of a mutation sample can be separated. By comparing the electropherogram patterns of the wild-type control with mutation samples, mutation positives can be readily identified (Fig. 27.1).

■ THE REVEAL MUTATION DISCOVERY SYSTEM

With the Reveal Mutation Discovery System, up to 96 samples can be screened for mutations simultaneously with a fraction of the cost of other methods with greater sample throughput. The ability of Reveal to analyze large number of mutation samples simultaneously comes from the innovative design that sweeps capillary array through a wide range of temperature during electrophoresis, covering all possible melting temperatures of the samples.

Guide to Mutation Detection, Edited by Graham R. Taylor and Ian N. M. Day.
ISBN 0-471-23444-3

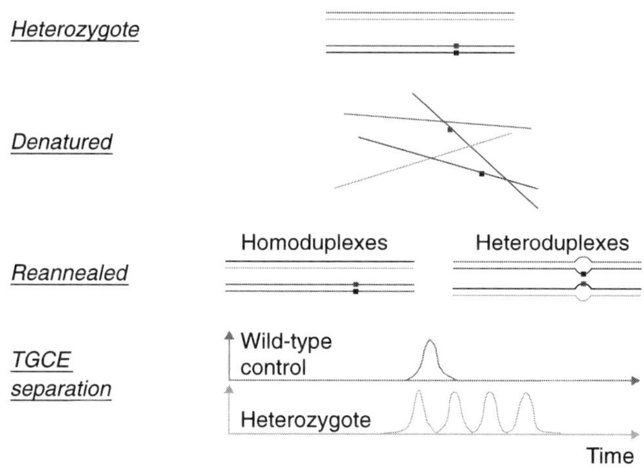

FIGURE 27.1. Scheme of sample preparation and mutation detection by capillary electrophoresis.

Modulated capillary cartridge design in the Reveal System allows precise and fast-response temperature control of the capillary array. Each sample stays at the vicinity of its T_m for a fraction of the total electrophoresis time. Therefore, the method does not require prior knowledge of the type and location of the SNP in a target amplicon (Fig. 27.2).

This novel approach provides two distinctive advantages:

1. Eliminates the need of calibration runs usually required with other technologies to determine the optimum melting temperature for a particular sample
2. Allows simultaneous screening of large number of samples with different T_m in a single run. The technique requires laser-induced fluorescence detection and SpectruMedix screening gel. The use of SpectruMedix screening gel eliminates the need of costly fluorescent dye-tagged reagents. It also differentiates double-stranded fragments from single-stranded DNA for clearer pattern recognition.

■ SAMPLE PREPARATION: HETERODUPLEX AND HOMODUPLEX FORMATION

In this procedure, DNA heteroduplexes and homoduplexes are formed from PCR-amplified products for mutation screening analysis. It starts with PCR amplification of DNA templates from homozygous or heterozygous samples and wild-type controls, which is followed by a denaturing and annealing step to produce heteroduplex and homoduplex species for temperature-gradient capillary electrophoresis.

Materials

1. DNA templates (25–40 ng/μl): homozygous or heterozygous samples and wild-type control
2. A thermal cycler and PCR-related reagents with a proofreading polymerase (e.g. *Pfu* DNA polymerase from Stratagene, La Jolla, CA)
3. 1× TE buffer: 10 mM Tris-Cl, 1 mM EDTA

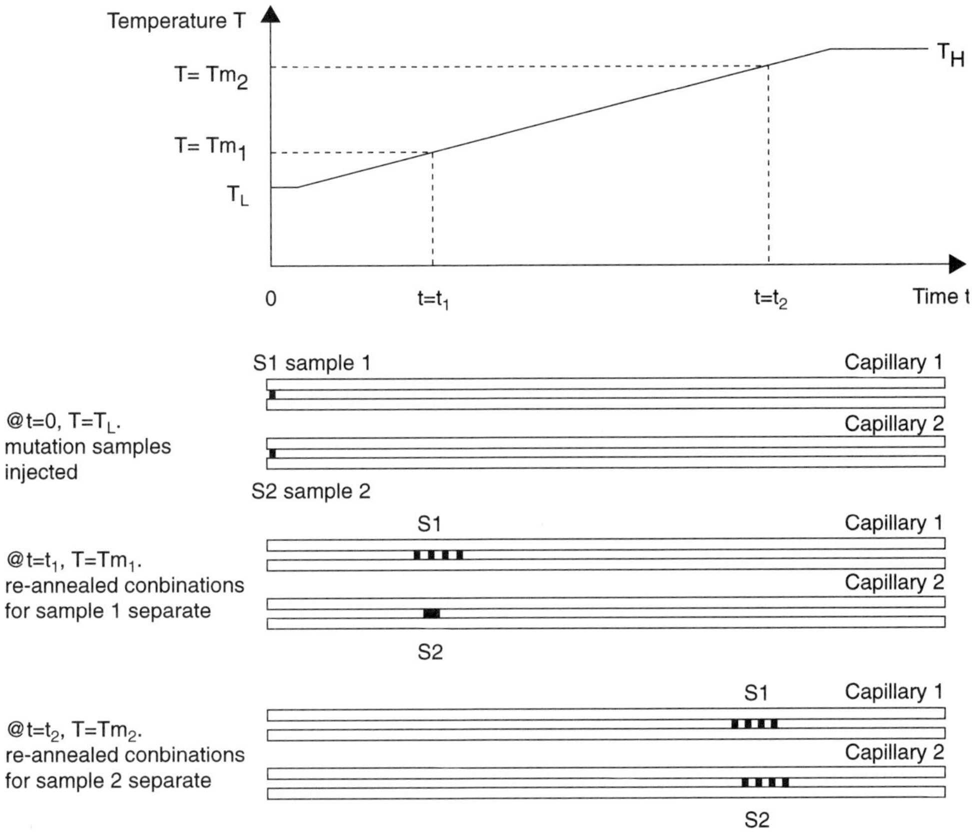

FIGURE 27.2. Parallel TGCE for high-throughput mutation screening.

Procedure Outline

1. Perform PCR reactions as described by the thermocycler manufacturer.
2. Hold 2 min at 94°C.
   a. 30 cycles of 94°C 30 s, 56°C 20 s, 72°C 40 s
   b. Hold at 72°C 10 min.
   c. Hold at 4°C indefinitely.
3. Analyze PCR products to validate purity and concentration.
4. Determine dilution factors for heteroduplex analysis. Crude PCR products can be used without purification. Dilution buffer is recommended for dilution if necessary. Typically, 1:10–1:50 dilutions are recommended. The minimum detectable concentration for dsDNA with SpectruMedix screening gel is ~200 pg/μl.

Prepare Heteroduplex Samples

a. For mutation heterozygotes, use PCR reactions directly.
b. For mutation homozygotes, mix the mutant DNA with the wild type at a 1-to-1 ratio.

c. Add a minimum of 4 µl of the diluted PCR product into each well in a 96-well tray. Overlay each sample with ~10 µl of mineral oil.

d. Prepare heteroduplex samples with a thermal cycler. A typical temperature profile is described as follows:

Hold 3 min at 95°C.

Decrease from 95°C to 80°C at 3°C/min.

Decrease from 80°C to 50°C at 1°C/min.

Hold 20 min at 50°C.

Decrease from 50°C to 45°C at 1°C/min.

Decrease from 45°C to 25°C at 2°C/min.

Samples can be stored at 4°C for a week or at room temperature for 24 h. Samples can be injected up to five times.

MUTATION SCREENING WITH THE REVEAL SYSTEM

Operation Parameters

Wetware (50-µm cartridge)

Bottle 1, Bottle 2, Syringe Bottle 3, Bottle 4

Purge 20, 5 0,0 12 ml 10,3 2 ml

(ml/min, min)

Cap. Flow 22.5,5 25,5 28 µl/cap. 6,6 n/a

(ml/min, min) (V# = 2)

Voltage

PreRun Inject* Current Monitoring Data Acquire*

V (kV) 10 3-6 9 9

Time 6 min 20–60 s 15 min 35–60 min

Temperature Settings

45–55°C, 21-min ramp

50–60°C, 21-min ramp

55–65°C, 21-min ramp

Note: A 21-min ramping is ideal for a sample tray containing fragments between 200 and 500 bp. A shorter ramp time is suggested for shorter fragments, and a longer ramp time is suggested for larger fragments. For example, if all fragments are over 500 bp use a 28-minute ramp.

Note: For TGCE applications the "Detection" (110 V) power cord must be disengaged from the heater controller to allow the DNA to reanneal at the detection window.

* Sample dependent (i.e., concentration and size).

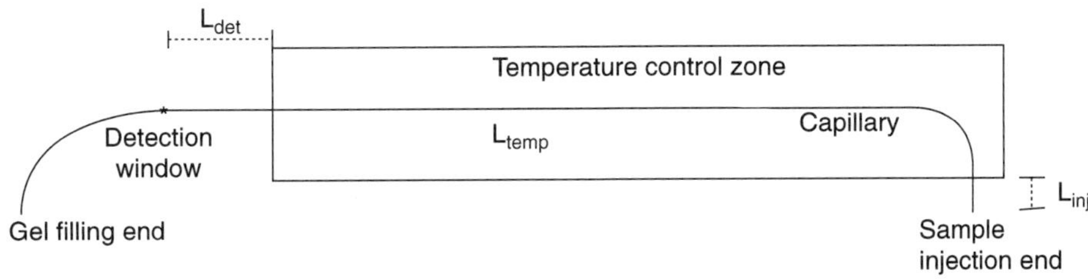

FIGURE 27.3. Schematic of capillaries in a temperature-regulated cartridge. L_{inj} = length of capillary before temperature control zone at sample injection end; L_{temp} = length of capillary in temperature control zone; L_{det} = length of capillary after temperature control zone up to detection window.

ESTABLISHING A NEW TEMPERATURE RAMPING PROFILE

The standard temperature ramping and electrophoresis parameters are provided in the method file. If a user needs to change the running voltage, or temperature ramp, a new set of operation parameters should be established and a new running method file created. To generate reasonably accurate temperature ramping range and speed for all of the fragments in a sample tray, the configuration of the capillary cartridge layout has to be considered.

Figure 27.3 depicts the layout of capillaries in a temperature-regulated cartridge. There are two sections of a capillary that contribute to the separation of samples before detector but are not temperature regulated, namely L_{inj} and L_{det} in Figure 3. The temperature of these two sections of capillaries is in the vicinity of 35°C. By controlling the temperature of the heating box at 35°C and performing a constant-temperature run for a set of different lengths of DNA fragments such as molecular ladders, one can calculate the time required for a certain length of fragment to migrate through these two sections of capillaries.

The data can then be used to calculate the time for a fragment to migrate through the temperature-controlled zone.

A Temperature Ramping Profile is Defined by the Following Parameters

1. Electrophoresis running voltage
2. Temperature ramping range (low temperature T_L to high temperature T_H)
3. The time to start ramping from T_L
4. The rate to ramp to T_H

Procedures to Determine a Temperature Ramping Profile

1. Decide the running voltage.
2. Select a sample standard that includes fragments covering the size range of the samples to be analyzed. The sample standard can be molecular ladder and/or SpectruMedix mutation standards. The fragment size envelope is from the smallest fragment F_s to the largest fragment.

Procedure

Run mutation standard or molecular ladder at 35°C. Obtain the migration time $t_{35C,FL}$ for the high end of the size envelope F_L fragment. The time to start the temperature ramping after sample injection, t_L, can be calculated as

$$t_L = \frac{L_{inj}}{L} t_{35C,FL}$$

where $L = L_{inj} + L_{temp} + L_{det}$. At this time, the largest fragment in the size envelope has entered the temperature-controlled zone.

Run mutation standard and/or molecular ladder at the high end of temperature ramp, T_H. Obtain the migration time for the low end of the size envelope F_S fragment. The latest time for the smallest fragment to exit the temperature controlled region during a temperature ramp t_H can be estimated as:

$$t_H = t_{TH,FL} - \frac{L_{det}}{L} t_{35C,FL}$$

This will be the time for the temperature to reach the high end of the temperature ramp, T_H.

The temperature ramping rate is then:

$$r = \frac{T_H - T_L}{t_H - t_L}$$

Because temperature ramping starts right after injection or before DNA samples enter the controlled-temperature zone, a broader ramp is required to compensate for the temperature ramping occurring in the zone of L_{inj} and ensure the estimate for the L_{temp} zone. Thus the starting low end of the temperature ramp can be estimated as

$$T'_L = T_L - rt_L$$

where rt_L is the temperature increment required along L_{inj}.

■ EXAMPLE

Assume that the sizes of the DNA samples range from 200 to 500 bp. Total length of the capillary for the Version 4 cartridge used on Reveal is:

$$L = L_{inj} + L_{temp} + L_{det} = 4.5\,\text{cm} + 40.5\,\text{cm} + 10.0\,\text{cm} = 55.0\,\text{cm}$$

When run at 10 kV and 35°C constant temperature (see above for other conditions), the migration time for 200 and 500 bp is 36 and 55 min, respectively.

The migration time is the time for a fragment to migrate to the detection window after electrophoresis starts. It includes the time for current monitoring, if any, and the time for data acquisition. The time to start the temperature ramping for the controlled-temperature zone, t_L, can be estimated as:

$$t_L = \frac{L_{inj}}{L} t_{35C,FL} = \frac{4.5\,\text{cm}}{55.0\,\text{cm}} \times 55\,\text{min} = 4.5\,\text{min}$$

When run at 10 kV and 60°C, the migration time for the 200-bp DNA fragment is 27 min. The time for the 200-bp fragment to exit the controlled temperature zone can then be determined as:

$$t_H = t_{TH,FL} - \frac{L_{det}}{L} t_{35C,FS} = 27\,\text{min} - \frac{10.0\text{ cm}}{55.0\text{ cm}} \times 36\,\text{min} = 21\,\text{min}$$

The rate for temperature ramping from 57 to 65°C is then estimated.

Because the Reveal Mutation System begins temperature ramping at the sample injection stage, the actual starting temperature should be adjusted experimentally.

A temperature ramping profile of 55 to 65°C ramp in 21 min has been found to be an effective starting point for analysis. Users should develop and test their own methodology.

For further information contact: SpectruMedix Corp., 2124 Old Gatesburg Rd. State College, PA 16803 (814-867/8600).

28

CHEMICAL CLEAVAGE OF MISMATCH AND ENZYME MISMATCH CLEAVAGE

Richard G. H. Cotton
Genomic Disorders Research Centre, St. Vincent's Hospital, The University of Melbourne, Victoria, Australia

■ INTRODUCTION

Chemical cleavage (1) is used to detect either mismatched thymines or mismatched cytosines in a length of DNA. The mismatch is recognized and modified by potassium permanganate or hydroxylamine and subsequently cleaved by piperidine (see Fig. 28.1). Products are analyzed on an automated DNA sequencer for example, an ABI 377 or 3100. All the reactions are carried out with solid phase for ease of handling.

DNA is usually a PCR product, biotinylated and fluorescently labeled. DNA is cleaned (nucleotides and primers removed) up with the Qiagen PCR Clean up kit and eluted with 50 µl water, and then 50 µl 2× annealing buffer is added. It must be heteroduplexed, and if the patient is homozygous then mixed with equal amounts of wild-type DNA. Heteroduplexes are formed by heating to 95°C for 5 min and then cooling to 25°C at 1°C/min.

Enzyme cleavage is similar but simpler and analyzed in the same way (2–5).

■ REAGENTS

Hydroxylamine

0.556 g of hydroxylamine (on the shelf, room temp.) in a 15-ml tube. Add 1.25 ml of water, add 500 µl of diethylamine, and check the pH by taking 5 µδ into 2 ml of water

Guide to Mutation Detection, Edited by Graham R. Taylor and Ian N. M. Day.
ISBN 0-471-23444-3

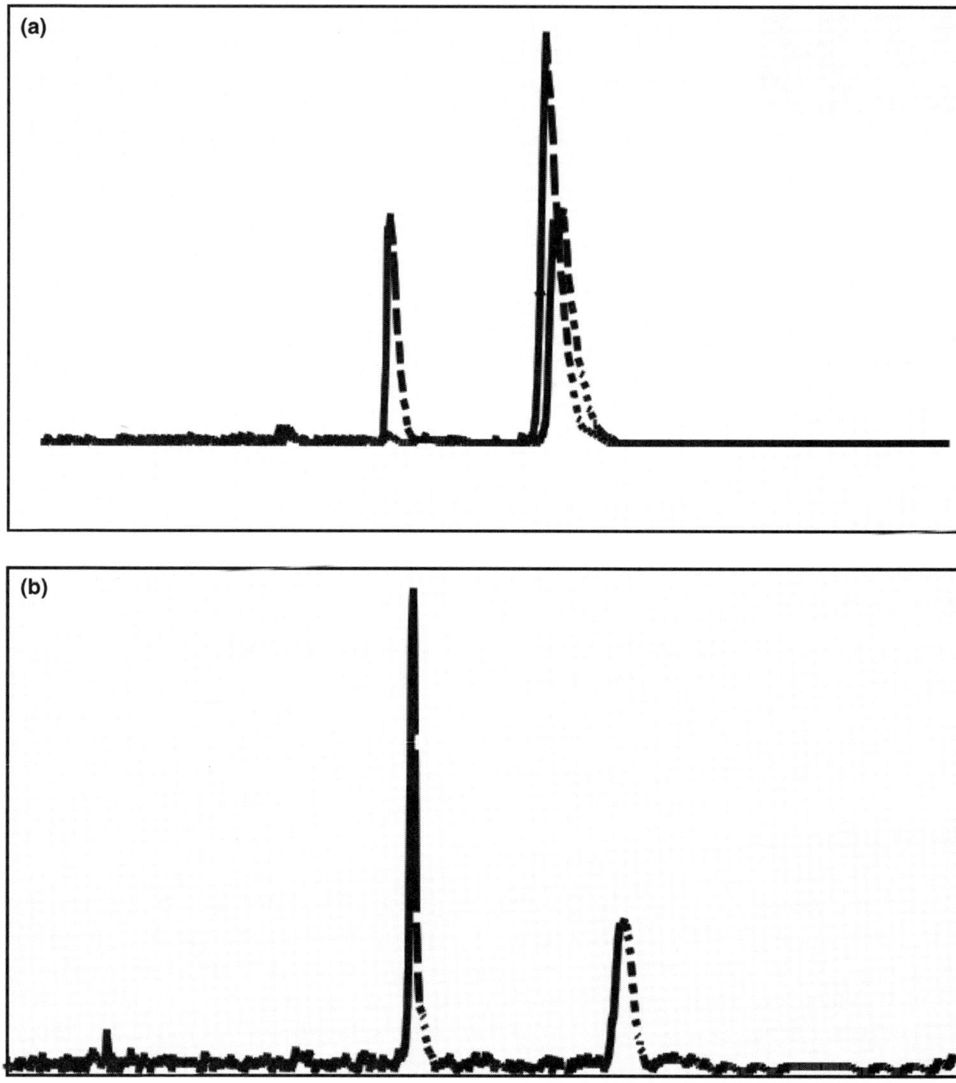

FIGURE 28.1. Cleavage of a 1200-bp heteroduplex by potassium permanganate (a) and T4 endonuclease 7 (b).

and checking with the pH meter. Add more diethylamine until the pH reaches 6; too much, and the pH will be too high and the efficiency will be reduced. With this, the concentration of hydroxylamine will be about 4M and the diethylamine about 2–3M. Make up the volume to 2ml, aliquot into 110µl, and freeze at –80°C. Keep for up to 3 months.

The diethylamine is used in the fume hood and waste discarded in an incineration bin.

Potassium Permanganate

0.16g in 10ml of water. Make this up fresh each time.

Tetraethylammonium Chloride

Make up tetraethylammonium chloride 0.5 g in 1 ml water (don't worry about adjusting the volume, just add 1 ml).

1× Annealing Buffer

For 20 ml

100 mM Tris·HCl pH 8.0	1 ml of 1 M
50 mM NaCl	100 µl of 5 M
40 mM $MgCl_2$	400 µl of 1 M

2× Binding and Wash Buffer (BWB)

For 10 ml

10 mM Tris pH 7.5	100 µl of 1 M
1 mM EDTA	20 µl of 0.5 M
2 M NaCl	4 ml of 5 M

TE

For 10 ml

10 mM Tris pH 8.0	100 µl of 1 M
1 mM EDTA	20 µl of 0.5 M

Piperidine Mix

100 µl deionized formamide
25 µl piperidine
12.5 µl TAMRA 2500 markers

Make this up and aliquot it in the fume hood.

■ METHOD

Hydroxylamine (mismatched Cs)

1. Amount of DNA per reaction : 50–100 ng (or the amount that can be just seen on an agarose gel, normal exposure) in 1µ annealing buffer. Final volume of 15 µl.
2. Wash (10 µl per reaction) streptavidin-coated magnetic beads (the PMPs from Promega are fine and they are much cheaper than Dynabeads) 2× with 1× BWB and resuspend in twice the volume of 2× BWB.
3. Mix DNA (in 15 µl) with 20 µl of washed beads. Leave on wheel for 15 min.

4. Remove supernatant with magnet and add 20 µl of hydroxylamine solution. Make sure beads are resuspended. 37°C for 1 h. Go to step 3.

Potassium Permanganate Reaction ($KMnO_4$)

5. Use the same amount of DNA as for HA reaction and the same amount of beads.
6. Remove the supernatant from the beads and add 20 µl of the reaction mix (5 µl of $KMnO_4$ in 500 µl of TEAC). Incubate at 25°C for 1h. Then go to step 3.
7. Remove supernatant and wash once with 50 µl of TE.
8. Remove TE and leave tubes open to dry while you make up the dyes. Add 5.5 µl of piperidine mix.
9. Make sure the temp block/PCR machine is up to 90°C and then add the tubes and incubate for 30 min.
10. Cool immediately on ice. Load 1.5 µl on the ABI 377 or 3100.

ENZYME MISMATCH DETECTION

Enzyme mismatch cleavage uses endonucleases (e.g. phage T4 endonuclease 7) to selectively cleave mismatched bases in heteroduplex molecules as an alternative to chemical cleavage of mismatches.

EMC Assay Protocols

Centricon/Dynabead Purification

1. Amplify sufficient PCR product to yield 100 ng DNA per EMC reaction (concentrations estimated by comparison with BRL 1-kb ladder on an agarose gel appear to be accurate enough).
2. Purify PCR reactions with a Centricon 100: make PCR reactions up to 2 ml with TE and centrifuge at 2000 rpm for 30 min. Repeat the wash with another 2 ml TE, then with 2 ml Avitech annealing buffer (50 mM Tris_HCl pH 8.0, 25 mM NaCl, 20 mM $MgCl_2$).
3. Invert the Centricon and elute DNA into the retentate cup by centrifuging for 5 min at 2000 rpm.
4. Make eluate up to an estimated concentration of 20 ng DNA/µl with annealing buffer.

Heteroduplex Formation

To make heteroduplex DNA, overlay with mineral oil, then heat to 95°C for 5 min and slowly cool to 25°C at 1°C/min. Leave samples at room temperature.

Dynabead Binding Reaction

1. While heteroduplexing DNA prepare Dynabeads. Aliquot 10 µlL Dynabeads per EMC reaction. Wash twice with an equal volume of freshly made PBS 0.1% BSA (8.0 g NaCl, 0.2 g KCl, 1.44 g Na_2HPO_4, 0.24 g KH_2PO_4 pH to 7.4 with HCl and make up to 1 liter with dH_2O) or 1× TE.

2. Wash Dynabeads once with 20 μl binding buffer (10 mM Tris·HCl, 1 mM EDTA, 1 M NaCl) for each sample to be bound, then with 15 μl. Aliquot 15 μl Dynabeads to each sample tube, magnetize, and remove the supernatant and add 15 μl fresh binding buffer.
3. Add 5 μl (100 ng) heteroduplex DNA to the Dynabeads and incubate at room temperature for 30 min on the rotator or gently agitating every 10 min to prevent Dynabeads settling out of suspension.
4. Magnetize beads and remove binding buffer.
5. Wash samples twice with 50 μl 1× T4 EVII reaction buffer (50 mM Tris·HCl pH 8.0, 10 mM $MgCl_2$, 5 mM DTT, 100 μg/ml BSA). Proceed with enzyme digestions.

T4 EVII Digestions

1. Use a 10-μl reaction volume. Dilute stock T4 EVII (1000 U/μl) 1:8 in dilution buffer (10 mM Tris·HCl pH 8.0, 5 mM DTT, 20% v/v glycerol, 100 μg/ml BSA) and add 1 μl of diluted enzyme to 1 μl 10× T4 EVII reaction buffer and μl dH_2O, i.e., 62.5 U enzyme per 10-μl reaction.
2. Add reaction mix to washed Dynabeads and incubate at 37°C for 1h, agitating every 10 min.
3. Stop reactions by adding an equal volume of deionized formamide/10 mM EDTA loading buffer.
4. Denature samples at 93°C for 3 min, then place on ice before loading on a gel.

Perseptive Particle Purification

1. Pipette 10 μl Perseptive magnetic particles per 50-μl PCR reaction and wash three times with 0.5 M EDTA pH 8.0.
2. To each aliquot of particles add a 50-μl PCR reaction and 50 μl hybridization buffer (2.5 M NaCl, 20% PEG 8000). Mix and leave at room temperature for 10 min.
3. Remove supernatant with a magnet and wash particles twice with 70% ethanol.
4. Allow particles to air dry for 2–10 min, then resuspend in 20 μl distilled water. Incubate at room temperature for 5 min, then magnetize the particles and remove eluate containing the DNA.
5. Spin eluate at top speed in a microfuge for 10 min, then carefully pipette off the supernatant without disturbing the remaining particles that have pelleted in the bottom of the tube.
6. **NB**: this is unnecessary if a sufficiently strong magnet is used.
7. Add an equal volume of 2× annealing buffer (100 mM Tris·HCl pH 8.0, 50 mM NaCl, 40 mM $MgCl_2$) to the eluate, adjust volume to an estimate concentration of 8 pmol/μl with 1× annealing buffer and heteroduplex as above
8. Repeat Perseptive purification after heteroduplex formation.

T4 EVII Digestions [Mashal et al. (2)]

Carry out reactions as detailed above.

Dilute stock enzyme 1 in 100 in dilution buffer (as detailed above) and use 2 μlL of this per 10-μl reaction.

T7 EI Digestions [Mashal et al. (2)]

Carry out reactions as above, adding DMSO to a final concentration of 7.5% in 1× T4 EVII reaction buffer. Use enzyme at an equivalent of 0.002 μl per 10-μl reaction.

REFERENCES

1. Ellis TP, Humphry KE, Smith MJ, Cotton RGH (1998) Chemical cleavage of mismatch: A new look at an established method/recent developments. Human Mutation 11: 345–353.
2. Mashal RD, Koontz J, Sklar J (1995) Detection of mutations by cleavage of DNA heteroduplexes with bacteriophage resolvases. Nat Genet 9(2): 177–183.
3. Babon JJ, McKenzie M, Cotton RG (2003) The use of resolvases T4 endonuclease VII and T7 endonuclease I in mutation detection. Mol Biotechnol 23(1): 73–81.
4. Babon JJ, McKenzie M, Cotton RG (1999) Mutation detection using fluorescent enzyme mismatch cleavage with T4 endonuclease VII. Electrophoresis 20(6): 1162–1170.
5. Youil R, Kemper B, Cotton RG (1996) Detection of 81 of 81 known mouse beta globin promoter mutations with T4 endonuclease VII—the EMC method. Genomics 32(3): 431–435.

29

AUTOMATED DYE TERMINATOR SEQUENCING

Ruth Charlton
Regional DNA Laboratory, St. James's University Hospital, Leeds, UK

■ INTRODUCTION

Most DNA sequencing needs can now be met with terminator chemistry (e.g., Big Dye, Applied Biosystems) and a sequencer such as Applied Biosystems 377 or 3100. Unlabeled primers can be used to sequence PCR products and to detect heterozygotes in fragments of at least 600 bases. PCR products are cleaned to remove excess primers and other material, a single primer and the PCR product are added to sequencing mix, and the reaction is run for about 2.5 h. The reaction product is cleaned by either ethanol precipitation or column purification and loaded onto a 36-cm sequencing gel for automated data collection. Sequencing data are archived on read-write optical disks. Each sequence run must specify the PCR sheet and sample number to enable an audit trail. Reliable calling of heterozygotes requires good signal-to-noise ratios. This can be assessed by comparison with a known sequence and by setting a lower limit on the base-calling accuracy over the region reported.

■ SEQUENCING WITH BIG DYE TERMINATORS ON THE ABI 377

Reagents

Stock Acrylamide Solution

Each week a stock of acrylamide in urea TBE (250 ml) should be prepared. If more than 5 runs per week take place, this may need to be increased. Use acrylamide/bis 19:1 (Bio-Rad) made up to 40% in distilled water.

Guide to Mutation Detection, Edited by Graham R. Taylor and Ian N. M. Day.
ISBN 0-471-23444-3

To Prepare 250-ml Acrylamide for 5 Gel Runs with 36-cm Plates

Add the following in a 500-ml beaker

Urea	90 g
40% Acrylamide/bis	25 ml
Distilled water	125 ml
Amberlite mixed bed resin	2.5 g

Stir until all the urea crystals have dissolved (about 15 min)

Filter 25 ml of 10× TBE through a 0.220-μm filter attached to a side arm flask.

Filter the urea-acrylamide through the same flask.

Important: The TBE must be filtered first so that it does not come into contact with the deionizing resin.

Allow the solution to degas for about 15 min.

Gently pour into a 250- or 500-ml measuring cylinder and adjust the volume to 250 ml with distilled water.

Decant gently into a bottle and store at 4°C for up to 1 week.

Agarose Gel Solutions

1–3% agarose depending on the fragment size, in TAE buffer.

Big Dye Terminator Mix

As supplied by Applied Biosystems

Half Big Dye

As supplied by Genetix

Primers

As supplied by Genosys, Oswel, or MWG Biotech.

95% Ethanol

As supplied by BDH; do not dilute down from absolute ethanol.

70% Ethanol

70 ml absolute ethanol
30 ml distilled water

Loading Buffer

Combine the following in a 5-to-1 ratio:
deionized formamide
25 mM EDTA (pH 8.0) with blue dextran (50 mg/ml)

Ammonium Persulfate

10% Ammonium Persulphate (make fresh)

TEMED

As supplied by Sigma

■ METHOD

Casting the Gel

Assemble the gel plates and spacers in the casting frame as described in the ABI Prism 377 users' manual, Chapter 2.

Take 50 ml of the urea-acrylamide, add 125 μl ammonium persulfate and 25 μl TEMED, and gently swirl to mix without creating air bubbles.

Pour the gel, using either a syringe or a peristaltic pump to introduce the gel solution at the top of the gel or using the syringe adapter from the base of the gel. Gently tap the plates at the gel front to prevent bubbles forming as the solution spreads down the plates by capillary action.

Make sure that there is surplus gel solution at the top of the plates and insert the shark-tooth comb flat side down, taking care not to introduce air bubbles. Allow the gel to set (30–90 min).

Template Preparation

PCR

1. Set up the PCR (50- to 100-μl reactions) with minimal concentrations of nucleotides and primer.
2. Run 5 μl on a 2% agarose gel test gel to check whether the reaction has been successful and to check product purity.

Purification of PCR Products with Centricon Columns

1. Assemble the Centricon-100 column according to manufacturer's recommendations.
2. Load 2 ml deionized water onto the column.
3. Add the entire PCR sample to the column.
4. Spin the column at 3000 *g* in fixed-angle centrifuge for 10 min.
5. Dispose of the filtrate and spin again at 3000 *g* for 5 min.
6. Remove the waste receptacle and attach the collection vial.
7. Invert the column and spin at 500 *g* for 5 min to collect the sample (40–60 μl).
8. Determine concentration of purified PCR products by running 5 μl on an 2% agarose gel against 1 μl 1-kb ladder (1.6-kb band, i.e., strongest band, contains 100 ng DNA).
9. Use 15–45 ng PCR product per sequencing reaction.

Sequencing Reaction

For each PCR sequencing reaction add the following reagents to a separate tube (200 µl)and mix well

2 µl	Big Dye Terminator ready reaction mix
2 µl	Half Big Dye
0.5 µl	Primer (3.2 pmol/µl)
15–45 ng	Purified PCR product
X µl	dH_2O
10 µl	Total volume

To set up control pGEM sequencing mix:

2 µl	Terminator ready reaction mixture
2 µl	Half Big Dye
1.5 µl	pGEM (supplied in kit)
2 µl	M13 primer (supplied in kit)
2.5 µl	dH_2O

Place the tubes in thermocycler (GeneAMP 9700, Applied Biosystems)

Cycle 25 times	96°C	10 s
	55°C	5 s
	60°C	4 min

Set volume to 10 µl.
Then hold at 4°C until ready to purify (or freeze overnight).

Purification of Sequencing Reactions

1. Pipette entire contents of sequencing reaction into 0.5-ml microcentrifuge tube.
2. To each tube add 1 µl 3 M NaAc and 25 µl nondenatured 95% ethanol. Mix well.
3. Leave the tubes at room temperature for 10–15 min to precipitate the extension products.
4. Place the tubes in a microcentrifuge and mark their orientations. Spin 15 min at maximum speed. Proceed immediately to the next step.
5. Carefully aspirate the supernatants (pellets not usually visible).
6. Add 100 µl of 70% ethanol and mix.
7. Spin tubes in microcentrifuge, in the same orientation as step 4, for 5 min at maximum speed.
8. Aspirate the supernatants carefully.
9. Dry the samples with lids open in thermal block for 1 min at 90°C.

Electrophoresis on ABI 377

1. Start up "data collection" software.
2. Create Run Folder (choose "New" from file menu and select the "Sequence Run" icon; this will present you with a worksheet from which you can plate check, prerun, and run the gel.)
3. To plate-check the gel place the cassette into the machine, close the door, and click the plate-check button on the run worksheet. The computer will automatically display a scan window containing 4 color traces. Clean any marks or dust by using a Kimwipe tissue dampened in dH_2O. Once the gel is clean, terminate the plate check.
4. Attach the top buffer chamber and pour in sufficient 1× TBE to cover the wells—**check for leaks**. If no leaks are evident, rinse the wells with a syringe/pipette.
5. Attach the heating/cooling plate and add 1× TBE to the bottom buffer chamber. Start the prerun by clicking the prerun button, prerun the gel for approximately 5 min, then pause the run; this will stop a current passing through the gel, but the gel will continue to be heated by the heating plate. The status of the run can be monitored via the status window that is opened from the "window" section at the top of the screen. Allow the gel to heat to 51°C before loading. While the gel is heating set up the sample sheet. Select "new" from the file dropdown menu at the top left of the screen. Select the sequence "sample icon"; this will present you with a sample worksheet.
6. Once the sample sheet is complete, the sheet can be saved either with its default name or by a unique identifier.
7. Prepare a loading buffer by combining the following in a 5-to-1 ratio: deionized formamide, 25 mM EDTA (pH 8.0) with blue dextran (50 mg/ml).
8. Resuspend each sequencing reaction in 4–6 μl loading buffer.
9. Heat to 95°C for 2 min, then chill on ice.
10. When the gel has reached 51°C, rinse wells and load the first set of samples in alternate wells.
11. Resume the prerun by selecting "Resume" from the "Instrument" dropdown menu.
12. Run for 5–10 min.
13. Pause the run, rinse the wells to be loaded, and load the second set of samples.
14. Cancel the prerun.
15. Choose the appropriate sample sheet from the dropdown menu on the run worksheet.
16. Start the run by clicking the run button. The run can be monitored by the Gel image window (in the "window" section at the top of the screen).

■ SEQUENCING WITH BIG DYE TERMINATORS ON THE ABI 3100

Introduction

Unlabeled primers can be used to sequence PCR products and to detect heterozygotes in fragments of at least 600 bases. Sequencing is usually carried out in both forward and reverse directions.

PCR products are purified with shrimp alkaline phosphatase (SAP) and exonuclease 1 (EXO), which remove excess primers. A single primer and the PCR template are added to the sequencing mix to perform the sequencing reaction. The reaction product is cleaned with an ethanol precipitation. The products are resuspended before loading onto the ABI 3100. 50-cm capillaries are used with POP6 polymer for sequencing.

Reagents

Capillary array 50 cm (Applied Biosystems part number 4315930)
96-Well tray (part number AB-0600)
96-Well septa (Applied Biosystems part number 4315933)
Shrimp alkaline phosphatase (SAP) (USB part number 70073Z)
Exonuclease 1 (EXO) (USB part number 70092Z)
Half Big Dye (Genetix)
DNA sequencing kit—Big Dye version 2 (Applied Biosystems part number 4303152)
10× EDTA buffer (Applied Biosystems part number 402824)
POP6 polymer (Applied Biosystems part number 4316357)
Deionized formamide (Applied Biosystems part number 4311320 or equivalent)

Sequencing in Microtiter 96-Well Plates

PCR Purification

1. Check that PCR has worked on agarose gel.
2. Set up the SAP/EXO reaction for PCR product.

Set up the following for each sample in a microtiter 96-well plate.

1 μl exonuclease I
2 μl shrimp alkaline phosphatase
5 μl PCR product

The reaction can be scaled down to half-volume when sequencing in one direction only.

1. Cover the plate with a heatproof sticky lid.
2. Centrifuge plate for a few seconds to get the samples to the bottom of the wells.
3. Run the following program on the ABI9700 thermal cycler.
   - 37°C for 30 min
   - 80°C for 15 min

Sequencing Reaction

3 μl of purified product is used as a template in a standard sequencing reaction. Keep the Big Dye sequencing mix and Half Big Dye on ice while setting up the reactions.

Note: For problem templates, omit Half Big Dye reagent and use 4 µl of Big Dye reaction mix.

Set up the following 10 µl reactions with primers in the forward and reverse directions. In a microtiter 96-well plate

dH_2O	2.5 µl
Primer	0.5 µl of 3.2 pmol/µl
Half Big Dye	2 µl
Big Dye ready mix	2 µl
PCR template	3 µl of SAP/EXO-treated PCR product

If the volume of PCR template is changed, for example, if the PCR yield is low and more template is added, adjust the volume of dH_2O accordingly.

Apply a heatproof adhesive lid, vortex briefly to mix, and then centrifuge to spin the samples to the bottom of the wells.

Run on an Applied Biosystems 9700 or MJ Tetrad thermal cycler, using the following program

96°C for 10 s	25 to 30 cycles
55°C or 60°C for 5 s	
60°C for 4 min	
4°C hold	

Ethanol Clean up of Sequencing Reactions in Microtiter 96-well Plates

1. Add 1 µl of 3 M sodium acetate and 25 µl of 95% ethanol to each well (can make a master mix and use the distriman repeat pipette).
2. Seal the plate with a sticky lid and vortex to mix.
3. Leave at room temperature for 15–20 min to precipitate.
4. Centrifuge at 2254 *g* for 30 min.
5. Remove lid and invert plate onto tissue paper, centrifuge at 700 *g* for 1 min (prog 2).
6. Add 75–100 µl of 70% ethanol and centrifuge at 2254 *g* for 10 min.
7. Invert plate onto tissue paper and centrifuge at 700 *g* for 1 min.
8. Dry the pellet by placing onto a heating block at 95°C for 1 min.
9. Cover with storage sticky lid and keep at –20°C until day of 3100 run.

Preparation of Samples for 3100 Run

1. Resuspend the sample in 20 µl of HDF (add HDF to the wells without samples to create full runs of 16).
2. Cover with a septa and vortex briefly.
3. Centrifuge for a few seconds to bring to bottom.
4. Heat the plate at 95°C for 1 min, then place immediately on ice.
5. Dab away any excess ice/liquid and place into a 3100 plate holder, ensuring that the orientation of the plate matches the holder.
6. Sample plate is now ready; check that the correct capillary (50 cm) and polymer (POP6) is installed. Check polymer levels and fill if required.

Preparing a Sample Sheet

1. Open the 3100 Data collection by double clicking the icon.
2. Select the sample sheet icon (2nd from left).
3. Type in the plate name as on the sample sheet, e.g., SEQ_1
4. Select sequencing and confirm plate is 96 well; add any comments and click finish.
5. The sequencing sample sheet will open; fill in as below:

Sample name = label each of the wells that contains a sample.
Position, pcr no.,sample, forward or reverse, e.g., A1 = 01_32_1F
B1 = 02_32_2F
A2 = 09_32_1R

Dye set = E
Mobility file = DT3100POP6{BD}v2.mob
bioLIMS project = e.g., 3100 project1
Run module = e.g., a) Stdseq50_600bases
or b) Stdseq50_600bases low inject
or c) Stdseq50_600bases high inject.
Analysis module = BC-3100POP6SR_seqoffFtoff.saz
Check that this sample sheet corresponds to your written sample sheet and plate.
Click **finish**, closing the sample sheet and placing it in the pending plate records box.

Note: Run modules are set up with the following conditions

	StdSeq50_ 600bases	StdSeq50_ 900bases	StdSeq50_ 600bases_ highinject	StdSeq50_ 600bases_ low inject
Run temperature	50	50	50	50
Cap fill volume	184	184	184	184
Current tolerance	100	100	100	100
Run current	100	100	100	100
Voltage tolerance	0.6	0.6	0.6	0.6
Prerun voltage	12.2	12.2	12.2	12.2
Prerun time	180	180	180	180
Injection voltage	1.5	1.5	7.5	1.5
Injection time	10	10	10	2
Run voltage	12.2	12.2	12.2	12.2
Data delay time	1200	1200	1200	1200
Run time	5500	10 000	5500	5500

Starting your 3100 Run

1. Push the tray button on the outside of the machine (make sure button does not stick).
2. Autosampler will move to front of machine; when it has stopped, open the doors.
3. Place the plate onto the autosampler, into position A or B, with cut-off corner of plate holder toward you.

4. Check that the buffer/water septas are laying flat with the plate.
5. Close the doors; autosampler will return automatically.
6. In plate view, click on your pending plate record and then onto the yellow plate linking the plate, which turns green.
7. Your sample sheet will now be in the linked plate records.
8. Select the start button on the toolbar (green arrow 4th from left)
9. You can view your run in array or capillary view, but do not leave like this for long.

Collecting and Analyzing the Data

1. Once a run has finished, "completed" appears in the status column in the run view tab.
2. The resulting data is placed in folders, identifiable by the plate number, within "extracted runs' (shortcut on main screen).
3. Double click on one of your runs and again on one of your samples; this will open Sequence Analysis.
4. Select **add files** and select your runs, click **finish**.
5. Open a sample, select the full view (5th button from left), identify the end point of your sequence, note data points, and close window.
6. Change the number in the stop point column to this; repeat for all samples.
7. Select **analyze** and fill down, deselect the print option.
8. Click **start**, analysis begins and becomes green when complete.
9. Open a sample to check peaks and end point, then close sequence analysis.

Transferring Data

1. Open the shortcut to extracted runs folder on the main screen and identify your runs.
2. The runs should be transferred into a 3100 Working sequencing folder on a network server.
3. Create a new folder, named according to plate name, e.g., Seq_1
4. Select your runs from the extracted runs folder and drag across to the appropriate folders.
5. When complete, drag the same runs to the copied folder, also in the extracted runs window.
6. Close extracted runs and other windows, leaving only the 3100 data collection window open.

Printing Analyzed Data

1. Open the sequence analysis icon.
2. Open **add samples**, find your samples, add them, and click finish.
3. Deselect analysis column and select and fill down print column.
4. Press start to print.
5. Close sequencing analysis when finished.

30

MUTATION DETECTION WITH THE STADEN PACKAGE

James Bonfield
Medical Research Council, Laboratory of Molecular Biology, Cambridge, UK

David Judge
Department of Genetics, University of Cambridge, Cambridge, UK

■ INTRODUCTION

The Staden package is a group of applications for preparing DNA sequence trace data for analysis or assembly, analysis, contig-assembly, trace viewing, and mutation detection (1–3). Full details can be found at: http://staden.sourceforge.net/overview.html.

Versions of the package are available that run in Windows, Unix (including Linux), and Macintosh OS-X. Most automated sequencing devices (including Applied Biosystems, MegaBace, ALF, and Beckman) can output files readable by the Staden package. Mutation detection functions are a relatively recent addition to the package, and they have considerable potential for increasing the throughput and accuracy of mutation calling in automated sequencing.

■ PROCEDURES

Pregap4

Obtaining the Course Data

Please visit http://sourceforge.net/project/showfiles.php?group_id=100316&package_id=108254 for data files used in this course.

Guide to Mutation Detection, Edited by Graham R. Taylor and Ian N. M. Day.
ISBN 0-471-23444-3

Starting Pregap4

Pregap4 is the tool for automating the various steps to take a binary sequence trace file and to convert it into something usable by the sequence assembly tools. Along the way it also performs the mutation analysis steps described below.

Start pregap4 by typing into your command line: **pregap4** (Unix), or selecting "pregap" from the Windows program menu.

Loading Your Files

Pregap4 has a series of menus at the top of the window and a large tabbed notebook underneath this (containing Files to Process, Configure Modules, and Textual Output). We will start with the first tab—Files to Process.

Click on the Add files button to bring up a file browser. Select all the .scf files and the HS14680.embl file by clicking on Any to view all files. Left click on the first file in the "Files:" panel and drag down to the end before letting go (as there are no other files in this directory).

Press OK to accept your selection, and Pregap4 will update to reveal the sequences you have selected.

Selecting the Modules

Now click on the Configure Modules tab to reveal the list of tasks that Pregap4 can perform. The left side of the window is the module name, and the right side shows the configuration parameters for the currently selected module. Select different modules by clicking on their name to the left. Click on the [ ] or [x] boxes on the far left to enable or disable a module. Modules with an [x] next to them will be used by Pregap4. There are a lot of options available in Pregap4, but for our case we can ignore most of them, so follow this simple walk-through. Any modules not listed below should be left disabled.

General Configuration (This module cannot be disabled.) Accept "No" to the Get entry names from trace files question. Pregap4 is offering to rename the files by extracting the "sample names" contained within them. This is useful if you filled out the sample sheet using the ABI sequencing software, but this is not the case for our data.

Phred Enable this. Phred is an alternative base caller to the ABI software. The biggest benefit in using Phred is that for every base called it assigns a probability of it being called correctly.

Estimate Base Accuracies Disable this. It simply measures signal-to-noise ratios as an alternative to re-calling the bases with Phred and assigning confidence values, but it conflicts with Phred, so it should be disabled. If your machine does not have Phred installed, select this option instead.

Trace Format Conversion
Initialize Experiment Files
Augment Experiment Files

Keep these three as they are—enabled. The Trace Format Conversion compresses the trace files to save some disk space. The Experiment Files are textual copies of the traces, much like EMBL files. The Augment step applies the "naming scheme," which we will come to later.

Quality Clip Keep this enabled. It identifies the poor-quality regions of the trace files and hides them from assembly (although the data are not deleted). The average confidence used to determine good data defaults to 15, which is approximately equivalent to a 3% error rate.

Sequencing Vector Clip
Screen for Unclipped Vector
Cloning Vector Clip

Disable all three of these modules: They are intended for shotgun-sequencing projects where a vector sequence (bluescript, pUC, etc) will be present at the start of every sequence. This is not true with our data.

Reference Traces Enable this.

Here we tell Pregap4 what our reference trace files are and the reference sequence is. There are three buttons labeled Browse. Click on each in turn and specify the following:

Reference Trace (+ve strand) = 000906_11cF.scf
Reference Trace (–ve strand) = 000906_11cR.scf
Reference Sequence = HS14680.embl

We shall see later on (in Gap4) how these files are used.

Trace Difference
Heterozygote Scanner

Enable both of these modules: They perform the primary task of automatically detecting mutations.

Gap4 Shotgun Assembly Enable this module. You will also need to specify the sequence database to align (assemble) your sequences into, and as this is does not yet exist we shall create a new one. So answer Gap4 database name with BRCA1 and click on the checkbox next to Create new database.

Assuming all has gone according to plan you should only have the modules enabled that Figure 30.1 shows. Here the Reference Traces module has been highlighted to show the reference traces and sequences used.

Naming Schemes How does Pregap4 know whether a sequence should be compared against the +ve strand or –ve strand reference sequence? We have to tell it. The easiest way of doing this is by careful naming of the files. Note that our files are always in pairs, for example, 000310_11cF.scf and 000310_11cR.scf.

This is exon 11 primer c for patient 000310 in the Forward and Reverse direction. Pregap4 has a choice of several simple naming conventions, including this one.

Pregap4

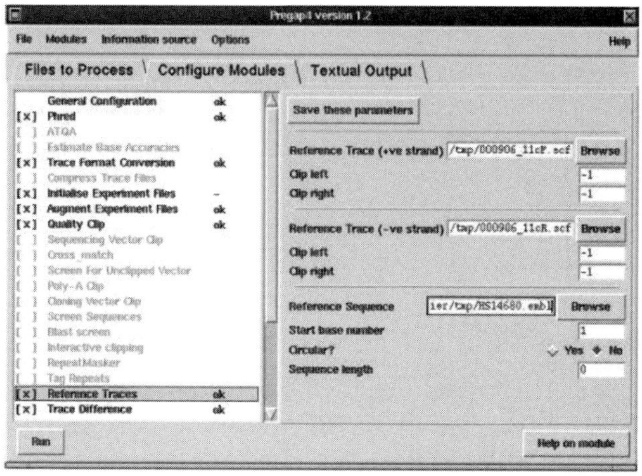

- Automates processing steps form ABI trace files to aligned database of sequences with annotated mutations.
- "Modules" to perform are selected. Example shown here is to specify the reference traces and sequence.

FIGURE 30.1. Pregap4 module selection.

Use the main File menu and select Load Naming Scheme. Press Browse and double click on the mutation_detection.p4t file. Then press OK to close the Load naming scheme window.

Running Pregap4

Before we run pregap4, it is wise to save all of our settings so far. Use the File menu and select Save all Parameters (in all modules). This creates a configuration file in this directory so that any future pregap4 sessions will be far quicker to set up.

Now click the Run button in the bottom left (or select Run from the Modules menu) to set pregap4 going. It will automatically switch to the Textual Output window and display the results and ultimately claim:

*** Processing finished ***

Switch back to your terminal window and list your files once more:

ls

You'll see lots of new files created, the most important being

pregap.passed
  a list of files that passed all processing steps
pregap.failed
  a list of files that failed at least one processing step
BRCA1.0
BRCA1.0.aux
  a gap4 assembly database containing your aligned sequences

Assuming all has worked correctly, exit pregap4.

Gap4

To start up gap4 type into your terminal window: **gap4**

We need to open the assembly database produced by Pregap4, so use the File menu and select ⟨open⟩. Double click on the BRCA1.0.aux file. A new window titled Contig Selector will pop up. A contig is a set of overlapping sequences—in this case our aligned patient and reference sequences. We only have the one aligned set, and so the contig selector window is showing one single horizontal line representing that set. Move the mouse cursor over it, and the line will turn white and it will report the length and number of sequences underneath.

With the mouse cursor on top of the contig line, click the right mouse button to bring up a menu and select ⟨Edit Contig⟩.

Contig Editor

A new window named Contig Editor: 1 HS14680 will now appear (Fig. 30.2).

At the top of the editor are a series of buttons, menus, and scroll bars. Underneath this we have the sequence names on the left and the DNA sequence on the right. The topmost sequence should be HS14680. This is our reference sequence, which can be seen by the S to the left of its name.

Scroll along so position 2200 is central. Note that the reverse strand sequences (those with names ending in R) all have negative numbers before their names (e.g., –19 001321_11cR) whereas the forward strand sequences have positive numbers. This indicates that the reverse strand sequences have been reversed and complemented. (Note that if this is the other way around then it implies that the entire contig needs reversing and complementing, which was one of the other options available from the contig selector popup menu we saw a moment ago.)

Click on the Cutoffs button at the top of the editor window. More sequence should appear, displayed in a light gray color. This is the poor-quality sequence that was hidden by Pregap4's Quality Clip module. We can see the confidence values assigned by the Phred base caller by using the editor Settings menu and selecting Show Reading Quality. The dark bases indicate low confidence, whereas the light bases indicate high confidence. Click on a base to see the confidence (and probability) shown at the bottom of the editor. Double click on a base to view the trace and confirm Phred's base-calling decisions. Tidy up your display by clicking on the Cutoffs button once more to hide the poor-quality data, deselecting Show Reading Quality, and closing any trace windows.

Try positioning your editor cursor somewhere by single clicking on a base call with the left mouse button. An inverse base (light text with a black background) represents the current position of the editing cursor. Try typing some new bases. and you will see that the default editing mode is to overtype. Initially this is all you are allowed to do, but the editor allows much more than this with the Edit Modes menu. We do not have time to discuss all these features here, so for now press the Undo button several times until it beeps, informing you that there are no more edits to undo.

Click on the editor Settings menu and select Highlight disagreements. Now only bases that differ from the consensus (which is the same as the reference sequence in this case) are shown as base calls; all others are displayed as dots. This is a quick way of identifying base calls that differ from the reference sequence (Fig. 30.2).

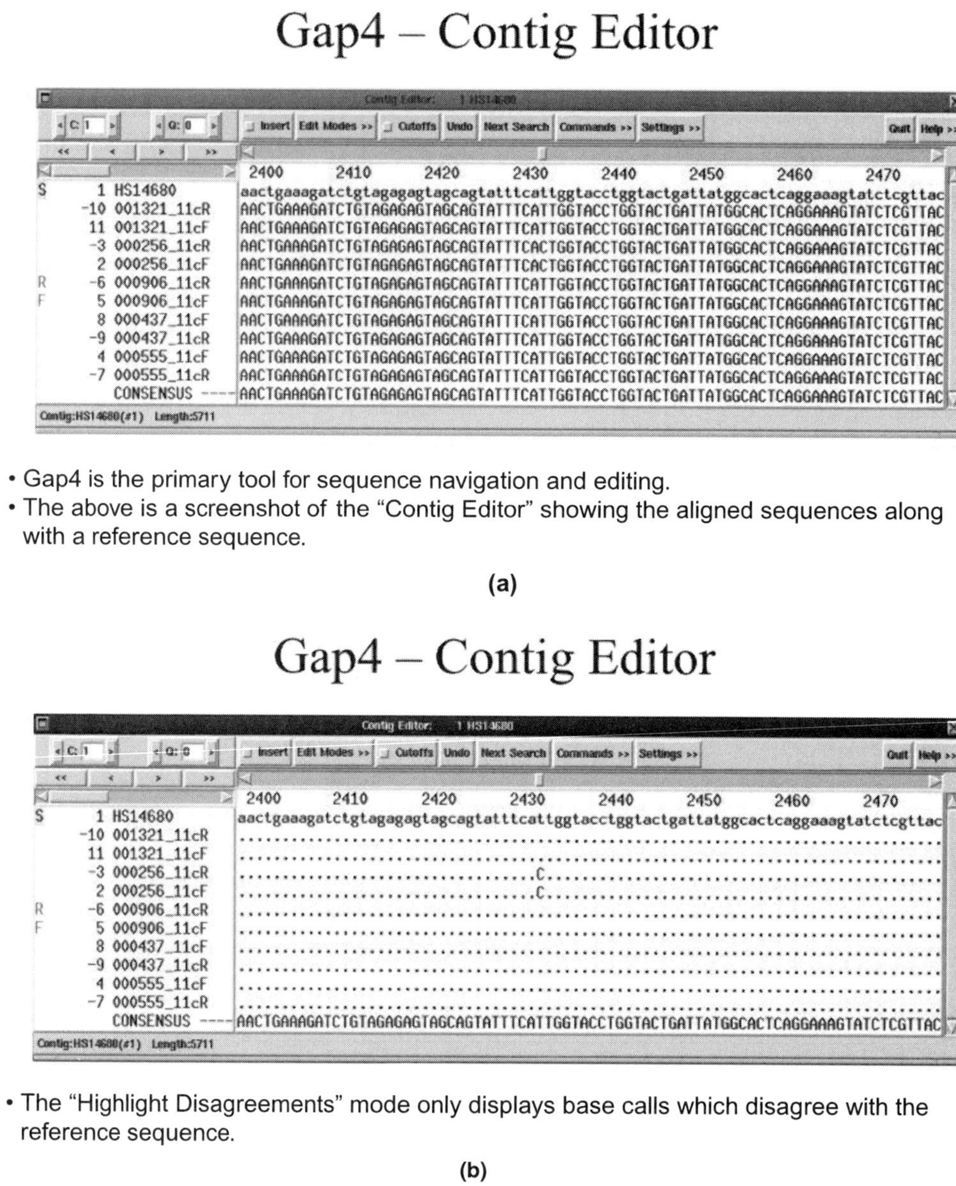

FIGURE 30.2. The gap4 contig editor. (a) Default display; (b) highlighting sequence differences.

Double click at editor base position 2196 for sequences 000256_11cR and 000906 11cR traces, to show their traces. You should be able to see some differences between the two traces. Right click on the top trace and select Trace Difference. Left click on the bottom trace. The differences trace is now visible showing two mutations, one of which is at a heterozygous position. The heterozygous base has been labeled as G by the base caller, which is the same as the reference base call. So looking for base call differences alone is clearly misleading and Highlight Dis-

Reference Sequence

- Load an EMBL sequence with features. CDS features used to determine exon locations.

```
FT   CDS             120..5711
FT                   /codon_start=1
FT                   /db_xref="SWISS-PROT:P38398"
FT                   /gene="BRCA1"
FT                   /product="breast and ovarian cancer
                     susceptibility"
FT                   /protein_id="AAA73985.1"
```

- Used to define a standard numbering for all mutations, regardless of missing base call or alignment issues.
- Gene named and amino acid translation shown underneath contig editor consensus sequence.

FIGURE 30.3. EMBL sequence feature tables.

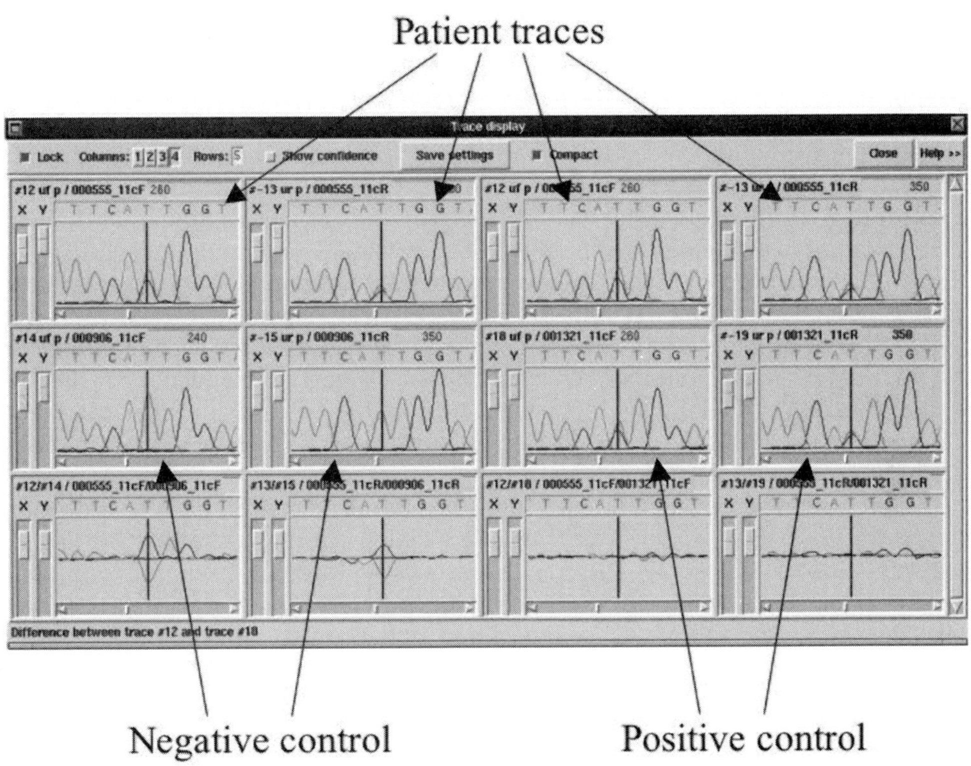

FIGURE 30.4. Viewing trace subtractions for rapid mutation scanning by eye.

agreements alone is not sufficient. Turn off Highlight Disagreements now (by going back into the Settings menu).

The difference plot is computed by aligning the two traces and subtracting one from the other, without any scaling in the *Y*-axis. Hence, when two traces are identical, we expect a flat line central to the trace display. A mutation can be seen as a double peak, one above and one below this central line—two are clearly present in the picture above. A peak in one direction implies a change in the signal strength for one dye only without a corresponding change for a second dye. This is not a mutation, but is typically caused by the presence of an adjacent mutation (as the peak heights are strongly dependent on the preceding bases). Searching for double peaks in the difference plot is precisely how the automatic methods within Pregap4 work. Double peaks sufficiently higher than the background noise will be automatically labeled as mutations and pregap4 then adds "tags" (the red- and orange-colored bases) the bases, which we see in the editor window.

Moving back to the trace display, drag the scroll bar on the bottom difference trace to the right until you see another obvious mutation. Note that the editor window has also been scrolling at the same time. You should be somewhere around editor base 2430 now. To make it easier to verify when mutations have been automatically detected on both strands we can display the forward and reverse sequences vertically adjacent to one another in the editor. (This is what the reading name convention in Pregap4 was used to determine.) Use the Settings menu to select Group Readings by Templates. Note that one pair have an F and R next to the far left of their names—these are the forward and reverse reference traces that we specified in Pregap4.

Now go back to the Settings menu, into the Trace Display submenu, and select Auto-diff Traces.

Try double clicking on a red or orange tag. You should now see 6 traces displayed in a 2 × 3 grid. The lefthand side shows the top (forward) strand, and the right side shows the bottom strand (Fig 30.4).

Next turn on Translate using Feature Tables in the Settings → Status Line menu. The bottom line of the editor should now be showing an amino acid translation of the BRCA1 gene. This information has been extracted from the CDS records in the EMBL reference sequence HS14680 (Fig 30.3).

This allows us to determine what the amino acid change is for each mutation, or whether it is a "silent" mutation. We could also use a genomic sequence for the reference (possibly containing multiple genes with multiple exons), in which case each exon is translated as appropriate. This then allows us to determine when mutations are in noncoding or splicing regions.

To get a summary of mutation positions use the Report Mutations command in the editor Commands menu. This brings up a dialog offering a choice reporting all tagged bases or simply all bases with base calls different from the reference sequence. We can also sort the output sequence by sequence or position by position, depending on whether we wish to know details about a particular patient or are looking for SNPs at a particular site. For now, just accept the defaults and press OK. It may appear that nothing has happened, but if you find your main Gap4 text output window (the first one that came up when you started Gap4) you should see the following:

```
000256_11cF 2196G>R (expressed D>[ND]) (strand–only)
000555_11cF 2196G>R (expressed D>[ND]) (strand–only)
000256_11cF 2201C>T (silent S) (strand–only)
000416_11cF 2201C>T (silent S) (strand–only)
000437_11cF 2201C>T (silent S) (strand–only)
000465_11cF 2201C>Y (silent S) (strand–only)
```

000555_11cF 2201C>T (silent S) (strand–only)
002461_11cF 2201C>T (silent S) (strand–only)
000256_11cF 2430T>C (silent L) (double stranded)
000416_11cF 2430T>Y (silent L) (double stranded)
000437_11cF 2430T>Y (silent L) (double stranded)
000465_11cF 2430T>Y (silent L) (double stranded)
000555_11cF 2430T>Y (silent L) (double stranded)
001321_11cF 2430T>Y (silent L) (double stranded)
002461_11cF 2430T>Y (silent L) (double stranded)
000256_11cF 2731C>T (expressed P>L) (double stranded)
000416_11cF 2731C>Y (expressed P>[PL]) (double stranded)
000437_11cF 2731C>Y (expressed P>[PL]) (double stranded)
000465_11cF 2731C>Y (expressed P>[PL]) (double stranded)
000555_11cF 2731C>Y (expressed P>[PL]) (double stranded)
001321_11cF 2731C>Y (expressed P>[PL]) (double stranded)
002461_11cF 2731C>Y (expressed P>[PL]) (double stranded)
000256_11cF 2788G>T (expressed G>V) (strand–only)

Let's look at the first line in detail.

000256_11cF 2196G>R (expressed D>[ND]) (strand–only)

000256 11cF is the patient sample (actually this represents both the F and R sequence as both are considered here). Position 2196 in the aligned reference sequence is base G, but in the patient sample it is a mix of bases A and G (DNA ambiguity code R). This is the first base of codon GAC, which codes for amino acid D. Mutating it to AAC yields amino acid N. So our mix of A and G gives either amino acid D or N, depending on the copy of the gene being expressed.

This is reported as expressed D>[ND]. Finally note that this mutation has only been observed on the reverse strand sequence.

Tags/Annotations

The Report Mutations function used the colored tags to identify the mutations. These tags were automatically added by the Pregap4 modules. Trace Difference produced the orange tags ("Mutation" tag—"MUTA" in short form) short) by computing the difference traces and detecting peaks above and below the baseline. Heterozygote Scanner produced the red tags ("Heterozygous" tag—"HETE" in short form) by searching for superimposed peaks.

Move back to the editor window and right-click on a red HETE tag and select the Edit Tag command from the popup menu. A new window appears showing the tag type and a text comment, which may be edited to include any details you wish. The very start of the tag will be two base calls, representing the mixture present (e.g., "CT").

Cancel this window and try creating a new tag by right clicking on another base call and selecting Create Tag. Click on the Type:comment field and scroll down the list to select HETE and press OK. Now type a comment, remembering to start your comment with two adjacent base calls representing the mixture (e.g., "AG My comment."). Then press Save and run Report Mutations once more to verify the new output.

Right click on an orange MUTA tag and select Edit Tag again. The format of the first line is slightly different (e.g., C → T) representing the reference sequence base call and the patient base call.

Positive Controls

We may be looking for a specific mutation and not just any difference from the reference sequence.

So far we have used the reference traces as a negative control; a trace in which we expect there to be no mutation. Gap4 also allows us to set sequences to be positive controls.

Right click on the sequence name 001321_11cR in the left side of the editor. You will get a popup menu containing, among other things, set as reference trace. Select this, select Positive control, and press OK. Do the same for 001321_11cF, too. You should now see a lowercase f and r next to these two sequence names. (Note that this menu also allows you to change the negative control traces should you wish to pick something different than 000906_11cR and 000906_11cF.)

Now try double clicking on another sequence at a site of a known mutation to view the traces, such as position 2731 in sequence 000555_11cF. After a short pause you will now see 12 traces arranged in a 4 × 3 grid. The leftmost two columns are the same as before (top and bottom strands compared against the wild-type/negative control reference traces). The rightmost two columns contain the top and bottom strand patient traces compared against the positive control. The difference traces for the positive control should be flat indicating that there is a match between the patient and the positive control.

Try the same again, but this time for sequence 000256_11cF. Here you can see that this patient is different from both our negative and positive controls.

Finally, before exiting the editor use the Settings menu and select Save Settings. This will now remember the various options (such as Group Readings by Templates) you have enabled. Exit the editor by pressing the Quit button in the top right and save any changes.

Template Display

To get an overview of your sequences try the Template Display. Right click on the contig line in the Contig Selector and select Template Display. You'll see a graphical representation of your sequences, starting with lots of short lines with arrows on their ends at the top (representing our patient sequences), a longer line underneath this representing the reference sequence, and finally at the bottom (with numbers next to it) is the consensus sequence.

Try zooming up by pressing the 50% button a few times until the patient data more or less fill the screen.

You'll see lots of colored markers on the sequences indicating the positions of the mutation tags. Try double clicking on one of these tags and you'll get a contig editor launched at that position. Try to rearrange the windows to see the template display and as much of the contig editor window as possible. You'll see that there's now a vertical cursor in the template display. This indicates the position of the editing cursor in the contig editor. Try clicking on the vertical line in the template display and dragging it left or right. You will see that this scrolls the editor. The template display is particularly useful when you are dealing with multiple exons and/or multiple sets of primers as it provides an easy visual overview of the sequence alignments against a genomic reference sequence.

Exiting

Finally, shut down gap4 by going to the main Gap4 text output window and selecting the File menu and Exit.

DISCUSSION OF MUTATION DATA PROCESSING METHODS

The original version of these methods was described by James K Bonfield, Cristina Rada and Rodger Staden in 1998 (2). More recent work has been done by Mark Jordan and James Bonfield.

At present, pregap4 and gap4 clearly show their primary usage in the field of genome assembly. The automatic mutation and heterozygote detection programs work well on all the test data we have, but now they require evaluation by external groups. Such analysis would enable us to improve the algorithms and to tune their parameters. At present, we know that sometimes a base will be declared both as a mutation and as a heterozygous position when visual inspection shows that it is one or the other.

There is still much that can be done overall to improve the methods, but the text above summarizes their status in July 2002. However, since then further development on these methods have ceased due to funding issues. The tools remain available (at http://staden.sourceforge.net/), but for clinical work they should be considered as a way to greatly reduce manual curation rather than a way to replace it.

REFERENCES

1. Bonfield JK, Smith K, Staden R (1995) A new DNA sequence assembly program. Nucleic Acids Res 23: 4992–4999.
2. Bonfield JK, Rada C, Staden R (1998) Automated detection of point mutations using fluorescent sequence trace subtraction. Nucleic Acids Res 26: 3404–3409.
3. Staden R, Judge DP, Bonfield JK (2001) Sequence assembly and finishing methods. Methods Biochem Anal 43: 303–322.

31

GENE DOSAGE BY SEMIQUANTITATIVE PCR

Kim Flintoff
Regional DNA Laboratory, St. James's University Hospital, Leeds, UK

■ INTRODUCTION

This method involves direct detection of duplications or deletions by quantitative fluorescent multiplex PCR (1, 2). To take the case of the APC gene, three fragments from APC (exon 8 and two exon 15 fragments) and three exons (exons 1, 2, and 3) from a control gene, myelin protein zero (P_0) on 1q22–23, are coamplified in a fluorescent multiplex PCR; 2–10 pmol of each primer and 100–200 ng of genomic DNA are used. Only 18–20 PCR cycles are used to ensure that the reaction is kept within the exponential phase.

The products are resolved on either an Applied Biosystems 373 or 377 sequence analyzer or a 310 sequence analyzer (capillary electrophoresis) according to manufacturer's instructions.

The results from the test sample are compared with those from a nondeleted, nonduplicated control sample. Dosage quotients for each STS with each control exon are calculated with peak heights or peak areas as described previously for dosage analysis of the dystrophin gene.

$$\frac{APC \text{ exon 8 (test sample)}/P_0 \text{ exon 1 (text sample)}}{APC \text{ exon 8 (control sample)}/P_0 \text{ exon 1 (control sample)}}$$

A total of 9 dosage quotients are calculated for each test sample. The mean and standard deviation of the 9 dosage quotients are calculated and a qualitative result, duplicated or not duplicated, is determined. If the standard deviation is greater than 10% of the mean, the result is rejected and the test repeated.

Guide to Mutation Detection, Edited by Graham R. Taylor and Ian N. M. Day.
ISBN 0-471-23444-3

TABLE 31.1. Primer Sequences

Primer	Sequence	Ref	Product Size (bp)	pmol per Reaction
APC exon 8 for	ACC TAT AGT CTA AAT TAT ACC ATC	28	184	5
APC exon 8 rev	GTC ATG GCA TTA GTG ACC AG	28		5
APC exon 15A for	GTT ACT GCA TAC ACA TTG TGA C	28	371	2.5
APC exon 15A rev	GCT TTT TGT TTC CTA ACA TGA AG	28		2.5
APC exon 15D for	CTG CCC ATA CAC ATT CAA ACA C	28	382	2.5
APC exon 15D rev	TGT TTG GGT CTT GCC CAT CTT	28		2.5
APC exon 15F for	AAG CCT ACC AAT TAT AGT GAA CG	28	435	7.5
APC exon 15F rev	AGC TGA TGA CAA AGA TGA TAA TG	28		7.5
MPZ exon 1 for	CAG TGG ACA CAA AGC CCT CTG TGT A	Yau (pers. commun.)	389	7.5
MPZ exon 1 rev	GAC ACC TGA GTC CCA AGA CTC CCA G	Yau (pers. commun.)		7.5
MPZ exon 2 for	CTC ACT TCC TCT GTA TCC CTT ACT G	Yau (pers. commun.)	393	10
MPZ exon 2 rev	GGA GGA CAA TGT AGT CAG GGT GAC A	Yau (pers. commun.)		10
MPZ exon 3 for	TGA CAG CTG TGT TCT CAT TAG GGT C	Yau (pers. commun.)	453	7.5
MPZ exon 3 rev	TCC GAG TGT ATG CCC TGC ATT GAG G	Yau (pers. commun.)		7.5

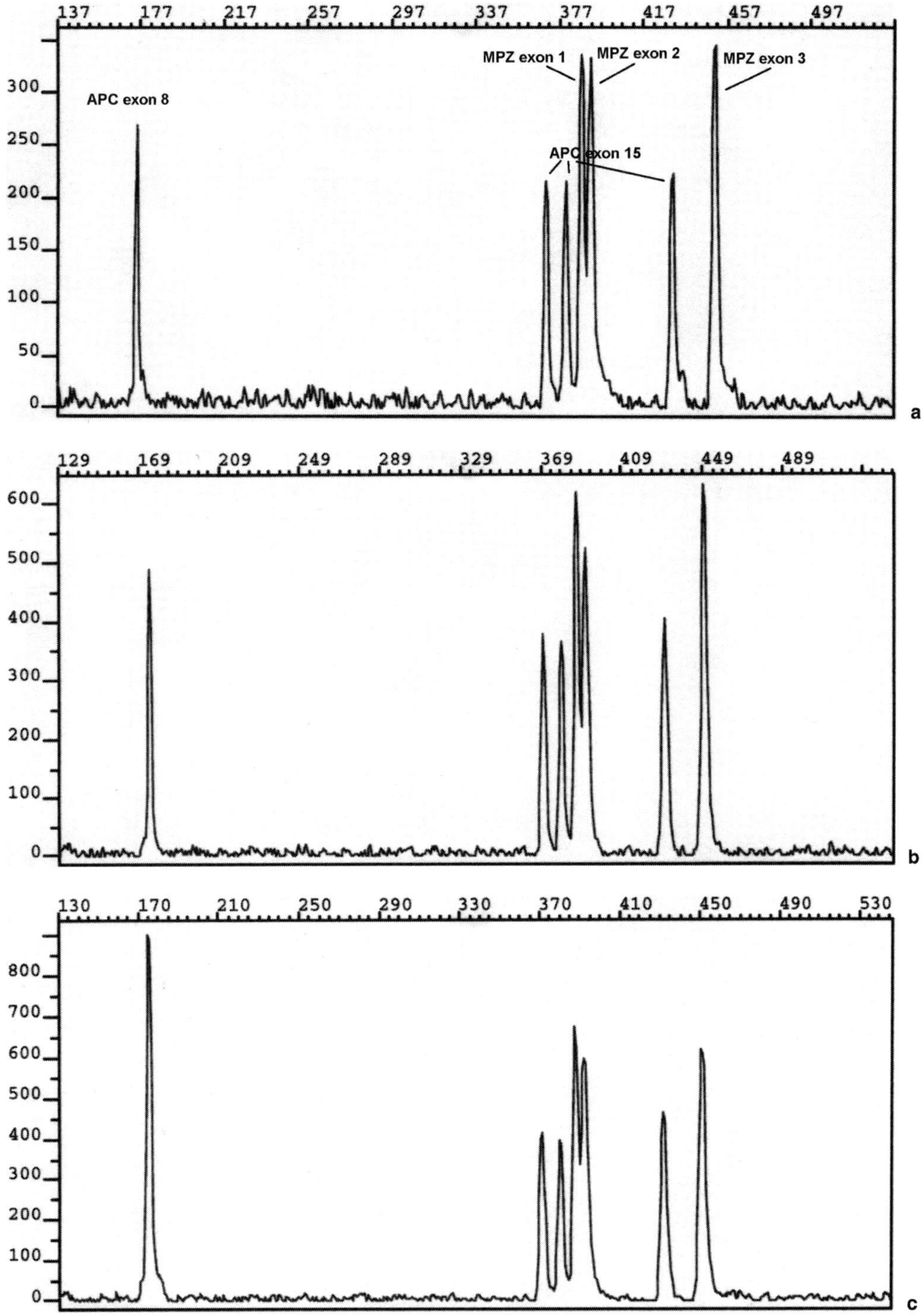

FIGURE 31.1. Quantitative fluorescent PCR of APC and MPZ. (a) Deleted for APC exons 8,15A, 15D, and 15F; (b) deleted for exons 15A, 15D, and 15F, but not exon 8; (c) wild type, no deletion.

TABLE 31.2. Dosage Quotients

A. Deleted for exons 8, 15A, 15D, and 15F

Exon	Sample Peak Height	Control Peak Height	DQs ex1	DQs ex2	DQs ex3
APC8	270	494	0.50	0.51	0.48
APC15A	217	359	0.55	0.56	0.53
APC15D	219	415	0.48	0.49	0.46
APC15F	234	452	0.47	0.48	0.45
PO1	340	309			
PO2	342	319			
PO3	353	307			
	mean DQ (exons 8 and 15)				
	0.50				
	SD				
	0.035				

B. Deleted for exons 15A, 15D, and 15F, but not exon 8

Exon	Sample Peak Height	Control Peak Height	DQs ex1	DQs ex2	DQs ex3
APC8	916	494	0.85	0.98	0.90
APC15A	414	359	0.53	0.61	0.56
APC15D	402	415	0.44	0.51	0.47
APC15F	465	452	0.47	0.54	0.50
PO1	675	309			
PO2	603	319			
PO3	634	307			
	Mean DQ (exon 8)			mean DQ (exon 15)	
	0.91			*0.51*	
	SD			SD	
	0.067			*0.052*	

C. Wild type, no deletion

Exon	Sample Peak Height	Control Peak Height	DQs ex1	DQs ex2	DQs ex3
APC8	375	494	1.00	1.15	1.10
APC15A	265	359	0.97	1.12	1.07
APC15D	249	415	0.79	0.91	0.87
APC15F	308	452	0.90	1.04	0.99
PO1	234	309			
PO2	210	319			
PO3	211	307			
	mean DQ (exons 8 and 15)				
	0.99				
	SD				
	0.110				

Dosage quotients (DQ) and standard deviation (SD) calculations for electropherograms shown in Figure 1.

METHOD

The quantitative multiplex PCR contained amplimers for exons 8, 15A, 15D, and 15F of *APC*, with exons 1, 2, and 3 of *myelin protein zero* (MPZ) (Ref. 2 and Yau SC, personal communication) used as a control (Table 31.1). The 5′ end of one primer from each pair was labeled with the fluorescent phosphoramidite 6-FAM. The reaction was carried out in a total volume of 10 µl and contained the following: 200–250 ng of template DNA, primers, 0.5 U of Promega *Taq* polymerase, 200 µM of each dNTP, 1.5 mM $MgCl_2$ in buffer (50 mM Tris, 15 mM $(NH_4)_2SO_4$, 50 mM KCl, 0.085 mg/ml BSA).

Thermocycling Parameters

95°C for 5 min, 18 cycles of 95°C for 30 s, 52°C for 30 s, 70°C for 1 min, and a final extension at 72°C for 5 min.

Analysis

Products were analyzed with GeneScan software (Applied Biosystems).

Figure 31.1 shows results from two cases and a individual with normal dosage for APC. The amplification of the test sample was compared with the amplification of a control sample with a known point mutation in a different part of the APC gene. The peak height of each amplified exon was used to calculate dosage quotients for all amplicons, for example, the dosage quotient (DQ) for exon 8 of APC and exon 1 of MPZ: (sample APC exon 8 peak height/sample MPZ exon 1 peak height)/(control APC exon 8 peak height/control MPZ exon 1 peak height). Each APC exon was compared with each MPZ exon. A dosage quotient close to 1.0 indicates two copies of the test gene are present; a dosage quotient close to 0.5 indicates only one copy. Dosage quotients and standard deviations of the mean dosage quotients were calculated with duplicate test and control sample loadings.

Dosage quotient and standard deviation results are shown in Table 31.2. The values are close to expected values, with standard deviations approximately 10% of the mean dosage quotient. Two clear groups of results emerge, with no overlap seen between values close to unity or values close to 0.5. Results indicating deletions (dosage quotients close to 0.5) were seen in each case with more than one amplicon, excluding allelic dropout as a cause of reduced amplification.

REFERENCES

1. Yau SC, Bobrow M, Mathew CG, Abbs SJ (1996) Accurate diagnosis of carriers of deletions and duplications in Duchenne/Becker muscular dystrophy by fluorescent dosage analysis. J Med Genet 33: 550–558.
2. Flintoff KJ, Sheridan E, Turner G, Chu CE, Taylor GR (2001) Submicroscopic deletions of the APC gene: a frequent cause of familial adenomatous polyposis that may be overlooked by conventional mutation scanning. J Med Genet 38: 129–132.

32

GENE DOSAGE BY MULTIPLEX AMPLIFIABLE PROBE HYBRIDIZATION

David Cockburn and Helen Snowden
Regional DNA Laboratory, St. James's University Hospital, Leeds, UK

■ INTRODUCTION

Multiplex amplifiable probe hybridization (MAPH) is a quantitative PCR method that involves binding of denatured genomic DNA to small pieces of nylon membrane, followed by hybridization of probes to the immobilized DNA (1–3) (see Fig. 32.1). After PCR amplification of these probes, the products are resolved, the peaks are identified, and the gene dosage of each amplicon can be calculated by comparison to a control. The use of common amplimers for each of the probes enables high-order multiplexing of up to 40 fragments of a size range from 100 to 450 bases.

■ METHOD

Solutions

Blocker Mix

20 μM each of blocking primers (these are same as the primers used for amplification, MAPH-F and MAPH-R)

Prehybridization Solution

280 μl 1.0 M NaH_2PO_4
720 μl 1.0 M Na_2HPO_4

Guide to Mutation Detection, Edited by Graham R. Taylor and Ian N. M. Day.
ISBN 0-471-23444-3

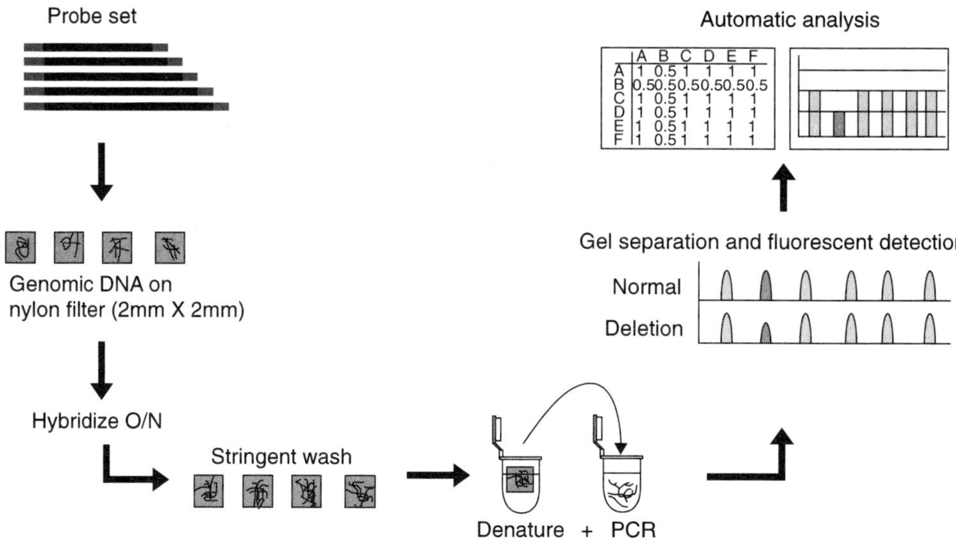

FIGURE 32.1. Schematic of the MAPH process (adapted from www.dmd.nl/DMD_MAPH.html).

700 µl 20% SDS
276 µl H_2O
20 µl herring sperm DNA (10 mg/ml)
4 µl 0.5 M EDTA (pH 8.0)

Wash Solutions

Wash Solution A (1%)

25 ml 20% SSC
50 ml 10% SDS
425 ml H_2O

Wash Solution B (0.1%)

2.5 ml 20% SSC
5 ml 10% SDS
492.5 ml H_2O

Preparation of Filters with Genomic DNA

Where the number of probes exceeds around 40, probes may be divided into 2 sets for hybridization, sets A and B.

A set	Name	B set
1	Sample 1	4
2	Sample 2	5
3	Sample 3	6

1. A list of samples is drawn up and numbered as shown above.
2. For each sample, 2 µl genomic DNA (@ 500 µg/ml) is added to each of 2 tubes (or however many sets of probes are being used) containing 1 µl 1 M NaOH.
3. The DNA is immobilized on small pieces of nylon filter. Using Hybond N membrane, cut a strip approximately of 3-mm width and then cut into 2 mm × 3 mm pieces. Before cutting the pieces, number and underline them with a very sharp pencil as your worksheet (*1*, *2*, *3*, for example). Arrange these pieces on a sheet of Saran film on the bench. Mark the number of each filter next to it on the Saran film.
4. Spot each genomic DNA/NaOH mix onto the filters, 1 µl at a time, allowing the filter to dry between applications (the filter pieces become nontranslucent, pencil labels on the reverse side become nonvisible).
5. When the entire DNA has been applied to the filters, allow to dry and wrap up in the Saran film.
6. Cross-link the DNA to the filters by UV irradiation (Stratalinker @ 500 mJ/cm^2, both sides).

Prehybridization

1. Solutions of prehybridization solution can be made up in advance and frozen (1-ml aliquots and 300-µl aliquots).
2. The A probe filters are prehybridized together in an Eppendorf tube containing 1 ml prehybridization solution. The B probe filters are hybridized together in another Eppendorf tube containing 1 ml prehybridization solution. These are then incubated at 60°C (for 2 hours to overnight).

Hybridization

1. Defrost two 300-µl aliquots of pre-prepared prehybridization solution; to each add 3 µl human Cot-1 DNA and boil for 2 min in the Eppendorf.
2. Remove the prehybridization solution from filters and add 200 µl of the boiled Cot-1/ prehybridization solution. Incubate for 30–60 min at 60°C in the oven or the water bath.
3. Prepare hybridization solution (one for A and one for B tubes)

   1 µl Cot-1

   1 µl herring sperm

   1 µl blocker mix (unlabeled MAPH forward and reverse primer)

   3 µl water

   2 µl probe mix A to one tube, probe mix B to the other tube

   2 µl 1.0 M NaOH
4. Incubate this mixture for 1 min at 37°C, then place on ice.
5. Add 3 µl 1.0 M NaH_2PO_4 and mix thoroughly.
6. Add this mixture to the Eppendorf tubes with the filters (the A probe mixture to the A filters and the B probe mixture to the B tubes).
7. Incubate overnight at 60°C.

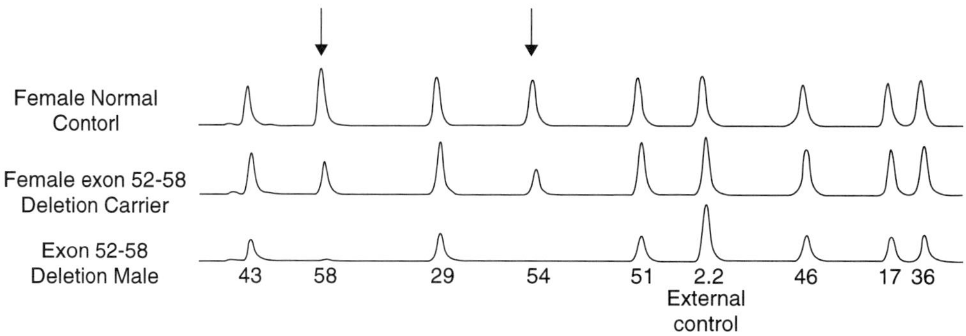

FIGURE 32.2. Example of unprocessed MAPH traces from three individuals probed with dystrophin exons. The relative dosage of each exon (numbered) is determined by examination of peak height or area.

Posthybridization Washes

1. Transfer the filters to a 50-ml tube containing 25 ml prewarmed (60°C) wash solution A, filters from tube A in one 50-ml tube, filters from tube B in a another 50-ml tube.
2. Wash for a total of 45–60 min in the oven:
3. 5 times at 60°C in 25 ml prewarmed (60°C) wash solution A
4. 5 times at 60°C in 25 ml prewarmed (60°C) wash solution B
5. When washing is complete, tip out the filters into a petri dish, identify each filter, and place into its own appropriately labeled PCR tube containing 50 µl PCR buffer.

PCR and Capillary Electrophoresis Fragment Run

1. Incubate the filter tubes at 95°C for 5 min and put on ice.
2. PCR reagents for 1 sample:

Water	4.3 µl
Buffer	1.0 µl
dNTPs	0.2 µl
Mg 0.5 mM	0.4 µl
Taq Polymerase	0.6 µl
Primer	0.8 µl
DNA	2.5 µl

3. PCRs are carried out for both the A probes and the B probes. Both are amplified under the same conditions: 95°C for 5 min, (94°C for 45 s, 63°C for 1 min, 68°C for 1 min) × 23, 68°C for 10 min, 60°C for 40 min.
4. To prepare for analyzing on an ABI 3100, add 9.5 µl HDF and 0.5 µl internal size standard (ROX MapMarker 1000, Cambio) into each well of a 96-well plate. Add 2 µl PCR product, vortex, then spin plate briefly. Heat for 5 min at 95°C and place on ice. The plate is then run on an ABI 3100 analyzer, and data analyzed with the Genescan program.

Data Analysis

Analyze the collected data on Applied Biosystems Genescan. When analysis is complete and all peaks are sized, this can then be imported into Genotyper. In Genotyper, open a premade template for MAPH (probe A or B) to import the Genescan data. This template can have macros set up to label peaks within predetermined categories and to set up a table of data. This table is then copied and pasted into an Excel template spread sheet for dosage analysis and graphical presentation of data.

■ REFERENCES

1. Armour JA, Sismani C, Patsalis PC, Cross G (2000) Measurement of locus copy number by hybridisation with amplifiable probes. Nucleic Acids Res 28: 605–609.
2. Armour JA, Barton DE, Cockburn DJ, Taylor GR (2002) The detection of large deletions or duplications in genomic DNA. Hum Mutat 20: 325–337.
3. Hollox EJ, Akrami SM, Armour JA (2002) DNA copy number analysis by MAPH: molecular diagnostic applications. Expert Rev Mol Diagn 2: 370–378.

33

GENE DOSAGE BY MULTIPLEX LIGATABLE PROBE AMPLIFICATION

Jan Schouten
MRC-Holland, Amsterdam, The Netherlands

Ruth Charlton
Regional DNA Laboratory, St. James's University Hospital, Leeds, UK

■ INTRODUCTION

Multiplex ligatable probe amplification (MLPA), like MAPH, works by comparative quantitation of specifically bound probes that are amplified by PCR with universal primers (1–3). In the MLPA technique (see Fig. 33.1), genomic DNA is hybridized in solution to probe sets, each of which consists of two halves. One half consists of a target-specific sequence (20–30 nucleotides) flanked by a universal primer sequence and can be generated synthetically. The other half also has a target-specific sequence at one end (25–43 nucleotides) and a universal primer sequence at the other, but it has a variable length random fragment in between (19–370 nucleotides) to generate the size differences in the probes. The two probe halves are designed such that the target-specific sequences bind adjacently to the target DNA and can then be joined by use of a ligase. This generates a contiguous probe flanked by universal primer binding sites that can then be amplified by PCR, whereas unbound probe halves cannot be amplified, and hence eliminates the need for removal of excess probe by washing (Fig. 33.1). Under optimal conditions, the yields of amplified, ligated probes will be proportional to the target copy number, so that the relative peak heights indicate gene dosage of target sequence (Fig. 33.2).

Guide to Mutation Detection, Edited by Graham R. Taylor and Ian N. M. Day.
ISBN 0-471-23444-3

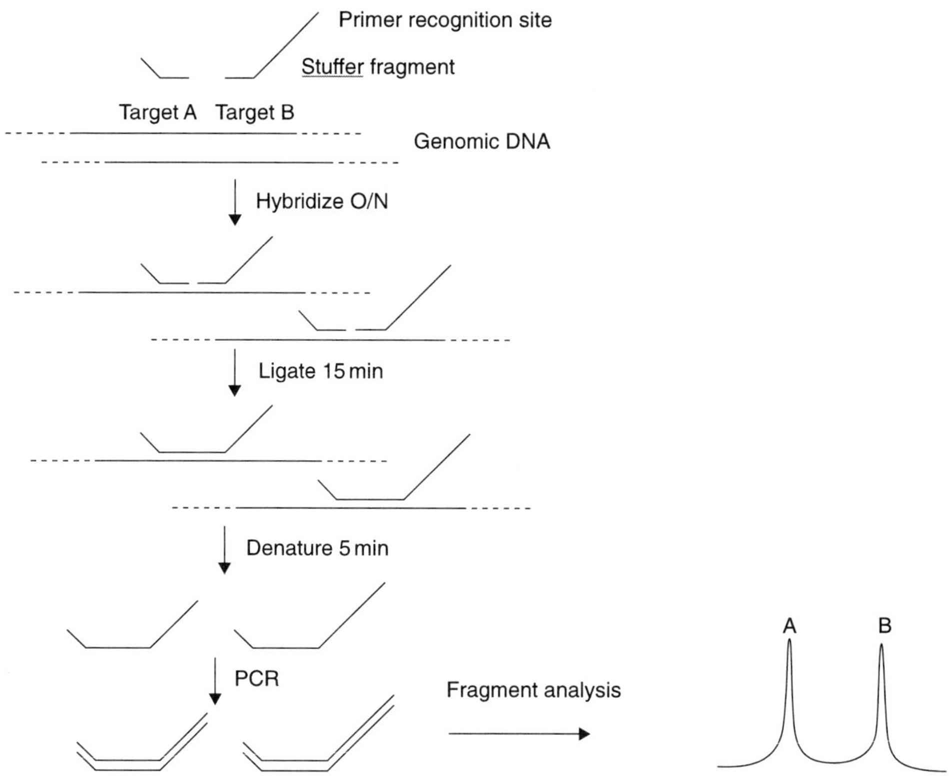

FIGURE 33.1. Outline of the MLPA process.

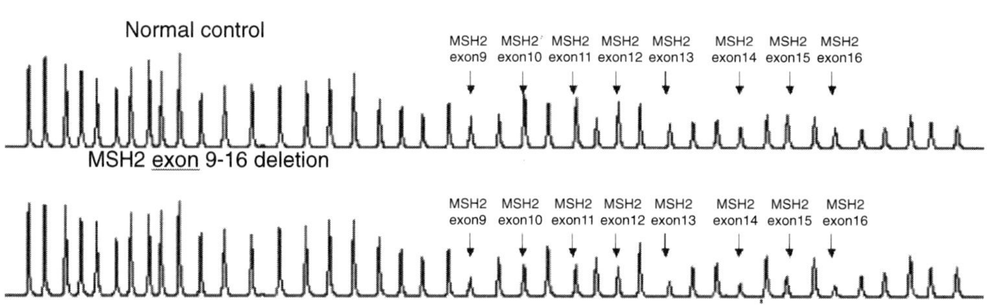

FIGURE 33.2. Example output from an MLPA experiment using the MSH2/MLH1 probe set.

■ REAGENTS

Contents of a Kit for 100 MLPA Reactions

1. SALSA probe mix: 150 μl (black cap).
2. SALSA MLP A buffer: 200 μl (yellow cap).
3. Ligase-65: 120 μl (brown cap).
4. Ligase-65 Buffer A: 400 μl (transparent cap).
5. Ligase-65 Buffer B: 400 μl (white cap).

6. SALSA PCR buffer: 1000 μl (red cap).
7. SALSA PCR primer mix: 250 μl (purple cap): mix of two PCR primers +dNTPs; note the fluorescent label
8. SALSA polymerase: 75 μl (orange cap).
9. SALSA enzyme dilution buffer: 250 μl (blue cap).

Store the kit in the dark at –15 to –25°C. All kit constituents are stable through the control date printed on top of the box when stored at –15 to –25°C.

Items 3 and 8 are enzymes purified from bacteria. Item 1 contains nucleic acids purified from bacteria. Item 7 contains synthetic nucleic acids. Items 2, 4, 5, and 6 contain nontoxic buffers, salts, and detergents. All enzymes, nucleic acids, and buffer constituents are nonhazardous.

■ METHOD

The use of a thermocycler with a heated lid (105°C) is essential. Use 0.2-ml PCR-tubes.

DNA Denaturation and Hybridization of the SALSA Probes

Dilute the DNA-sample (20–500 ng DNA) with TE to 5 μl.
Heat 5 min at 98°C; cool to 25°C before opening the thermocycler.

Add:
1.5 μl SALSA probe mix (black cap)
1.5 μl SALSA MLPA buffer (yellow cap)
Mix with care. Incubate 1 min at 95°C, followed by a 16-h incubation at 60°C.

Ligation reaction:
Mix for each reaction: 3 μl Ligase-65 buffer A (transparent cap)
3 μl Ligase-65 buffer B (white cap)
25 μl water. Mix
Add 1 μl Ligase-65 (brown cap) and mix again.
Reduce temperature of the thermal cycler to 54°C.
Add the ligase mix (32 μl) to the MLPA reaction while at 54°C and mix.
Incubate 15 min at 54°C, than heat 5 min at 98°C.
Proceed with the PCR.

PCR

Mix:
4 μl 10× SALSA PCR buffer (red cap)
26 μl water
10 μl MLPA ligation sample
While tubes are in the thermal cycler at 60°C, add to each tube 10 μl polymerase mix and start PCR reaction.

Polymerase mix (made less than 1 h before use and stored on ice):
2 μl SALSA PCR-primers (purple cap)
2 μl SALSA enzyme dilution buffer (blue cap)
5.5 μl water
Mix and add while on ice: 0.5 μl SALSA polymerase (orange cap)
Mix well but do not vortex.

Incomplete mixing of the viscous 50% glycerol enzyme solutions with dilution buffers is a major source of errors.

PCR Conditions

1. 20 s 95°C
2. 30 s 60°C
3. 60 s 72°C, 33 cycles

End with 20-min incubation at 72°C.

If the use of heat-stable polymerases is still protected by patents in your country, the SALSA polymerase can be disposed of as being nonhazardous waste. Replacement by a licensed preparation of *Taq* polymerase (5 units/μl) will usually provide you with acceptable MLPA results.

Notes on the Denaturation/Hybridization Reaction

1. The use of a thermocycler with a heated lid (105°C) and 0.2-ml PCR-tubes is essential.
2. Use 20–500 ng of human chromosomal DNA dissolved in water or TE. The EDTA concentration in the DNA sample should not exceed 1 mM.
3. The volume of the reaction is important for the hybridization speed, which is probe- and salt concentration dependent.
4. SALSA probe mix and MLPA buffer can be mixed in advance.

The MLPA control mix included in most SALSA probe mixes will generate 4 fragments of 64, 70, 76, and 82 bp even when ligation is omitted. These bands will be small when more than 50 ng human DNA is used for MLPA. If these bands are stronger than the expected MLPA products, either the ligation reaction failed or the amount of sample DNA was much lower than the recommended minimum amount of 20 ng. In either case, the results obtained may not be reliable. A fifth band of 94 bp is generated by the control mix, which is ligation dependent. This is an amplification product of a synthetic MLPA probe specific for the human IL1B gene. This band should be of similar peak area as most of the expected MLPA products.

Recommended Thermal Cycler Program for the Complete MLPA Reaction

1. 5 min 98°C; 25°C pause
2. 1 min 95°C; 60°C pause
3. 54°C pause; 15 min 54°C
4. 5 min 98°C; 4°C pause
5. 60°C pause

6. 20 s 95°C; 30 s 60°C; 60 s 72°C, 33 cycles
7. 20 min 72°C; 4°C pause

Notes on the Ligation Reaction

1. After the 98°C ligase inactivation treatment, samples can be stored at 4°C for 48 h or at –20°C for longer periods. However, best results are obtained when the PCR reaction is started immediately.
2. The ligase mix can be added to the samples as soon as the temperature of the samples has dropped to 54°C.
3. Diluted Ligase-65 can be stored at 4°C for at least 1 h.

Notes on the PCR

1. Never use micropipettes that have been used for handling MLPA PCR products for setting up a MLPA reaction.
2. The manual hot start is not essential. Preparing the complete PCR on ice will result in almost identical results.
3. The number of PCR cycles can usually be reduced to 30. This will result in a slight improvement in the linearity of relative probe signal with target sequence copy number.
4. PCR products can be stored at 4°C for at least 48 h. As the fluorescent labels used are light sensitive, they should be stored in the dark.
5. The relative peak areas of some amplification products are sensitive to the KCl concentration during the PCR reaction. Especially during the PCR protocol, it is important to use identical amounts of MLPA buffer (containing most of the KCl) in each reaction.
6. The volume of the PCR reaction can be brought down to save on reagents.
7. Do not use chemically modified hot-start enzymes such as ABI AmpliTaq Gold Polymerase.

Separation/Relative Quantification of Amplification Products

Electrophoresis

The amount of the MLPA PCR reaction required for analysis by capillary electrophoresis or by sequence gel electrophoresis depends on the apparatus and fluorescent label used.

Agarose gels/Agilent DNA LabChips: (10–12 probes) SALSA unlabeled PCR primer mix contains 6 pM unlabeled M13–577 + 6 pM unlabeled M13–582 primer + 2 mM of each dNTP.

Use 10–15 μl for ethidium bromide-stained 2.5–3% agarose gels.

ABI-310: SALSA FAMPCRprimer mix contains 6 pM unlabeled M13–577 + 4 pMFAM-labeled M13–582 primer + 2 mM of each dNTP.

After the PCR reaction, mix: 0.75 μl of the PCR reaction +0.75 μl water +0.5 μl internal standard (TAMRA-500 Genescan, AB1401733) +12 μl deionized formamide.

This mix is heated for 2 min at 94°C and cooled on ice.

Capillaries: 5–47 cm 50 μm. AB1402839. POP-4 polymer. Settings: 5-s injection at 15 kV.

Run time: 30 min at 60°C and 15 kV.

ABI-3100 (16 capillaries):SALSA FAM PCR primer mix contains 6 pM unlabeled M 13–577 + 4 μM FAM-labeled M13–582 primer + 2 mM of each dNTP.

After the PCR reaction, mix: 1 μl of the PCR reaction + 0.5 μl ROX-labeled internal size standard (ROX-500) + 8.5 μl deionized formamide.

Capillaries: 36 cm polymer: pop 4 Run temperature: 60°C. Cap fill volume: 184 steps. Prerun voltage: 15 kV. Prerun time: 180 s. Injection voltage: 3.0 kV. Injection time: 10 s. Run voltage: 15 kV. Data delay time: 1 s. Run time: 1500 s.

Patient		Control											
9		4											
Category	Height	Category	Height		Ave	SD	5q31	10p11.2	11p13	17q21	4q25	10p14	11p12
C5q31	2371	C5q31	2597	**C5q31**	0.974	0.073	1.000	1.022	0.996	1.085	0.915	0.928	0.870
C10p11.2	2256	C10p11.2	2526	**C10p11.2**	0.952	0.072	0.978	1.000	0.974	1.061	0.895	0.907	0.851
MLH1.EX1	2207	MLH1.EX1	2338	MLH1.EX1	1.007	0.076	1.034	1.057	1.030	1.121	0.946	0.959	0.899
MSH2.EX1	1601	MSH2.EX1	1722	MSH2.EX1	0.991	0.075	1.018	1.041	1.014	1.104	0.932	0.945	0.886
MLH1.EX2	1828	MLH1.EX2	1841	MLH1.EX2	1.059	0.080	1.088	1.112	1.083	1.180	0.995	1.009	0.946
MSH2.EX2	1368	MSH2.EX2	1639	MSH2.EX2	0.890	0.067	0.914	0.935	0.910	0.992	0.836	0.848	0.795
MLH1.EX3	1784	MLH1.EX3	1863	MLH1.EX3	1.021	0.077	1.049	1.072	1.044	1.138	0.960	0.973	0.912
MSH2.EX3	2347	MSH2.EX3	2634	MSH2.EX3	0.950	0.071	0.976	0.998	0.972	1.059	0.893	0.905	0.849
MLH1.EX4	1754	MLH1.EX4	1698	MLH1.EX4	1.102	0.083	1.131	1.157	1.127	1.227	1.035	1.050	0.984
MSH2.EX4	2354	MSH2.EX4	2784	MSH2.EX4	0.902	0.068	0.926	0.947	0.922	1.004	0.847	0.859	0.806
C11p13	1675	C11p13	1827	**C11p13**	0.978	0.074	1.004	1.027	1.000	1.089	0.919	0.932	0.874
MLH1.EX5	1818	MLH1.EX5	2087	MLH1.EX5	0.929	0.070	0.954	0.975	0.950	1.035	0.873	0.885	0.830
MSH2.EX5	1493	MSH2.EX5	1682	MSH2.EX5	0.947	0.071	0.972	0.994	0.968	1.054	0.890	0.902	0.846
MLH1.EX6	1655	MLH1.EX6	1789	MLH1.EX6	0.987	0.074	1.013	1.036	1.009	1.099	0.927	0.940	0.881
MSH2.EX6	1867	MSH2.EX6	2245	MSH2.EX6	0.887	0.067	0.911	0.931	0.907	0.988	0.833	0.845	0.792
MLH1.EX7	2629	MLH1.EX7	2791	MLH1.EX7	1.004	0.076	1.032	1.055	1.027	1.119	0.944	0.957	0.897
MSH2.EX7	2485	MSH2.EX7	2771	MSH2.EX7	0.956	0.072	0.982	1.004	0.978	1.065	0.899	0.911	0.854
MLH1.EX8	1544	MLH1.EX8	1610	MLH1.EX8	1.023	0.077	1.050	1.074	1.046	1.139	0.961	0.974	0.914
MSH2.EX8	1830	MSH2.EX8	1758	MSH2.EX8	1.110	0.084	1.140	1.166	1.135	1.237	1.043	1.058	0.992
MLH1.EX9	1623	MLH1.EX9	1465	MLH1.EX9	1.181	0.089	1.213	1.240	1.208	1.316	1.110	1.126	1.056
C17q21	947	C17q21	1125	**C17q21**	0.898	0.068	0.922	0.943	0.918	1.000	0.844	0.855	0.802
MSH2.EX9	562	MSH2.EX9	1099	MSH2.EX9	[illegible]	0.041	0.560	0.573	0.558	0.607	0.512	0.520	0.487
MLH1.EX10	1268	MLH1.EX10	1327	MLH1.EX10	1.019	0.077	1.047	1.070	1.042	1.135	0.958	0.971	0.910
MSH2.EX10	802	MSH2.EX10	1542	MSH2.EX10	[illegible]	0.042	0.570	0.582	0.567	0.618	0.521	0.528	0.496
MLH1.EX11	1555	MLH1.EX11	1612	MLH1.EX11	1.029	0.077	1.057	1.080	1.052	1.146	0.967	0.980	0.919
MSH2.EX11	984	MSH2.EX11	2349	MSH2.EX11	[illegible]	0.034	0.459	0.469	0.457	0.498	0.420	0.426	0.399
MLH1.EX12	1558	MLH1.EX12	1438	MLH1.EX12	1.155	0.087	1.187	1.213	1.182	1.287	1.086	1.101	1.032
MSH2.EX12	943	MSH2.EX12	1984	MSH2.EX12	[illegible]	0.038	0.521	0.532	0.518	0.565	0.476	0.483	0.453
MLH1.EX13	2003	MLH1.EX13	1886	MLH1.EX13	1.133	0.085	1.163	1.189	1.158	1.262	1.064	1.079	1.012
MSH2.EX13	573	MSH2.EX13	1255	MSH2.EX13	[illegible]	0.037	0.500	0.511	0.498	0.542	0.458	0.464	0.435
C4q25	930	C4q25	932	**C4q25**	1.064	0.080	1.093	1.117	1.088	1.185	1.000	1.014	0.951
MLH1.EX14	1044	MLH1.EX14	1102	MLH1.EX14	1.010	0.076	1.038	1.061	1.033	1.125	0.949	0.963	0.903
MSH2.EX14	533	MSH2.EX14	1042	MSH2.EX14	[illegible]	0.041	0.560	0.573	0.558	0.608	0.513	0.520	0.487
MLH1.EX15	1306	MLH1.EX15	1493	MLH1.EX15	0.933	0.070	0.958	0.979	0.954	1.039	0.877	0.889	0.833
MSH2.EX15	758	MSH2.EX15	1426	MSH2.EX15	[illegible]	0.043	0.582	0.595	0.580	0.631	0.533	0.540	0.506
MLH1.EX16	1472	MLH1.EX16	1428	MLH1.EX16	1.099	0.083	1.129	1.154	1.124	1.225	1.033	1.047	0.982
MSH2.EX16	406	MSH2.EX16	797	MSH2.EX16	[illegible]	0.041	0.558	0.570	0.556	0.605	0.511	0.518	0.485
MLH1.EX17	1028	MLH1.EX17	901	MLH1.EX17	1.217	0.092	1.250	1.278	1.244	1.355	1.143	1.159	1.087
MLH1.EX18	1085	MLH1.EX18	1115	MLH1.EX18	1.038	0.078	1.066	1.090	1.061	1.156	0.975	0.989	0.927
MLH1.EX19	1296	MLH1.EX19	1223	MLH1.EX19	1.130	0.085	1.161	1.187	1.156	1.259	1.062	1.077	1.010
C10p14	871	C10p14	885	**C10p14**	1.050	0.079	1.078	1.102	1.073	1.169	0.986	1.000	0.938
C11p12	1652	C11p12	1574	**C11p12**	1.119	0.084	1.150	1.175	1.145	1.247	1.052	1.066	1.000

FIGURE 33.3. Spreadsheet of MLPA results.

Sequence of the PCR Primers

SALSA PCR 582 primer (labeled): 5′-GGGTTCCCTAAGGGTTGGA-3′

SALSA PCR 577 primer (unlabeled): 5′-GTGCCAGCAAGATCCAATCTAGA-3′

Data Analysis. Analysis of a limited number of samples or rapid quality inspection can be done by visual examination of the capillary electrophoresis peak profiles. For the analysis of large numbers of samples, export of peak heights or areas to excel files is necessary.

Export from Genotyper to Excel

Check that probe name, probe size, and peak height are exported correctly.

Paste into analysis template.

Read the results.

Ready made Genotyper templates can have macros set up to label peaks within predetermined categories and to set up a table of data. This table can then be copied and pasted into an Excel template spread sheet for dosage analysis and graphical presentation of data. The dosage analysis is usually a high-order dosage quotient as developed by Yau et al. (4) for QF PCR, but with many more data points. Macros are set to produce red backgrounds for gene deletions, blue for duplications, and green for standard gene dosage in the "mean" dosage quotient columns (Fig. 33.3).

■ REFERENCES

1. Gille JJ, Hogervorst FB, Pals G, Wijnen J, Van Schooten RJ, Dommering CJ, Meijer GA, Craanen ME, Nederlof PM, De Jong D, McElgunn CJ, Schouten JP, Menko FH (2002) Genomic deletions of MSH2 and MLH1 in colorectal cancer families detected by a novel mutation detection approach. Br J Cancer 87: 892–897.
2. Schouten JP, McElgunn CJ, Waaijer R, Zwijnenburg D, Diepvens F, Pals G (2002) Relative quantification of 40 nucleic acid sequences by multiplex ligation-dependent probe amplification. Nucleic Acids Res 30: e57.
3. Hogervorst FB, Nederlof PM, Gille JJ, McElgunn CJ, Grippeling M, Pruntel R, Regnerus R, van Welsem T, van Spaendonk R, Menko FH, Kluijt I, Dommering C, Verhoef S, Schouten JP, van't Veer LJ, Pals G (2003) Large genomic deletions and duplications in the BRCA1 gene identified by a novel quantitative method. Cancer Res 63: 1449–1453.
4. Yau SC, Bobrow M, Mathew CG, Abbs SJ (1996). Accurate diagnosis of carriers of deletions and duplications in Duchenne/Becker muscular dystrophy by fluorescent dosage analysis. J Med Genet 33: 550–558.

34

ARRAY CGH

Jan P. Dumanski and Patrick G. Buckley
Department of Genetics and Pathology, Uppsala University, Uppsala, Sweden

■ INTRODUCTION

Comparative genomic hybridization (CGH) was originally performed by labeling a control sample with one fluorochrome and a reference sample with a different label. Hybridization of these probes with metaphase chromosomal preparations enabled genome-wide measurement of gene dosage (1). Array-based adaptations replace the metaphase spread with an array of DNA from genomic clones or PCR products, enabling much higher resolution, down to 40 kb for an entire genome or less for individual chromosomes (2–4).

■ METHODS

Preparation of Amino Silane-Coated Slides

The slide racks used for the coating procedure are standard staining jars used in histology labs (see Appendix). Make sure you have containers and slide holders for washing, coating, and rinsing of the slides and that the slide holders fit in ALL containers.

1. Fill the slide racks with 19 slides according to Figure 34.1. Look at each slide and discard those that are visibly stained with streaks of dirt.

Guide to Mutation Detection, Edited by Graham R. Taylor and Ian N. M. Day.
ISBN 0-471-23444-3

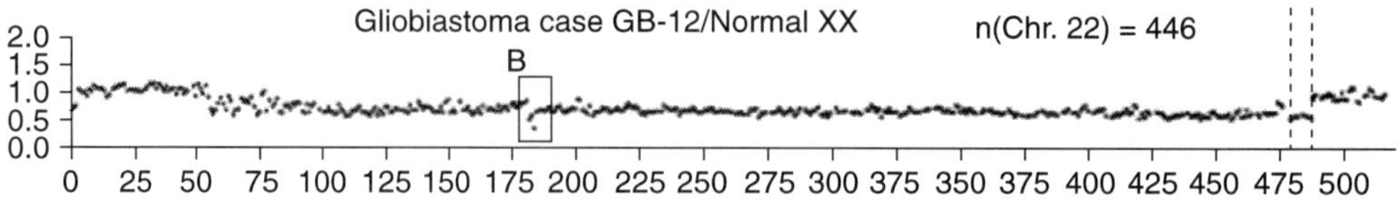

FIGURE 34.1. Detection of a homozygous deletion in a glioblastoma. Adapted from Buckley et al. (2002) Hum Mol Genet 11: 3221–3229.

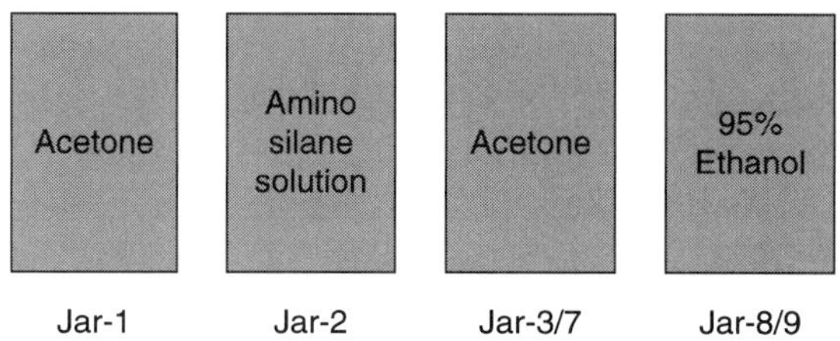

FIGURE 34.2. Diagrammatic scheme of the coating procedure.

2. Prepare wash solution for the slides. Use 200 ml wash solution per slide jar. The wash solution is made of 20-g NaOH pellets dissolved in 80 ml ddH_2O and 120 ml EtOH per jar. This solution should be made fresh.
3. Submerge the holder into the NaOH solution and shake gently for at least 2 h. From now on, keep slides submerged to avoid dust.
4. Rinse off the NaOH cleaning solution from the slides with water. Wash each slide holder for at least 3 min. After rinsing, keep them submerged in water.
5. Prepare the coating procedure by placing 7 of the slide jars in a fume hood. Two additional jars for the final 95% ethanol wash should also be prepared. Figure 34.2 outlines the washing scheme.
6. Prepare the amino silane coating solution by adding 220 µl 3-aminopropyltrimethoxysilane to 200 ml of HPLC-grade acetone.
7. Fill the remaining jars with 200 ml laboratory-grade acetone.
    1. Submerge a slide rack into jar 1 to remove water. Let it sit for ~30 s while gently pulling it up and down out of the jar.
    2. Move the rack to jar 2. Shake gently for 2 min.
    3. Move the rack to jar 3 for 2 min.
    4. Move the rack to jar 4 for 2 min.
    5. Move the rack to jar 5 for 2 min.
    6. Move the rack to jar 6 for 2 min.
    7. Move the rack to jar 7 for 2 min.
    8. Move the rack to a final jar of 95 % EtOH.
    9. When 2 racks have been completed (i.e., after EtOH washes), place them into a plate-spinning adaptor for the Beckman J6 centrifuge and spin at 1400 rpm for 1 min.
    10. Put the racks into a 45°C incubator for 30 min.
    11. Allow the slides to cool and check the surface quality of the coated slides.
    12. Store the slides in a slide box until use (slides can be stored for up to a month).

Sonication of Target DNA for Printing

1. Plug in the sonicator and set the timer on the desired time plus one extra second. **Note**: Sonication time will vary according to DNA size and concentration.
2. Using a squeeze-bottle, rinse the sonicator probe thoroughly with water.
3. Wear ear protection from now on!
4. Fill a 50-ml Falcon 2053 tube with water and sonicate under the same conditions as for the real DNA sample. (e.g., 20% power, cycle 1)
5. Dry the probe with paper tissue.
6. Place the probe as deep as possible into the DNA sample, making sure not to touch the tube. Start sonication by pressing the start button on the timer for at least 1 s.
7. After sonicating a sample, the timer will beep. Briefly press Start/Stop button to stop beeping. This also sets the timer back to the previously adjusted value.
8. Wash the probe with distilled water and wipe clean with a clean tissue.
9. When finished set the timer on "00:00" by pressing simultaneously the "seconds" and "minutes" buttons.
10. Clean the probe thoroughly and unplug the sonicator.
11. After sonication, remove 2 µl for size analysis on 1% agarose minigel (40 ml agarose), with the smallest possible width of combs. Use Lambda-*Hind*III and 123-bp ladder as size markers. The size of your sonicated DNA should be between 5 and 20 kb. If you have a large band at the top of the gel, you may need to resonicate. It is therefore important not to start precipitation before you have the results from the agarose gel.

Double Precipitation and Dissolving of DNA in DMSO

1. Decrease the volume of sonicated DNA in 1.5-ml microfuge tube to approximately 75 µl by spinning the tube in a vacuum centrifuge. For approximation, compare the decreased volume with standard tubes.
2. Move the DNA to a clearly labeled 0.5-ml Eppendorf tube.
3. Precipitate the DNA. (add 1/8 volume of 3 M sodium acetate pH 5.2, mix 30 s, briefly spin, add 1.1 volume of isopropanol, mix slowly, 15-min freeze at –70°C, 15-min centrifuge, remove supernatant, and allow to air dry for 15 min). At this stage it is not necessary to wash with 80% ethanol. Add 20 µl water to the DNA pellet.
4. Reprecipitate the 20-µl DNA solution again as described above. However, wash the pellet with 200 µl 80% ethanol (–20°C) and centrifuge for 7 min at 13,000 before drying.
5. After reprecipitation, open the tube and allow the pellet to air dry for 15 min at room temperature (assuming that most of the ethanol has been removed with smallest pipette tip). There should be no traces of EtOH left in the tube; however, overdrying causes the pellet to crisp, thus making it very difficult to dissolve in the printing solution.
6. Hydrate the pellet in 2 µl ddH_2O. Allow the DNA to absorb the water for 3–5 min.
7. Spin briefly and add 8 µl of DMSO.
8. **Note**: The concentration of DMSO may vary from 50% to 80%.

9. Place the tube on an IKA-Vibrax or Eppendorf shaker overnight to ensure a homogenous concentration of DNA.
10. Prepare an Excel sheet with the full names of all the clones to be pipetted into the printing plate.
11. Pipette 4 µl of target DNA into a well of the 384-well printing plate.

Cleaning of Microarray Printing Pins

Note: SMP3 pins are cleaned before and after the printing process. Extreme care should be taken in handling the printing pins.

1. Prewarm a 2% microcleaning solution to 65°C.
2. Dip the SMP3 pins in the microcleaning solution for 30 min.
3. Sonicate the pins in distilled H_2O in a low-amplitude sonicator for 2 min.
4. Dry the pins with dust-free pressurized air.
5. Observe the cleanliness of the pins under light microscope.
6. If the pins are not clean, repeat steps 2 through 5.

Microarray Printing

AxySys software, supplied with the Cartesian printing robot, is used for programming the printing procedure.

1. Before placing the slides in the printing robot, ensure the uniformity and cleanliness of the slides. Any fibrous material on the slide might clog the pins and result in bad quality of printing.
2. Spin the printing plate(s) at 1400 rpm for 2 min in Eppendorf plate centrifuge.
3. The printing is done at 35% humidity and 22°C. Distilled water at 70°C is used for the intermittent washes between the loading of DNA into the pins.
4. The printing of 80 chromosome 22 array takes approximately 5 h 30 min.
5. After printing, the slides are cross-linked in ultraviolet light and tagged with bar-coded strips.

UV Cross-Linking of Slides

1. Place the printed slides on a cardboard tray.
2. Put the tray into the cross-linking machine.
3. Press "energy set 5000" (5000 µJ).
4. Start.
5. Store the cross-linked slides in a black slide box at room temperature.

Labeling of DNA with Random Priming

1. Sonicate the DNA at 80% amplitude for 30–60 seconds in a volume of 200 µl.
2. Check the size of the sonicated DNA by electrophoresing 1 µl (add 1 µl dye and 8 µl water) of sonicated DNA on a small 1% agarose gel (size of DNA should range from 300 to 2000 bp). Run a 123-bp ladder and Lambda-*Hind*III marker on the gel for size comparison. Resonicate if the DNA fragments are too large.

■**TABLE 34.1. Cy3 and Cy5 dNTP Fluorescent Mix**

Cy3 labeled dCTP, Final Concentration	Stock conc.	Volume for 10 Reactions-20 µl
1 mM of dGTP, dTTP, dATP	10 mM	2 µl of each
0.65 mM of dCTP	10 mM	1.3 µl
0.35 mM of CY3-dCTP	1 mM	7 µl
1 mM EDTA	50 mM	0.4 µl
10 mM Tris, pH 8.0	0.1M	2 µl
ddH_2O		3.3 µl
Cy5 labeled dCTP, Final Concentration	**Stock conc.**	**Volume for 10 Reactions-20 µl**
1 mM of dGTP, dTTP, dATP	10 mM	2 µl of each
0.65 mM of dCTP	10 mM	1.3 µl
0.35 mM of CY5-dCTP	1 mM	7 µl
1 mM EDTA	50 mM	0.4 µl
10 mM Tris, pH 8.0	0.1 M	2 µl
ddH_2Or		3.3 µl

3. Precipitate the DNA. (add 1/8 volume of 3 M sodium acetate pH 5.2, vortex for 30 s, briefly spin, add 1.1 volume of isopropanol, mix gently, 15-min freeze at –70°C, 15-min centrifuge, remove supernatant).
4. Wash with 200 µl of 80% ethanol (–20°C) and centrifuge for 7 min at 13,000 rpm. Allow the pellet to air dry for 15 min at room temperature.
5. Dissolve the pellet in ddH_2O, ensuring that the final concentration of DNA is 250 ng/µl, and place on the IKA-VIBRAX or Eppendorf shaker for 15 min.

The following refers to the labeling of 6 µg DNA (three hybridizations):

6. Add 24 µl 2.5× random priming buffer, vortex, and briefly centrifuge.
7. Boil for 5 min on a heat block at 95°C and then snap cool on ice for 10 min.
8. Prepare the 10× fluorescent dNTP mix. When cooled, add 6 µl 10× fluorescent dNTP mix (see Table 34.1).
   1. Add 3 µl Klenow fragment enzyme (bioprime labeling kit).
   2. Incubate at 37°C for 2–6 h.
   3. Stop the labeling reaction by adding 15 µl 0.5 M EDTA pH 8.0.
   4. Add water to a final volume of 450 µl and transfer the solution into a microcon column.
   5. Spin at 10,000 rpm for 8 min. To elute the DNA turn the microcon column upside down, place it into a new tube, and spin at 10,000 rpm for 2 min. If there are still traces of the dye on the microcon column, add 30 µl of water to dissolve the remaining part and repeat the "upside down" spinning step once again.

Slide Prehybridization

1. Add 400 µl of bovine serum albumin (1 aliquot) to a Coplin jar.
2. Fill the Coplin jar 3/4 full with prehybridization solution (see Appendix) and place in the 45°C incubator for 45 min.

■TABLE 34.2. Quantity of DNA Used in One Hybridization Reaction

Test, Cy3 labeled	~2 μg
Reference, Cy5 labeled	~2 μg
Cot-1 DNA	~100–150 μg*

* Amount of Cot-1 DNA depends on the array type and the quality of Cot-1 DNA.

■TABLE 34.3. Components of the Hybridization Reaction

Reagent	5 μl Hybemix	10 μl Hybemix	15 μl Hybemix
1% SDS	1.0 μl	2 μl	3 μl
ddH_2O	0.5 μl	1 μl	1.5 μl
Master mix	3.5 μl	7 μl	10.5 μl

3. When the prehybridization solution has reached 45°C, place the microarray slide in the jar and incubate for 1 h at 45°C.
4. After incubation, move the slide to a new Coplin jar containing ddH_2O. Wash the slide by emptying the jar of water and refilling it with fresh ddH_2O.
5. Carefully blow dry the slide with pressurized air.

Probe Prehybridization

1. Add the labeled DNA and the blocking DNA into a brown Eppendorf tube. Use the amounts shown in Table 34.2 for probe prehybridization:
2. Precipitate the DNA mix (add 1/8 volume of 3 M sodium acetate pH 5.2, vortex for 30 s, briefly spin, add 1.1 volume of isopropanol, mix gently, 15-min freeze at –70°C, 15-minute centrifuge at 13,000 rpm, remove supernatant).
3. Add 200 μl of 80% EtOH to the pellet (wash the walls of the tube a number of times to remove the DNA that is smeared on the tube wall). Centrifuge for 7 min at 13,000 rpm and allow to air dry for approximately 15 min in the dark.
4. Redissolve the pellet in 1/10 ddH_2O, 2/10 1% SDS, and 7/10 master mix (see Table 34.3 and Appendix).
5. Denature the probe for 20 min at 72°C, centrifuge briefly, and incubate the tube at 45°C for 2 h (the probeprehybe period).

Microarray Hybridization

1. Place the dry prehybridized microarray on the hybridization template (a black box that corresponds to the size of your array).
2. Carefully pipette the hybridization solution onto the array. If any bubbles appear, try to remove them with a syringe needle.
3. When bubble free, apply the coverslip carefully with tweezers.
4. Put the slide into a Corning hybridization chamber; add 20 μl water into the small wells in the hybe chamber to prevent drying.
5. Put aluminum foil over the chambers, and hybridize for 16–18 h at 45°C.

Posthybridization Washes

Note: washes 1 and 2 should be carried out in the dark

1. Fill a Coplin jar with wash solution 1 and heat to 65°C.
2. Remove the coverslip by rinsing the array slide with PBS (wash solution 2).
3. Put the slide into the 65°C wash solution 1 for 20 min.
4. Transfer the slide to a new Coplin jar containing PBS solution for 20 min at room temperature.
5. Soak the slide in wash buffer 3 for 20 s and rinse it quickly with ddH_2O. Blow the slide dry with pressurized dust-free air.
6. Scan for results with the GenePix 4000b scanner.

MATERIALS

Reagent	Supplier	Cat No.
dGTP	Invitrogen AB	18252-011
dCTP	Invitrogen AB	18252-013
dATP	Invitrogen AB	18252-015
dTTP	Invitrogen AB	18252-018
Bioprime DNA labeling kit	Invitrogen AB	18094-011
Lambda DNA	Invitrogen AB	25250-010
Sodium hydroxide	Sigma-Aldrich	E-7023
Formamide	Sigma-Aldrich	F-7503
Dimethyl sulfoxide (DMSO)	Sigma-Aldrich	D-8906
Bovine serum albumin	Sigma-Aldrich	A-2153
Coverslips	Sigma-Aldrich	Z36,590-4
HPLC-grade acetone	Sigma-Aldrich	27,072-5
Acetone	Sigma-Aldrich	17997-3
Amino silane	Sigma-Aldrich	281778
Dextran sulfate	Sigma-Aldrich	D-8906
Tough tags	Sigma-Aldrich	Z35,912-2
Microscope slides	Corning life sciences	Z16270-1
Microcon columns	Millipore AB	42410
Cy-5 dCTP	Amersham Biosciences	210735
Cy-3 dCTP	Amersham Biosciences	210734
Dextran sulfate sodium salt	Amersham Biosciences	17-0340-01
Microseal "F" foil	MJ Research Inc	MSF-1001
Microcleaning solution	TeleChem International	MCS
Human Cot-1 DNA	Roche	1581074

EQUIPMENT

Equipment	Supplier
Smp3 split pins	TeleChem International
Prosys 5510-A microarray printing robot	Cartesian
384-Well storage plate (Costar 3656)	Corning
Hybridization chambers	Corning
Sonicator—UP 50H Ultrasonic processor	Dr. hielscher, GmbH
Staining jar, with rack and handle	VWR
Speed Vac concentrator centrifuge	Savant
Thermomixer compact	Eppendorf
Genepix 4000b array scanner	Axon instruments

APPENDIX

Master Mix

Chemical	Amount
Formamide	5 ml
Dextran sulfate (Ref. 1)	1 g
20× sodium citrate buffer (SSC) (Ref. 1)	1 ml
ddH_2O	to 7 ml

Suspend the dextran sulfate in the formamide and quickly add the SSC and the water. Vortex vigorously. Heat the solution if it doesn't dissolve. Make aliquots of the solution and freeze.

Prehybridization Solution

Chemical	Final concentration	Volume in 1000 ml
20× SSC[1]	20%	250 ml
Bovine serum albumin (BSA)*	1%	10 g
10% SDS[1]	0.1%	10 ml
ddH_2O	Fill to 1000 ml	

* When preparing the prehybridization solution it is recommended to make a stock solution of BSA, aliquot it, and freeze. Thaw an aliquot and add it to the solution for each prehybridization.1Sambrook and Russell, Molecular Cloning: A Laboratory Manual, 3rd edition, Vol. 3, Appendix A1. New York: Cold Spring Harbor Laboratory Press, 2001.

Wash Solution 1

Chemical	Concentration	Volume in 1000 ml
20× SSC	2×	100 ml
10% SDS	0.1%	10 ml
ddH_2O		Fill to 1000 ml

Wash Solution 2

1× phosphate-buffered saline (PBS)

Wash Solution 3

0.2× SSC

REFERENCES

1. Kallioniemi A, Kallioniemi OP, Sudar D, Rutovitz D, Gray JW, Waldman F, Pinkel D (1992) Comparative genomic hybridization for molecular cytogenetic analysis of solid tumors. Science 258: 818–821.
2. Bruder CE, Hirvela C, Tapia-Paez I, Fransson I, Segraves R, Hamilton G, Zhang XX, Evans DG, Wallace AJ, Baser ME, Zucman-Rossi J, Hergersberg M, Boltshauser E, Papi L,

Rouleau GA, Poptodorov G, Jordanova A, Rask-Andersen H, Kluwe L, Mautner V, Sainio M, Hung G, Mathiesen T, Moller C, Pulst SM, Harder H, Heiberg A, Honda M, Niimura M, Sahlen S, Blennow E, Albertson DG, Pinkel D, Dumanski JP (2001) High resolution deletion analysis of constitutional DNA from neurofibromatosis type 2 (NF2) patients using microarray-CGH. Hum Mol Genet 10: 271–282.

3. Buckley PG, Mantripragada KK, Benetkiewicz M, Tapia-Paez I, Diaz De Stahl T, Rosenquist M, Ali H, Jarbo C, De Bustos C, Hirvela C, Sinder Wilen B, Fransson I, Thyr C, Johnsson BI, Bruder CE, Menzel U, Hergersberg M, Mandahl N, Blennow E, Wedell A, Beare DM, Collins JE, Dunham I, Albertson D, Pinkel D, Bastian BC, Faruqi AF, Lasken RS, Ichimura K, Collins VP, Dumanski JP (2002) A full-coverage, high-resolution human chromosome 22 genomic microarray for clinical and research applications. Hum Mol Genet 11: 3221–3229.
4. Mantripragada KK, Buckley PG, Benetkiewicz M, De Bustos C, Hirvela C, Jarbo C, Bruder CE, Wensman H, Mathiesen T, Nyberg G, Papi L, Collins VP, Ichimura K, Evans G, Dumanski JP (2003) High-resolution profiling of an 11 Mb segment of human chromosome 22 in sporadic schwannoma using array-CGH. Int J Oncol 22: 615–622.

35

MICROPLATE ARRAY DIAGONAL GEL ELECTROPHORESIS (MADGE)

Ian N. M. Day
Division of Human Genetics, Southampton University Hospital, Southampton, UK

■ INTRODUCTION

Microplate array diagonal gel electrophoresis (MADGE) (patent GB2284484) was invented by Day in 1994 (1) to enable compatibility between microplate-based liquid phase reactions and polyacrylamide or agarose gel electrophoretic analysis. In MADGE, a microplate-compatible array of wells is turned on a diagonal relative to the line of electrophoresis so samples can then run on a long track length between the wells (Fig. 1). Thus maintenance of a 96-well format can be coupled to high sample resolution. Microplates occupy an analytical "middle ground" in throughput between test tubes and microarrays, and MADGE can now meet the associated electrophoretic requirements at this level. The natural scalability of MADGE enables high throughput in large laboratories, while its simplicity facilitates distributability between smaller laboratories. The set-up and running costs of MADGE compare favorably with other technologies, and most electrophoresis-based analysis should be feasible with MADGE. The invention of MADGE was driven by higher throughput requirements in human single nucleotide polymorphism (SNP) genotyping, and similar requirements for de novo mutation scanning have further spawned the invention of meltMADGE. This set of core protocols reflects the origins of MADGE, but the scope of MADGE extends in principle to any higher-throughput genomic analysis.

Guide to Mutation Detection, Edited by Graham R. Taylor and Ian N. M. Day.
ISBN 0-471-23444-3

THE USE OF MADGE IN SIMPLE PCR CHECKING AND SNP GENOTYPING REACTIONS

MADGE gels can be used to check any array-based PCR reaction, and the MADGE system lends itself particularly well to SNP genotyping. MADGE gels can be used in genotyping reactions with restriction fragment length polymorphisms (PCR–RFLP/MADGE) (Fig. 2). However, as restriction enzyme costs are often prohibitive for genotyping "scale-ups," MADGE is generally used in conjunction with allele-specific amplification assays such as ARMS (amplification refractory mutation system) (2). ARMS-MADGE is usually set up as a two-reaction system; both reactions have a "common" primer near the SNP of interest, but each contains a different ARMS primer specific for one of the two SNP alleles. PCR amplification is verified by inclusion of control primer pairs in the reactions. Alternatively, both reactions can be performed in the same mix as a tetraprimer PCR.

STANDARDIZATION OF PCR FOR SUBSEQUENT MADGE ANALYSIS

For this protocol it is assumed that readers have optimized PCR genotyping reactions and now wish to develop their assays to allow high-throughput screening. To this end, we recommend standardization of assays to the following format:

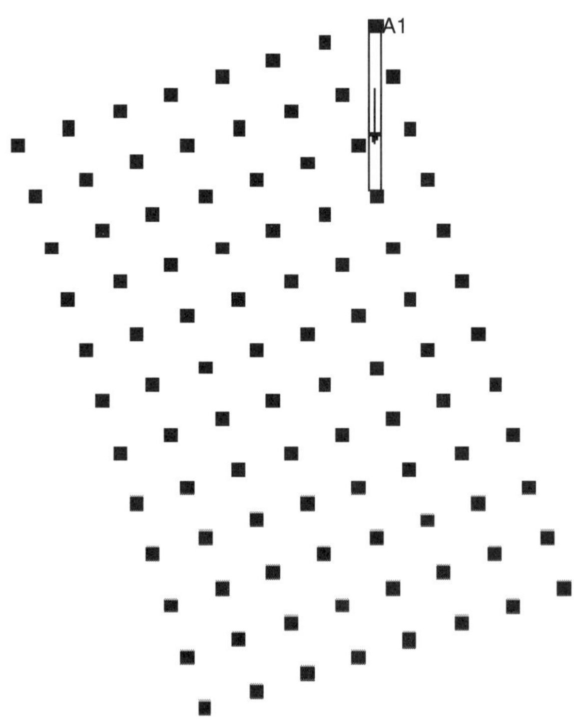

FIGURE 35.1. 96-Well MADGE with track A1 showing direction of electrophoresis.

1. Reactions set up in a volume of 10 μl
2. Assays set up in 96-well polypropylene plates (e.g., Thermoquest, Cat. no. PC1090, sealing with "rubber mats" (Hybaid, Cat no. HB TD MTS RS5)
3. Reactions run "oil free" with the cycling program to include a heated lid setting
4. Use of high-performance Peltier Thermal Cyclers (e.g., PTC-225 DNA Engine TETRAD, MJ Research Inc.)

MADGE PRINCIPLES AND PROCEDURES

As many assays are set in a 96-well format, MADGE was designed to preserve the exact configuration and 9-mm pitch of an 8 × 12 microtiter plate array. However, in the standard MADGE gel format, the array is set on a diagonal of 71.6°, giving a final track length of 26.5 mm. Thus the advantages of maintaining a 96-well format (such as 96-channel transfer of samples and information) can be coupled to high sample resolution.

MADGE gels can be made with a two-piece kit (MadgeBio, Cat no. M0011) consisting of a plastic "former" with 96 2-mm cubic "teeth" and a plain glass plate. The former molds a 2-mm-deep, 100 × 160 mm rectangular gel within which there are the 96 wells.

PREPARATION OF MADGE GELS

First coat one surface of the plain glass plate with "sticky silane" (99% ethanol, 0.5% glacial acetic acid, 0.5% v/v gamma-methacryloxypropyltrimethoxysilane). Use the edge of another glass plate to spread 1 ml of sticky silane as evenly as possible. When dry, wipe the surface with a damp tissue.

MADGE Gels can be Poured in Two Different Ways

1. Lay the MADGE gel former level, with the teeth upward, on paper towels or in a flat tray. Ensure that the broad border is nearest to you, with the array closest to this short edge.
2. Mix 35 ml of the required gel mix (see Table 35.1) and pour this mix into the nearest end of the MADGE gel former.

TABLE 35.1

% Acrylamide	5%	7.5%	10%
Effective separation range (bp)	80–500	70–400	50–300
For 50 ml Gel Mix:			
10× TBE	5 ml	5 ml	5 ml
30% Acrylamide:Bis (19:1)	8.3 ml	12.5 ml	16.7 ml
Distilled water	36.7 ml	32.5 ml	28.3 ml
25% Ammonium persulfate	150 μl	150 μl	150 μl
TEMED	150 μl	150 μl	150 μl

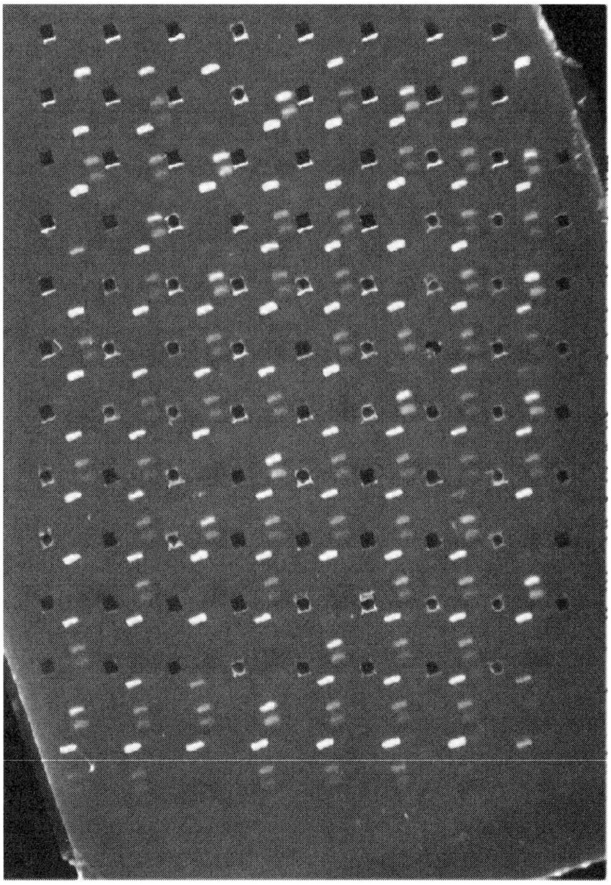

FIGURE 35.2. A 96-well MADGE genotyping gel showing easily defined band patterns.

3. Carefully place the glass plate (sticky silane side down) over the former. It is easiest to first place one end of the plate over the broad edge nearest to you, then to use one smooth, continuous action to lower the other end down.

Alternatively,

1. Place the glass plate over the former, leaving a small gap next to one of the broad borders.
2. Hold the plates together with one hand and tip the plates to a slight incline so the gap is at the top end.
3. Carefully pour the gel mix into the gap, so that it runs down in between the two plates and fills the mold up from the bottom. With this method gels can be poured without any acrylamide overspill.
4. Align the glass with the outside of the former; this will give a gap at the far end ready to prise away the former when set. A weight (such as a 50- to 100-g bottle) can be placed centrally on the glass to ensure uniform gel thickness.
5. Once the MADGE gel has set, take a flat-ended spatula and place under the edge of the glass, where there is a 2-mm gap. The glass plate with gel attached can then be prised away from the former.

It is a frequent practice in our laboratory to pour batches of gels, because making the gel mix takes longer than pouring the gels. Gels can be piled together, cling film wrapped, and stored in a fridge for days to weeks before use. It should be noted that ethidium bromide is unstable during prolonged storage, in which case prestaining immediately before use is often convenient.

ELECTROPHORESIS OF MADGE GELS

1. Submerge the gel and glass plate in a standard gel tank (such as Kodak BioMax MP-1015 Multipurpose tank, Sigma, Cat no. Z35,604-2), abutting the glass to one side of the tank. Well position A1 is placed nearest the cathode.
2. Add 5 µl PCR sample to 2 µl loading dye, load 5 µl using an 8-, 12-, or 96-channel pipette.
3. Typical running conditions would be 10 V/cm for 30 min in 7.5% PAGE, 1× TBE for a 200-bp PCR product, but conditions can be adapted according to experimental needs.

VISUALIZATION AND ANALYSIS OF MADGE GELS

Gels are usually prestained with ethidium bromide at a concentration of 1 µg/ml by gently agitating gels on a rotary shaker for 30 min. Prestaining aids tight band patterns, but poststaining does give acceptable results. Given that the human eye/brain is extremely good at pattern recognition, small-scale gels can be read manually, to reasonable resolution. However, use of digital imaging systems (such as a Fluorimager 595, Molecular Dynamics) and suitable software greatly facilitates analysis. MADGE-specific software is commercially available (web link-http:www.phoretix.com/1d.htm) where specific requirements for analyzing MADGE gels incorporate or replace the features and facilities for standard gel analysis. Phoretix 1D analysis software can take TIFF files from either a standard CCD camera or a flatbed scanner such as the Fluorimager 595, and selection of "MADGE mode" changes the appropriate dialogs, allowing identification of the 96 gel tracks. The standard gel analysis facilities can then be used for densitometry, molecular sizing, Rf adjustments, and band pattern matching. Results are presented in spreadsheet format, so data can readily be exported for genotype analysis. A step-by-step guide for used of MADGE software and the Fluorimager 595 is presented at www.sgel.humgen.soton.ac.uk. Dyes such as Vistra Green (Amersham Pharmacia, Cat no. RPN 5786) are more sensitive than ethidium bromide, allowing detection down to 20 pg per band. Other dyes (see http://www.probes.com/) create the potential for RNA, ssDNA, oligonucleotide, and multiplexing applications.

ARMS AND 192-WELL MADGE

Setting up ARMS assays from 96-well arrays generates 192 reactions. A 96-well MADGE can be repeat loaded with the second allele reactions after a few minutes of electrophoresis. Alternatively, 192-well MADGE gels can be set up with the appropriate formers (MadgeBio, Cat no. M0141). Here a second set of wells are located halfway along the tracks of the first and samples, so the two allele-specific reactions can be run in series. Although the gel tracks are shorter in length than the conventional MADGE system, resolution is sufficient for SNP genotyping reactions. For computer analysis of 192-well MADGE gels the two allele-specific tracks are treated as one lane. Phoretix 1D Advanced software can then be used to create and name lanes, and the MADGE lane classification option can be used to assign allele calls.

■ REFERENCES

1. Day IN, Humphries SE (1994) Electrophoresis for genotyping: microtiter array diagonal gel electrophoresis on horizontal polyacrylamide gels, hydrolink, or agarose. Anal Biochem 222: 389–395.
2. Newton CR, Graham A, Heptinstall LE, Powell SJ, Summers C, Kalsheker N, Smith JC, Markham AF (1989) Analysis of any point mutation in DNA. The amplification refractory mutation system (ARMS). Nucleic Acids Res 17: 2503–2516.

INDEX

Guide to Mutation Detection, Edited by Graham R. Taylor and Ian N. M. Day.
ISBN 0-471-23444-3